New Perspectives on Risk Communication

That risk communication ranks high on the policymaking agenda is beyond discussion today. The field is a point of intersection of social communication, practical management and policy making. It covers such diverse activities as to inform and educate the public about risk, and risk management in order to influence attitudes and behaviour, to act in situations of emergency or crises, to aid in decision-making and to assist in conflict resolution. Communication has grown into a major concern in current risk governance based on network co-ordinated management of public affairs conducted by authorities and companies and is recognized as a key component in the government of risk. This is especially salient in policy fields relating to environmental planning and resource management, urban planning, chemical and food regulation, or infrastructure planning, development and maintenance. This book explores risk communication research with a focus on new theoretical perspectives, research findings, and applied goals. It reflects on a broad range of innovative theoretical perspectives, methodological approaches and empirical areas.

This book was previously published as a special issue of the *Journal of Risk Research*.

Åsa Boholm is Professor of Social Anthropology at the School of Public Administration, University of Gothenburg. Research areas include cultural and organizational dimensions of risk, the communication and management of technological risks in public policy, land use planning, the role of science and technology in public administration and decision making.

New Perspectives on Risk Communication

Uncertainty in a Complex Society

Edited by
Åsa Boholm

Routledge
Taylor & Francis Group
LONDON AND NEW YORK

First published 2011
by Routledge
4 Park Square, Milton Park, Abingdon, Oxon OX14 4RN

Simultaneously published in the USA and Canada
by Routledge
605 Third Avenue, New York, NY 10017

Routledge is an imprint of the Taylor & Francis Group, an informa business

This book is a reproduction of the *Journal of Risk Research*, 11, 1-2. The Publisher requests that those citing this book use the bibliographical details of the journal issue on which the book is based.

Typeset in Times New Roman by Taylor & Francis Books

British Library Cataloguing in Publication Data
A catalogue record for this book is available from the British Library

ISBN13: 978-0-415-61787-1 (hbk)

Disclaimer
The publisher would like to make readers aware that the chapters in this book are referred to as articles as they had been in the special issue. The publisher accepts responsibility for any inconsistencies that may have arisen in the course of preparing this volume for print.

Contents

CONTENTS

Introduction

Risk communication is a point of intersection of social communication, practical management and policy-making. It covers such diverse activities as to inform and educate the public about risk, and risk management in order to influence attitudes and behaviour, to act in situations of emergency or crises, to aid in decision-making and to assist in conflict resolution. Communication has grown into a major concern in current risk governance based on network co-ordinated management of public affairs conducted by authorities and companies (Swyngedouw 2005) and is recognized as a key component in the government of risk (Hood, Rothstein, and Baldwin 2004). This is especially salient in policy fields relating to environmental planning and resource management, urban planning, chemical and food regulation, or infrastructure planning, development and maintenance. Polarization between the perceptions of lay publics on the one hand, and regulators, scientific experts and project proponents on the other has been a key issue in risk communication research (Petts and Brooks 2006). The field involves several distinctions, including that between expert and laymen, between those affected by decision and those who make the decisions, between conflict and co-operation, between facts and values, and between inclusion and exclusion in decision processes (for a recent overview of the field see Palencher and Heath 2007).

When the risk communication field was new, in the late 1960s and early 1970s, inspiration came from psychological research on how people assess risk information (Fischhoff 1995). It was shown that people when they make judgements about risk do not compute statistical information. They rely on heuristics – short cuts of information processing that simplify information – which often makes some risks more salient than others (Slovic 2000). The early rationale for risk communication research derived from the identified divide between the scientific way to assess risk (based on calculations of probability and estimated 'loss') and the lay people approach which tended to over- or underestimate risk. Hence risk communication initially adopted a pedagogical mission to teach the public about real risk so that they can act 'rationally' and make informed choices about what risks to take or not to take (see Leiss 1996, for an overview).

Over the last 15 to 20 years this technocratic approach has been gradually abandoned and today it is widely recognized that public values and preferences must be included in risk assessment and management (Renn 1998). Focus has moved to an emphasis on deliberation and dialogue processes, often with a normative element. Risk communication researchers have argued that communicators and audience must listen to each other and learn from one another (Petts 2001). Emphasis has shifted from education, framed as the monitoring of behaviour and attitude change induced from technocratic expertise, to consensus building and conflict resolution.

A dominant paradigm in risk communication has been a technical sender-receiver model stating that a message travels from a sender via some kind of channel or medium to a receiver. A condition for communication is that a message can be packaged into a signal according to a code and that both sender and receiver use the same coding device for packaging and unpacking messages. Communication failure occurs when the message at the receiver's end differs significantly from the message dispatched from the sender either due to distortion from 'noise' during transmission or due to discrepant coding and decoding. The Shannon and Weaver (1949) model originally focused on communication in a technical sense, including machines, but the model has also been widely applied to social communication in human interaction. The social amplification of risk framework which has been influential in risk communication research (see Pidgeon, Kasperson, and Slovic 2003) can be seen as a development of the Shannon-Weaver model. The distortion of risk messages is attributed to 'noise' in the channelling of a risk signal by means of various socially induced attenuations or amplifications of the signal (for example by the media, industry, interest groups, academics, or stakeholders).

In risk communication studies Jürgen Habermas's (1985) normative theory of communicative rationality has been advocated as an alterative to the Shannon-Weaver model. Following Habermas, it has been argued that communication should be a dialogue between actors who are willing to listen to each other and who are open to change their minds and positions on a certain issue depending on how the deliberative process unfolds (see for example Renn 2004). By means of dialogue, pluralities of viewpoints, evaluations and prioritizations can be considered which promotes sensible decision-making on collective and often controversial matters imbued by risk and uncertainty. Trust between participants is a crucial condition for dialogue (Petts 2001). In reality however social communication does not adhere to this ideal. Human communication is part of social interaction and can be co-operative as well as conflictive depending on many things such as social conventions, expectations, social roles and identities, power relations, and interpretations of meanings (Allwood 1978).

That risk communication ranks high on the policy-making agenda is beyond discussion today. But what current developments can be discerned with regard to risk communication as a research field? What is the theoretical status and use of the sender-receiver model and the dialogue model for communication? Are there other theoretical trends, and in such case, what are they? In order to find answers to those questions a research conference with the title 'New perspectives on risk communication: uncertainty in a complex society' was arranged in Göteborg, Sweden in August–September 2006. The aim was to stimulate a broad inter-disciplinary discussion on risk communication as a research field with a focus on new theoretical perspectives, research findings, and applied goals.

The conference which was funded by the Swedish Research Council for Environment, Agricultural Sciences and Spatial Planning, the Swedish Emergency Management Agency, the Swedish Research Council, the School of Public Administration, Gothenburg University, and the Centre for Public Sector Research, Gothenburg University, attracted 90 participants from 14 countries representing 58 institutional affiliations. In all, 60 academic papers were presented. This special issue of the *Journal of Risk Research* includes a selection of these papers together with a viewpoint paper on risk communication with commentaries. This volume reflects the broad range of theoretical perspectives, methodological and empirical foci that risk communication can incorporate and which makes it such a challenging and exciting field.

References

Allwood, J. 1978. On the analysis of communicative action. *Gothenburg Papers in Theoretical Linguistics* 38, Dept of Linguistics, Göteborg University. [Also in *The Structure of Action*, ed. P. Brenner, 168–91.] Oxford: Basil Blackwell.

Fischhoff, B. 1995. Risk perception and communication unplugged: twenty years of process. *Risk Analysis* 15, no. 2: 137–45.

Habermas, J. 1985. *The theory of communicative action*. Cambridge: Beacon Press.

Hood, C., H. Rothstein, and R. Baldwin. 2004. *The government of risk. Understanding risk regulation regimes*. Oxford: Oxford University Press.

Leiss, W. 1996. Three phases in the evolution of risk communication practice. *Annals of the American Academy of Political and Social Science* 545: 85–94.

Palenchar, M.J., and R.L. Heath. 2007. Strategic risk communication: Adding value to society. *Public Relations Review* 33: 120–9.

Petts, J. 2001. Evaluating the effectiveness of deliberative processes: waste management case-studies. *Journal of Environmental Planning and Management* 44, no. 2: 207–26.

Petts, J., and C. Brooks. 2006. Expert conceptualisations of the role of lay knowledge in environmental decision making: Challenges for deliberative democracy. *Environment and Planning (A)* 38: 1045–59.

Pidgeon, N., R.E. Kasperson, and P. Slovic, eds. 2003. *The social amplification of risk*. Cambridge: Cambridge University Press.

Renn, O. 1998. Three decades of risk research: Accomplishments and new challenges. *Journal of Risk Research* 1, no. 1: 49–71.

———. 2004. Participatory processes for designing environmental policies. *Land Use Policy* 23: 34–43.

Shannon, C.E., and W. Weaver. 1949. *The mathematical theory of communication*. Urbana, IL: University of Illinois Press.

Slovic, P. 2000. *The perception of risk*. London: Earthscan.

Swyngedouw, E. 2005. Governance innovation and the citizen: The Janus face of governance-beyond-the-state. *Urban Studies* 42, no. 11: 1991–2006.

Åsa Boholm
CEFOS, Gothenburg University, Sweden

Risk communication: world creation through collective learning under complex contingent conditions

Piet Strydom

Department of Sociology, University College Cork, Ireland

Risk communication, in the full sense of the word, is a discursive event in which speakers advance claims in the face of other responding participants before a general public. The presence of the public leads the participants to evaluate what happens in moral terms, with the result that their claims obtain an unavoidable normative quality and the discursive event takes the form of a public controversy which puts pressure on the participants to coordinate their disagreements. Proceeding from the assumption of socially distributed and shared cognition, the core argument of this paper is that risk communication, in the final analysis, is a cooperative learning process in and through which a communication community constructively arrives at a diagnostic interpretation of its common situation, the challenge it faces, and possible ways of dealing with it. Since such learning is possible only under conditions of relatively high complexity and contingency, however, its characteristic non-linear dynamic development makes uncertainty both many-sided and unavoidable. Often, however, such a collective achievement is put beyond reach, not simply because of complexity, contingency and uncertainty, but rather because the agents or groups involved follow one or other of a number of strategies which effectively block learning. Were the social sciences to contribute to the enhancement of risk communication (e.g., by facilitating value- and will-formation in the face of concrete problems), they should study the multi-levelled process of risk communication in the different communicative-discursive contexts within which it takes place with a view to clarifying the learning processes and potentials they harbour. Crucial here are the normative standards appealed to and the degree of legitimacy they allow. Not merely the management of uncertainty depends on this, but also the very world brought into being through risk communication.

Introduction

By ‘risk communication’, to begin with, I understand a broad societal phenomenon which is of a historically specific nature. It made its appearance in the late twentieth century risk society as part of the public discourse about risk and responsibility. And, in the meantime, it has become the characteristic feature, the signature as it were, of early twenty-first-century society. Historically, therefore, it is comparable to the ‘poverty communication’ of the nineteenth century and the ‘violence communication’ of the early modern period (Strydom 1999a, 1999b, 2000, 2002). This means that risk communication is a multi-levelled process that includes all communication, all the different streams or strands of communication, about the central issue of our time: the issue of risk which has increased communication in the past number of decades to such an unprecedented degree.

An adequate understanding of risk communication is possible, therefore, only in the context of the risk society and, more specifically, the public discourse about risk and responsibility. What such contextualization allows one to do, is to appreciate the complexity of the conditions under which risk communication takes place and, hence, also the multi-levelled or differentiated nature of risk communication itself. The unfolding of this discourse during the second half of the twentieth century revealed the different streams or strands which go into the makeup of risk communication together with their respective vehicles (Strydom 2002). Whereas risk communication at the outset took the form of closed elite communication about the level of acceptable safety, it later progressively broadened into a form of public communication about the societal production and constitution of risk. In the 1950s, experts in the nuclear industry and regulators were both senders and receivers of risk communication. By the 1970s, however, social movement organizations have entered the ranks of risk communicators and started to address ordinary members of society. Since then, a significant proportion of the citizenry has become sufficiently concerned and involved to provide the spawning ground of both ideas and organizations feeding into risk communication.

It is remarkable that whereas the discursive logic generated by risk communication for decades sharply divided those affected, even leading to destructive conflict, it has given rise in the course of time to discursive means for the potential coordination of divergent orientations and actions. This suggests that risk communication, in the final analysis, is a cooperative learning process in and through which a communication community arrives at a constructive diagnostic interpretation of its common situation, the challenge it faces, and possible ways of dealing with it, thus creating a form of life for its members.

The social scientific study of risk communication, then, must take into account a number of things: the historically coloured societal conditions of complexity under which it emerged and takes place; the multi-levelled or differentiated nature of risk communication and the specific character of each of the forms it takes; and, finally, the relations among these different dimensions and the direction taken by their interrelation. To comply with this requirement, I devote this paper to a brief treatment of three topics: first, the complex conditions of risk communication as a multi-levelled process dealt with in terms of the discursive constitution and organization of contemporary society; secondly, considering the different forms of risk communication from the perspective of the learning processes they harbour such that it becomes possible to determine typical configurations of relations between them and the potential such relations unlock; and, finally, reflections on the status and role of normative claims and criteria in the coordination of the different levels of risk communication.

Complexity conditions of risk communication

The complexity of the conditions under which risk communication takes place in contemporary society, including the uncertainty accompanying it, is best clarified by considering the mode of constitution of society and the contribution communication makes to it.

Internal differentiation of the multi-levelled process of risk communication

Risk communication is a central component of the communicative-discursive process whereby society – in this case, the 'risk society' (Beck 1992) – is constituted and organized. This process of constitution is characterized by a non-linear social dynamic. The latter rests on the generation of variety by different sources and the selective combination of these contributions into a not fully anticipated or expected outcome. This non-linearity draws our attention to the multi-levelled and internally differentiated nature of risk communication, since the latter is a principal generative source of variety. It is also here that we find an important part of the explanation of the high degree of contingency and uncertainty surrounding risk communication.

Although it may be referring to one and the same object (i.e., a risk reality), risk communication draws on a 'culture of contradictions' (Eder 1993, 194) and is carried and promoted by socially or institutionally distinct agents. For example, state and corporate institutions, on the one hand, and associations, NGOs and social movements, on the other, while attending to a mutually recognized risk, embody different risk perception and risk communication cultures.[1] They thus entertain different communicative and informative intentions in accordance with their distinct modes of engagement with the world.[2] This provides a basis for introducing a fundamental distinction between institutional and civic risk communication. Related to these different forms of communication, and reinforcing them, are furthermore corresponding resonance structures[3] to which they are oriented and which make possible both responses to such communications and the forging of connections between them. These structures stretch from institutional resonance to extra-institutional or civil society resonance, both of which implicate cultural resonance. Later, more attention will be given to resonance, at which stage it will be necessary to refer also to public resonance.

From a cognitive perspective,[4] accordingly, risk communication can be said to involve a whole range of related yet different competing and even conflicting cognitive devices, from schemata, operations and scripts – that is, modes of classifying a given risk, making sense of and acting upon it – to cognitive formats for selecting, condensing and generalizing relevant information, such as for instance statistics, reports, political analysis and ordinary language. This competitive and potentially conflicting diversity is mirrored, on the other side, by the corresponding resonance structures. Considering these various dimensions, risk communication must thus be regarded in terms of a multi-levelled network in and through which incoming information about a given risk reality is processed in a socially distributed way. Occupying different positions, yet parallel to one another, each of the participating agents frames and communicates the mutually recognized risk in its own particular way and thereby activates resonance structures which allow responses in the form of feedback of one kind or another as well as connections to be forged. And in doing so, each of the participants makes a contribution to the way in which the risk becomes collectively classified, understood and dealt with. Risk communication is thus socially shared, creative cognition or a creative collective cognitive process whereby a community forms and shapes a world for itself.

In contradistinction to contemporary approaches (e.g., Clark 1998; Wertsch 1998) to distributed cognition in networks of situated action which, notwithstanding their commendable highlighting of differences and interdependencies, tend to flatten out different levels and make disjunctions, tensions, contradictions and conflicts unrecognizable, it is vital for an adequate understanding of risk communication not only to stress complexity, but also to keep the concomitant contingency and uncertainty in mind. For this reason, it is imperative to acknowledge that risk communication is not simply a matter of representation, but at the same time also one of signification;[5] not simply one of communicating a position or content in keeping with a communicative and informative intention, but also of contextual references generating meanings going far beyond all intentions; not simply of problem-solving, but also of creating and bringing a world into being.[6]

This brings us to the context or larger intersubjective framework of risk communication – and that means something that duplicates complexity and multiplies uncertainty by imposing a communicatively and discursively mediated yet non-linear social dynamic on risk communication itself.

The public communicative context of risk communication

By attracting attention and providing a compelling object of reference, a risk opens a reflexive and communicative moment. Indeed, in conjunction with the schemata, operations and cognitive formats of those affected, risk gives rise to an intersubjective experiential context filled with concern and developing social relations. Individuals, organizations and institutions recognize the consequences of others' decisions and actions potentially affecting them, attribute those consequences to the agents involved, and relate to one another in an effort to collectively define the problem and decide how it could be dealt with. Risk and the cognitive structures[7] corresponding to it not only call forth communication, which in turn activates resonance structures, but also help to forge connections and intensify communication to the point that a larger communicative event takes place and overarching communicative structures come into operation, take effect and develop. Risk communication thus leads to the emergence of a discourse and the establishment of what is called the 'public sphere'.[8] In fact, each of the different forms of risk communication, together with its set of resonance structures, gives rise to a corresponding type of partial public sphere. At a higher level, a more embracing public sphere emerges at the intersection of the different forms of risk communication and resonance structures.

Depending on the conjunction of these strands of risk communication and resonance structures, the public sphere takes on different forms, with four basic types being distinguishable.[9]

Public sphere at rest

In Western type societies where certain fundamental arrangements are in place, the backdrop against which activated forms of public sphere appear could be conceived as a 'public sphere at rest' (Habermas 1996, 379).[10] It is characterized by a low level of constricted risk communication and hence a dearth of resonance. There is indeed an awareness of potential or real risks, but there is no cause for a systematic engagement in risk communication. However, once the perception and recognition of a risk reaches a critical threshold, once risk communication increases and flares up, any of a number of types of public spheres could emerge. This is illustrated by the historical unfolding of the discourse about risk and responsibility since the 1950s and '60s through four phases (Strydom 2002, 11–35) – from a restricted expert and regulator debate about risk calculation and assessment, eventually to a fully fledged public discourse about the societal constitution of risk and its implications for the creation and organization of society.

Institutional public sphere

One active type is an elite or institutional public sphere. It is generated by a form of risk communication engaged in by experts such as for example managers, planners, economists, engineers, lawyers, psychologists and regulators representing institutions such as private firms and state administration. Business promotion and political campaigning aimed at mobilizing consumers, clients or the citizenry to support some service, product or decision would count as instances of such risk communication. Corresponding to this form of risk communication, institutional resonance is a characteristic structural feature of this type of public sphere. It makes its appearance through the response of such institutions to the perceived needs, demands and criticisms of consumers, clients and citizens.

Liberal or mass public sphere

Another type is the liberal or mass public sphere. The form of risk communication generating this public sphere is borne and promoted by active members of society who are typically organized into consumer groups, neighbourhood groups, voluntary associations, NGOs and social movement organizations. To such civic risk communication corresponds civil society resonance which feeds on the response of active associations, organizations and movements belonging to civil society to political or economic encroachments or problems their members or others they identify with face in their private life-worlds. Central to this type of public sphere, therefore, are political, social or cultural critique and different forms of protest.

Discursive public sphere

A final active type is a mediated or discursive public sphere. This type of public sphere is more general, but also more complex, than the preceding institutional and liberal or mass public spheres. Depending on the degree of development of its defining feature, viz. discursive mediation, it could take two related yet distinct forms.

In its first deficient form, it is characterized by the presence of both institutions and civil society. Connections are established and even elaborated between institutional risk communication and resonance, on the one hand, and civic risk communication and resonance, on the other. Although both are affected by sharing a situation in which they face each other, the differences between them are maintained. In some sense or another, institutional and civic risk communication stand over against one another and even get embroiled in contestation and conflict. None of the parties expects and therefore exerts itself to achieve more general agreement and acceptance of the claims it advances. Rather than allowing the full institutionalization of discourse,[11] the structures of this public sphere facilitate the forgetting or suppression of the bridging potential of communication.

In the case of the enhanced and still more complex form of the discursive public sphere, the crucial difference is made by the incorporation of the public and hence public resonance – which means to say, the introduction of the third point of view over and above the ego and alter perspectives of the institutional and civil society actors.[12] In such a fully developed public sphere, the participating institutions as well as associations, organizations and movements relate to each other in and through the discursive medium only via a reference to the presence of the non-participating yet nevertheless involved observing, evaluating, judging and commenting public. The epistemic authority coming into play here unequivocally gives priority to the public interest. The publicly embodied third point of view therefore has a sobering structuring effect on the self-understanding, self-presentation, mutual sensitivity, mutual understanding and mutual recognition of the active participants and their respective forms of risk communication. This instantiation of the public sphere thus possesses the potential of refining risk communication to the highest degree of legitimacy and hence efficacy possible. But this is possible only under the contingent, uncertain and fragile conditions of the non-linear social dynamics leading to a collective outcome.

This raises the question of what precisely is at stake in risk communication.[13]

Risk communication as learning under contingent conditions

The identification of different types of public spheres in which distinct forms of risk communication take place underscores the non-linear dynamic nature of the multi-levelled process of risk communication. Far from simply the transfer of information, risk communication involves the processing of information through its division and recombination, implicating variable configurations of speakers, audiences and others. In turn, such an event implies that learning processes are at work in it.[14] As it takes its course, cognitive structures of different levels and scope, from individual self-understanding to cultural models, undergo a self-organizing transformation resulting in a new classification, reframing and reconstitution of the world. Considering risk communication from this perspective should throw some light on persistent problems and challenges faced by risk communicators, particularly perennial uncertainty, but also on potential opportunities and possibilities for the enhancement of risk communication.

Sociologically, learning processes can take on a variety of different forms, depending on the structure and complexity of the social relations, intersubjective experiential or communicative contexts presupposed. Different configurations of public spheres allow distinct learning processes. The multi-levelled process of risk communication, particularly the impetus driving it forward, cannot be adequately understood unless it is related to these learning processes. For this purpose, some basic as well as composite types of social learning are distinguished below and related to the levels of the risk communication process.[15]

Aggregative learning

First, aggregative learning at best is possible in a public sphere at rest representing an intersubjective framework characterized by a high level of structural openness and a low level of symbolic complexity. Such learning involves the aggregation of achievements gained through trial and error at the individual level, whether persons or legal personalities such as institutions. At issue here, therefore, are the individual learning processes of business executives, corporations, politicians and bureaucracies that take place without collective will-formation and, hence, to the exclusion of civil society and the public. It leads to configurations of socio-economic and political or administrative power which give rise to cleavages and class structures as well as associated dynamics of individualization and social exposure to risk. For example, this type of learning and its consequences were typical of the period prior to the emergence in the 1970s of the concern of social movements and the United Nations with the problem of risk.

Institutional learning

Second, institutional learning requires an intersubjective experiential context possessing a high level of structural closure and a low level of symbolic complexity such as an elite or institutional public sphere. It involves the kind of learning that different kinds of institutions have to undergo to be able to fulfil their particular missions – for instance, economic institutions (e.g., the corporate sector) to safeguard and further production, distribution and consumption, and political institutions (e.g., the state, the EU) to take care of the population and collective goods. This type of learning assumes the indifference, acquiescence or compliance on the part of the relevant other such as for instance labour, consumers and the citizenry. Thus it typically leads to institutional practices which also have a hierarchical, authoritarian or paternalistic side. In the risk communication of private firms and state institutions, this type of learning is closely allied with and complementary to aggregative learning. Here a form of institutional risk communication prevails which is articulated through a variety of contradictory cultural models. While corporations let themselves be guided by such models as entrepreneurship, profit, smartness, bullishness, risk-taking, customer service and need-fulfilment, and government agencies by such models as reason of state, the common good and public service, both tend to be paternalistic and secretive and therefore readily revert to practices of concealment, denial and cover-up.

Associational learning

Third, associational learning is made possible by a liberal or mass public sphere which presupposes a configuration of social relations displaying a low level of structural openness and a high level of symbolic complexity. It involves the coordination and synthesis of the cognitive structures of associating individuals and groups in civil society.[16] Such coordination and synthesis stretch from the construction of a collective identity, via the formation of a collective actor such as a voluntary association, NGOs or social movement, to collective mobilization and eventually collective action.[17] Involving little or no communication with relevant others, this kind of learning proceeds from the assumption of the opposition of corporations and the state or, at best, their indifference or acquiescence, while hope is pinned to the support of the citizens. Needless to say, this is typical of the contemporary environmental, anti-biotechnology, genetic rights and related movements engaging in civic risk communication. This form of communication, impelled as it is by associational learning, is guided and given direction by a range of cultural models, including from human and planetary health, through rights, citizenship, sustainability, responsibility, victimology and justice, to global civil society, democracy, cosmopolitanism and ecology.

This brings us to the kind of learning that takes place in a mediated or discursive public sphere which, as we have seen, takes two forms depending on the configuration of the relations involved. In the first case, both institutional and associational learning are present, while in the second the added reference to the public's third point of view gives rise to yet another type of composite learning process well beyond the first. In both contexts, however, we witness the socially or discursively constructed nature of risk communication under conditions of variable contingency which is at the core of the uncertainty enveloping it.

Double contingency learning

It is characteristic of institutional learning and associational learning that both involve the erection of boundaries to the exclusion of the other. Where both are present together, we are dealing with what may be called 'collective learning in situations of double contingency' (Trenz and Eder 2004, 12).[18] In such double contingency learning, the parties have a self-referential orientation and are willing and able only to adjust their preferences, accommodate themselves to a certain degree or accept integration up to a certain point under the conditions arising from a confrontation with the other. Given the difference between them yet their sharing of a situation facing each other, institutional and associational learning processes represent what can be regarded as two distinct 'logics of collective action' (Offe and Wiesenthal 1980) which are rather difficult to reconcile. While they could co-evolve, these two logics often do come to stand hard by one another. This involves contestation, resistance and conflict between, for instance, economic and/or political agents, on the one hand, and social movement actors, on the other. It is often unproductive, such as the stalemate between the biotech industry and the genetic rights movement, or it could even be destructive, as in the case of the demise of Monsanto or of the environmental opposition in France. At times, however, it is also productive, as Lewis Coser (1956) argued many years ago with reference to Simmel regarding the integrative function of conflict.[19] The latter is the most positive interpretation of double contingency learning. Through contestation and conflict the parties learn to accommodate to one another and to allow a low degree of coordination or even integration.

This kind of learning is obviously central to risk communication in so far as it involves a process of social or discursive construction which is impelled forward by the competition, contestation and conflict of opposing forces. It marks out the field of experiences familiar to those involved in risk communication. Many of the typical problems, obstacles and frustrations go back to the limits imposed on the relation between institutional and civic risk communication. Often the agents are not merely unable, but in fact obstinately unwilling to generalize their own peculiar frames, cultural models and cognitive formats. Predictably, they thus patently fail to anticipate the resonances elicited by their communications. Consequently, they fall in the trap of one or other of a number of strategies blocking learning (Miller 1986, 2002): defensively avoiding disagreement; an authoritarian insistence on a predefined consensus; ideologically insisting on an endemic and insurmountable antagonism; and regressively rejecting arguments by disqualifying the people who advance them.

Negative phenomena such as paternalism, scientism, technocracy and public relations spin, on the one side, and idealization, fundamentalism, over-dramatization and noise, on the other, are well-known manifestations of blocked learning in the risk communication field.[20] Once these blockages are removed or overcome, however, the way is open toward a more demanding, indeed less probable, yet much more promising learning process making possible more efficacious and legitimate risk communication.

Triple contingency learning

Societal learning, finally, like the previously discussed composite learning process, presupposes a discursively mediated public sphere in the sense of an intersubjective experiential context characterized by a high level of both structural openness and symbolic complexity. But it is distinct from double contingency learning in that it results from a confrontation and interrelation of the participants before an attentive public. Societal learning is thus a form of discursive learning in which both institutional actors and civil society actors take each other into account via a reference to the observing, evaluating, judging, commenting and thus monitoring public. It is only by passing through the coordinated system of standard setting and 'scorekeeping' (Brandom 2001, 180) perspectives that the participants are at all able to arrive at an understanding of what they have in common and, especially, of what they disagree about. It is this configuration of communicative relations that I have elsewhere called 'triple contingency' (Strydom 1999c) in distinction to the classical sociological concept of double contingency. And by relating learning to it, Trenz and Eder were able to take it yet a step further, conceiving of 'societal learning in situations of triple contingency' or 'collective learning by triple contingency' (2004, 12).

Triple contingency learning represents the collective learning mechanism which brings us closest to the communicative self-constitution and self-organization of society.[21] The threefold set of contingently related communicative relations increases and intensifies public communication and resonance. It draws more participants and observers into the situation and renders cultural models reflexive, autonomous and discursively available, thus activating a plethora of cultural resources for self-reflection, self-problematization, translation, connectivity and the transformation of self-understanding. And it compels both participants and observers to resonate by making unequivocal options and communicating those options effectively, while either accommodating themselves to the public or anticipating its response. It harbours the 'cosmopolitan' promise of a transformative moment in which a 'creative combination of different forces' (Delanty 2006, 38; also Beck 2005) occurs. The outcome of the multi-levelled process of risk communication, which depends on social construction guided by such discursive learning, is a collective achievement.[22] Risk communication is thus a cooperative collective learning process in which all those involved are required to reach beyond their particular perspectives so as to link up with something larger than themselves and sharable with others.[23]

Such learning is made possible by the emergence of cognitive forms, discourses, public spheres and a public possessing the epistemic authority of observing, evaluating, judging and forming opinion. In turn, public communication – in this case, risk communication – provides the medium for the spatial, temporal and structural expansion of the public sphere as an intersubjective context of learning. In proportion as this context is expanded together with its constitutive public, an improvement in the conditions of both participation and monitoring results. Such improvement facilitates, yet also makes more complex, the achievement of collectively constructed definitions and decisions as well as the enhancement of the conditions of discourse – or 'condensation' as well as 'confirmation', to use Brandom's (2001) terminology. Risk communication in the proper sense of collective learning contributes to both these achievements. Under contemporary conditions, it is crucial not only to the solution of collective problems, but also to the bringing into being of a new world.

The normative problem of risk communication

The normative problem connected with risk communication relates to the fact that the latter is the practice of a communication community. The normative aspect concerns the collective evaluative self-definition which a risk community constructs in the guise of a diagnostic interpretation of its common situation, the risk problem it faces, and possible ways of dealing with it. The self-definition expresses the members' agreement or acceptance of that evaluation – i.e., agreement or acceptance not in the conventional sense of uniformity of belief and evaluation, but in the post-conventional sense of allowing differences. In turn, the degree of agreement or acceptance gives an indication of the legitimacy of risk communication.

Now, considering the multi-levelled nature of risk communication under complex contemporary conditions, together with the different learning processes and hence rationalities at play in it, it should be readily apparent that there is no one single normative code that could secure agreement and legitimacy without difficulty. Those involved in risk communication distinguish themselves from one another by their distinct engagements with the world, each of which is guided and given direction by a particular conception of the good. Besides the vast difference between institutional and civic risk communication, institutional risk communication alone harbours a variety of conceptions of the good or normative codes, such as those of science, politics, economics, law and so forth. For example, science pursues the truth, but even within science itself there are different and potentially conflicting conceptions of the good, such as microbiology led by rDNA knowledge and ecology by environmental knowledge (Schomberg 1992). And politics or public policy, typically employing a rational decision-making model, seeks to attain the common good. The civic risk communication of associations, NGOs and social movements is likewise guided and directed by the common good, but, by contrast with politics' top-down version, rather a bottom-up one to which moral concern is central.

Due to their being partial, none of these conceptions of the good or normative codes is able to resolve the normative problem entailed by risk communication. Against the background of existing cleavages and class structures, the prevalence of tension, competition, contestation and conflict surrounding communication in the risk society bears this out. Endless scientific controversies, irresolvable disagreements among experts, and contested policy options, political decisions and industrial programmes, all demonstrate that the adoption of a partial normative code is fraught with dangers and is therefore unacceptable up to the point of lacking legitimacy. By accepting the truth as expressed by a particular scientific discipline (e.g., microbiology in the case of the release of genetically modified organisms to the exclusion of ecology), we might well diminish our ability to anticipate a future catastrophe or we might even open ourselves to one. However rational, political decisions based on scientific knowledge, while ignoring scientific uncertainty or allowing experts simultaneously to act as advocates, would have a comparable result. But settling for the opposite socially rational arguments of associations, NGOs or social movements alone would also place us in a vulnerable position. We could unwittingly render ourselves vulnerable to collective fictions or illusions which bear little or no relation to reality.

It has become a familiar strategy to criticize the idealism of moral points of view deriving from moral philosophers and articulated by associations, NGOs and social movements. In so far as such agents' implied normative codes are partial and, even more so, if they are reified into unrealizable absolutist or fundamentalist claims, it would indeed be proper to reject them as idealistic. But it is no less idealistic to think that the normative code of science or of rational political decision-making is going to resolve the normative problem raised by risk communication. None of these normative codes by itself is capable of mobilizing sufficient support to establish a legitimate self-definition for a risk community. Such a structural deadlock can be transcended only by fluidizing the normative problem communicatively. The tendency of the past decade and a half for public administration to enter into processes of acting out and negotiating with social groups is a response to this requirement. Many see the solution in participation – that is, admitting all those affected to take part in a process of social exchange or mediation of social consensus. That it more often than not eventuates in sub-rational zero-sum games of power and interest conflict, however, suggests that the quality of the outcome of mere participation leaves much to be desired. In situations like this, uncertainty is left to proliferate in an uncontrolled manner.

A major source of uncertainty in the context of risk controversies is the lack of an articulated form staking out who should participate in what way in joint or collective decision-making processes. A promising way of dealing with this deficiency is a discursive one – i.e., discursivization through the incorporation of partial rationalities and their associated normative codes in a critical discourse in which situational definitions and norms are jointly or collectively constructed. Discursivization is indeed necessary, but it is by no means sufficient. A further step towards the proceduralization of such discourse is required (e.g., Apel 1988; Habermas 1996; Dryzek 1999; Eder 2000b). Over and above institutional and associational learning processes and their sporadic conjunction in situations of double contingency, this would encourage the emergence of more systematic discursive learning processes in which all those involved (both ego and alter) have a rightful place and relate to one another through a reference to the third point of view (the Other). In such a situation of triple contingency, all the normative codes are in a justified[24] way incorporated into a larger moral-ethical framework which, in turn, defines and structures the situation in terms of the public interest. Without the reference to the public interest, the participants cannot develop a shared knowledge of their differences, and hence a critical discourse in which the participants collectively learn through the mutual exposure of biases, distortions, half-truths, illusions and rationalizations cannot come about.[25]

It should be noted that the discursively relevant public normative criterion, the public interest,[26] is not simply equivalent to the abstract moral point of view of the moral philosopher, but one emerging from and brought to bear on the particular situation.[27] The alternative to identifying with the normative code of a preferred or favoured agent (e.g., science, the state or the environmental movement) is not to adopt an absolute or abstract moral ideal indicating what ought to be and to hold it up so that participants can see how far short they are falling. It is much rather a matter of mobilizing and scrutinizing all the normative codes – those of ego, alter and the Other – for the purpose of locating starting points for a new constructive learning process which could possibly move us closer to an intact and inclusive communication community.

Even if in this way an evaluative self-definition is attained which is supported by a majority of the risk community's members and, hence, enjoys a high degree of legitimacy, the costs – both rational and moral costs – must be reflected on, acknowledged and justified. Since better grounds or reasons[28] than those underpinning the justification could conceivably later emerge, openness and a possibility of revision allowing a reconfiguring or minimizing of such costs must be retained. Although uncertainty can thus never be eliminated, we do have means to deal with it. Risk communication in the sense of a cooperative collective learning process, as it turns out, is one of the most eminent among the available contemporary means for managing uncertainty. And to the degree that such uncertainty management proves successful and meaningful, risk communication simultaneously helps to bring into being a world possessing its own broadly accepted evaluative self-definition.

Acknowledgements

I wish to put on record my gratitude towards Ragnar Löfstedt and Åsa Boholm as well as Vicki Johansson for their consideration, cordiality and support.

Notes

1. Douglas and Wildavsky 1982 made an important theoretical contribution to the conceptualisation of social and institutional differences, but, having done so from an unhistorical structuralist Durkheimian perspective, they neglected the crucial pragmatic-communicative dimension (e.g., Apel 1988; Habermas 1996; Boltanski and Thévenot 1991; Eder 1996; Strydom 2002, 2005, 2006a). In addition, also the differentiation of modern society into function systems (Luhmann 1995) has to be taken into account.
2. Habermas (1979, 41–2) identifies the 'double structure of speech', embracing the 'level of intersubjectivity' and the 'propositional content', to which corresponds Sperber and Wilson's (1986) conceptual pair of 'communicative intention' and 'informative intention'. Thévenot (1998, 2001, 2007) distinguishes different modes of engagement with the world and corresponding cognitive formats.
3. Different authors focus on distinct aspect of resonance – e.g., Gamson 1988: cultural resonance; Luhmann 1989: system resonance; Habermas 1996: civil society resonance; Eder 2000a: institutional resonance and extra-institutional resonance. An overview leading to a first attempt at systematization is offered in Strydom 2003.
4. Some relevant cognitivist literature would include: Cicourel 1990; Conein 2005; D'Andrade 1995; Eder 1996; Saferstein 1998, 2007; Strydom 2002, 2006a; Thévenot 1998, 2001, 2007.

5. See, e.g., Sinha 1988 for a discussion of the distinction between representation and signification.
6. Drawing on phenomenological and non-Western ideas, but obviously lacking an adequate knowledge of the social sciences, Varela, Thompson, and Rosch 1991 interestingly attempt to introduce the perspective of collective world creation or 'enactivism' into the cognitive sciences.
7. Elsewhere I proposed to consider the currently characteristic cognitive order and organization of society in terms of the cognitive scheme of safety and danger (Strydom 2002, 77–8).
8. On two occasions, Habermas critically reflected on this concept which he originally introduced in 1962: a conference held upon the publication of *The Structural Transformation of the Public Sphere* (1989) in English and the preparation of a new German edition (1990). Shortly after, however, he returned to the topic to update it more thoroughly (1996).
9. The following analysis builds on Habermas 1989, 1990, 1996; Trenz and Eder 2004; and Eder 2006.
10. To allow for the dynamic circulation of power, Habermas (1996) draws a basic distinction between two states of the public sphere. While a public sphere at rest is characterized by a normal or routine mode of problem-solving, an activated public sphere proceeds by an extraordinary mode of posing and solving problems, fuelled by heightened public attention, crisis consciousness, and an intensified search for solutions. It should be added that it is the latter state in which the counterfactual force of communicative reason enters actual social relations which allows world creation. Risk is an issue which typically gives rise to crisis consciousness and activates the public sphere, the implication being that risk communication is a lightening rod directing latent counterfactual considerations into everyday communication.
11. On the institutionalization of discourse, and hence self-legislation, and the role of communicative reason in such a configuration, see Brunkhorst 1993.
12. An awareness of the importance of the third point of view or third party reference goes back, among others, to Adam Smith's 'sympathy' or 'impartial spectator' (1982, I.i.1, I.i.5.8), Kant's 'sensus communis' (1972, § 40), Mead's 'taking the attitudes of others toward oneself' leading to the 'generalised other' (1974, 90), Piaget's 'placing oneself in reciprocal relation' leading to 'ideal norms at the back of all rules' (1983, 383, 385) and Habermas' 'public…constitutive for…the public sphere' (1996, 364) as well as 'the moral point of view' (1999, 84), and it is shared by various contemporary authors, including e.g., Miller 1986, 1992; Giesen 1991; Boltanski and Thévenot 1991; Chazel 1997; Thévenot 2001; and Eder et al. 2002. The elemental insight preserved in this tradition suffers from a lack of clarity, however, due to its implicit or explicit attachment to the traditional social theoretical concept of 'double contingency' and its interpretation from one-sided or cross-cutting objectivist, conventionalist and cognitivist perspectives. For some analysis, see Strydom 1999c.
13. Philosophically, we are encouraged to search for what is at stake in communication in the rationality structures transported in it (e.g., Apel and Kettner 1996; Wingert and Günther 2001). Without by any means denying the relevance of this, I propose more social scientifically to focus on the closely related cognitive structures and learning processes. As the later discussion of normative criteria suggests, however, it is impossible nevertheless to circumvent the question of rationality.
14. A cognitivist theory according to which learning is conceived as the self-organizing transformation of cognitive structures of different levels and scope which, in favourable cases, brings forth something structurally new is to be found in, e.g., Miller 1986, 2002; Eder 1996, 1999a; and Strydom 2000. At the intersection between cognitive structures and social or communicative-discursive processes, a rule-governed transformation takes place

which leads from a previous state of the world to a newly constructed one, involving the reorganization of cognitive classifications of reality and thus the reframing of the world. A good example is Eder's (1996) analysis of the communication and framing or the making of the environment into an issue and the consequent institutionalization of the environmental movement. In the course of such a learning process, the cognitive structures of social actors – e.g., social movement spokespersons, business executives, politicians – undergo change at the same time as the transformation of organizational – e.g., social movement, corporate, governmental – frames and, at a still higher level, cultural models of very wide scope – e.g., ecology taking the place of industrialism. The outcome of the process was thus the establishment of a new cultural model, a new kind of politics and a new kind of institutional order – in the form, respectively, of ecology, the politics of nature, and post-corporatism. Simultaneously, of course, all these levels imply a transformed self-understanding on the part of the actors involved.

15. I do so by drawing from and elaborating on Miller 1986, 1992, 2002 as well as Trenz and Eder 2004, who take up the concept of 'triple contingency' I introduced elsewhere (Strydom 1999c). See also Eder et al.'s (2002, 104–8) theory of 'communicative situations'.
16. This is the theme of the wide-ranging literature on collective identity formation and the constitution of collective action in social movements. On the problem of coordination and synthesis as such, see e.g., Eder 1993; Miller 2002; and Strydom 2006b. Also of relevance is Engel's (1997) distinction among collective representations, collective beliefs and collective judgements.
17. In his original conception, Habermas stressed this type, as when he focused on 'social movements as learning processes' (1979, 125). There he also envisaged that what was thus learned could be transposed to the level of society.
18. Parsons formalized this classical idea by introducing the 'proposition of double contingency', circumscribing it as follows: 'There is a *double contingency* inherent in interaction. On the one hand, ego's gratifications are contingent on his selection among available alternatives. But in turn, alter's reaction will be contingent on ego's selection and will result from a complementary selection on alter's part' (Parsons and Shils 1951, 16). Luhmann (1995) and even Habermas (1996) adopt the concept as being central to sociology. Strydom (1999c, 2001) offers a detailed overview and critique.
19. For similar arguments, see also Miller 1992 and Eder 1999b.
20. Blockages in learning go back largely to power. On the basis of bringing together the complementary approaches of Habermas and Foucault, I have elsewhere analysed different ways in which power enters into discourse (Strydom 2000, particularly 50–52). See also Eder 2006.
21. The core insight encapsulated here is the one from which both Apel and Habermas single-mindedly proceed throughout their writings, yet inconsistencies are apparent. For critical analyses, see e.g., Wellmer 1986; Brunkhorst 1993; Miller 2002; Strydom 2001, 2006b.
22. This collective achievement is a socio-cognitive construct which brings together convincing, credible, justified and therefore acceptable claims with elements of qualified authority and normatively recognized power in a way that inspires trust and legitimacy (Honneth 1991; Boltanski and Thévenot 1991; Habermas 1996; Engel 1997; Strydom 2000, 2002).
23. Turner 1994 has severely criticized the widely accepted social scientific assumption of sharing, but what is meant by 'sharable' here, by contrast, Piaget captured years ago: rather than subjecting 'each personality to rules that have a common content, it simply obliges individuals to place themselves in reciprocal relation with each other without letting the laws of perspective resultant upon this reciprocity destroy their individual points of view' (1983, 383).
24. The problem of justification is central to Apel and Habermas' discourse ethics and related work, but it has been taken up also into the new French 'pragmatic sociology' (Boltanski and Thévenot 1991).

25. See Strydom 2000, 2002 for the related concept of socio-cognitive critique.
26. On the common good or public interest, see e.g., Boltanski and Thévenot 1991 and Offe 2001. In a penetrating analysis of the contemporary renaissance of the concept, Offe submits that the field of public interest evaluation, judgement and argumentation lies between the formal sphere of legally sanctioned obligation and the informal sphere of legally allowed freedom, with the implication that it represents a third kind of criterion or norm. When he asks after the social reference of the concept of public interest, however, he does not raise the possibility of the third point of view beyond organized corporate actors – who, according to him, are tempted to misuse it for authoritarian-paternalistic purposes – and unorganized associations – whose micro-politics, he thinks, appeals to it without contradiction.
27. The key insight here is Habermas's insistence right up to his latest publication on 'the *factual* role of performatively presupposed *counterfactual* assumptions' (2005, 30, my translation). See e.g., Eder's 2006 defence of it against criticisms. One rationale for the introduction the concept of 'triple contingency' (Strydom 1999c) was to attempt to capture the relation between the counterfactual and the factual in a way amenable to empirical research. In the present case, the public interest is a counterfactual assumption which takes actual effect in social life through the orientations and actions of the participants and, in particular, through the impact its embodiment in the public's reaction to the participants has on the conditions of their actions.
28. On grounds or reasons, see Kettner 1996.

References

Apel, K.-O. 1988. *Diskurs und Verantwortung*. Frankfurt: Suhrkamp.

Apel, K.-O., and M. Kettner, eds. 1992. *Zur Anwendung der Diskursethik in Politik, Recht und Wissenschaft*. Frankfurt: Suhrkamp.

———, eds. 1996. *Die eine Vernunft und die vielen Rationalitäten*. Frankfurt: Suhrkamp.

Beck, U. 1992. *Risk society: Towards a new modernity*. London: Sage.

Beck, U., and E. Grande. 2005. *Das kosmopolitische Europa*. Frankfurt: Suhrkamp.

Boltanski, L., and L. Thévenot. 1991. *De la justification: Les economies de la grandeur*. Paris: Gallimard.

Brandom, R.B. 2001. *Making it explicit: Reasoning, representing, and discursive commitment*. Cambridge, MA: Harvard University Press.

Brunkhorst, H. 1993. Zur Dialektik von realer und idealer Kommunikationsgemeinschaft. In *Transzendalpragmatik*, ed. A. Dorschel, M. Kettner, W. Kuhlmann, and M. Niquet, 342–58. Frankfurt: Suhrkamp.

Chazel, F. 1997. Les ajustements cognitifs dans les mobilisations collectives. In *Cognition et sciences sociales*, ed. R. Boudon, A. Bouvier, and F. Chazel, 193–206. Paris: Presses Universitaires de France.

Cicourel, A.V. 1990. The integration of distributed knowledge in collaborative medical diagnosis. In *Intellectual teamwork*, ed. J. Galegher, R.E. Kraut, and C. Egido. Hillsdale, NJ: Erlbaum.

Clark, A. 1998. Embodied, situated, and distributed cognition. In *A companion to cognitive science*, ed. W. Bechtel and G. Graham, 506–17. Oxford: Blackwell.

Conein, B. 2005. *Les sens sociaux: Trois essays de sociologie cognitive*. Paris: Economica.

Coser, L.A. 1956. *the functions of social conflict*. Glencoe, NJ: Free Press.

D'Andrade, R. 1995. *The development of cognitive anthropology*. Cambridge: Cambridge University Press.

Delanty, G. 2006. The cosmopolitan imagination: Critical cosmopolitanism and social theory. *The British Journal of Sociology* 57(1): 25–47.

Douglas, M., and A. Wildavsky. 1982. *Risk and Culture: An Essay on the Selection of Technical and Environmental Dangers*. Berkeley, CA: University of California Press.

Dryzek, J.S. 1999. *Discursive Democracy: Politics, Policy, and Political Science*. Cambridge: Cambridge University Press.

Eder, K. 1993. *The New Politics of Class: Social Movements and Cultural Dynamics in Advanced Societies*. London: Sage.

———. 1999a. Societies learn and yet the world is hard to change. *European Journal of Social Theory* 2(2): 195–215.

———. 1999b. Integration durch Kultur? Das Paradox der Suche nach einer europäischen Identität. In *Kultur, Identität, Europa*, ed. R. Viehoff and R.T. Segers, 147–179. Frankfurt: Suhrkamp.

———. 2000a. Zur Transformation nationalstaatlicher Öffentlichkeit in Europa: von der Sprachgemeinschaft zur issuespezifischen Kommunikationsgemeinschaft. *Berliner Journal für Soziologie* 10, no. 2: 167–84.

———. 2000b. Taming risks through dialogues: The rationality and functionality of discursive institutions in risk society. In *Risk in the modern age: Social theory, science and environmental decision-making*, ed. M.J. Cohen, 225–48. Houndsmills: Palgrave.

———. 2006. Making sense of the public sphere. In *Handbook of contemporary European social theory*, ed. G. Delanty, 333–46. London: Routledge.

Eder, K., B. Giesen, O. Schmidtke, and D. Tambini. 2002. *Collective identities in action*. Aldershot: Ashgate.

Gamson, W.A. 1988. A constructionist approach to mass media and public opinion. *Symbolic Interaction* 11, no. 2: 161–74.

Engel, P. 1997. Croyances collectives et acceptations collectives. In *Cognition et sciences sociales*, ed. R. Boudon, A. Bouvier, and F. Chazel, 155–73. Paris: Presses Universitaires de France.

Giesen, B. 1991. *Die Entdinglichung des Sozialen*. Frankfurt: Suhrkamp.

Habermas, J. 1979. *Communication and the evolution of society*. London: Heinemann.

———. 1989. *The structural transformation of the public sphere*. Cambridge: Polity.

———. 1990. *Strukturwandel der Öffentlichkeit*. Frankfurt: Suhrkamp.

———. 1996. *Between facts and norms: Contributions to a discourse theory of law and democracy*. Cambridge: Polity.

———. 1999. *The inclusion of the other: Studies in political theory*. Cambridge: Polity.

———. 2005. *Zwischen Naturalismus und Religion: Philosophische Aufsätze*. Frankfurt: Suhrkamp.

Honneth, A. 1991. *Critique of power*. Cambridge, MA: MIT.

Kant, I. 1972. *Critique of judgement*. New York: Hafner.

Kettner, M. 1992. Über einige Dilemmata angewandter Ethik. In *Zur Anwendung der Diskursethik in Politik, Recht und Wissenschaft*, ed. K.-O. Apel and M. Kettner, 9–28. Frankfurt: Suhrkamp.

———. 1996. Gute Gründe: Thesen zur diskursiven Vernunft. In *Die eine Vernunft und die vielen Rationalitäten*, ed. K.-O. Apel and M. Kettner, 424–64. Frankfurt: Suhrkamp.

Luhmann, N. 1989. *Ecological communication*. Cambridge: Polity.

———. 1995. *Social systems*. Stanford, CA: Stanford University Press.

Mead, G.H. 1974. *Mind, self and society*. Chicago, IL: University of Chicago Press.

Miller, M. 1986. *Kollektive Lernprozesse: Studien zur Grundlegung einer soziologischen Lerntheorie*. Frankfurt: Suhrkamp.

———. 1992. Discourse and morality: Two case studies of social conflicts in a segmentary and a functionally differentiated society. *Archives Européennes de Sociologie* 33, no. 1: 3–38.

———. 2002. Some theoretical aspects of systemic learning. *Sozialer Sinn* 3: 379–421.

Offe, C. 2001. Wessen Wohl ist das Gemeinwohl?. In *Die Öffentlichkeit der Vernunft und die Vernunft der Öffentlichkeit*, ed. L. Wingert and K. Günther, 459–88. Frankfurt: Suhrkamp.

Offe, C., and H. Wiesenthal. 1980. Two logics of collective action. In *Political power and social theory*, ed. M. Zeitlin, 112–28. Greenwich, CT: JAI Press.

Parsons, T., and E.A. Shils, eds. 1951. *Towards a general theory of action*. Cambridge, MA: Harvard University Press.

Piaget, J. 1983. *The moral judgement of the child*. Harmondsworth: Penguin.

Saferstein, B. 1998. Ethnomethodology. In *A companion to cognitive science*, ed. W. Bechtel and G. Graham, 391–401. Oxford: Blackwell.

Saferstein, B. 2007. Process narratives, grey boxes, and discourse frameworks: Cognition, interaction, and constraint in understanding genetics and medicine. *European Journal of Social Theory* 10, no. 3: 423–46.

Schomberg, R. von. 1992. Argumentation im Kontext wissenschaftlicher Kontroversen. In *Zur Anwendung der Diskursethik in Politik, Recht und Wissenschaft*, ed. K.-O. Apel and M. Kettner, 260–77. Frankfurt: Suhrkamp.

Sinha, C. 1988. *Language and representation: A socio-naturalistic approach to human development*. Hemel Hempstead: Harvester Wheatsheaf.

Smith, A. 1982. *The theory of moral sentiments*. Indianapolis: Liberty Fund.

Sperber, D., and D. Wilson. 1986. *Relevance: Communication and cognition*. Oxford: Blackwell.

Strydom, P. 1999a. The civilisation of the gene: Biotechnological risk framed in the responsibility discourse. In *Nature, risk and responsibility*, ed. P. O'Mahony, 21–36. London: Macmillan.

———. 1999b. The challenge of responsibility for sociology. *Current Sociology* 74, no. 3: 65–82.

———. 1999c. Triple contingency: The theoretical problem of the public in communication societies. *Philosophy and Social Criticism* 25, no. 2: 1–25.

———. 2000. *Discourse and knowledge*. Liverpool: Liverpool University Press.

———. 2001. The problem of triple contingency in Habermas. *Sociological Theory* 19, no. 2: 165–86.

———. 2002. *Risk, environment and society*. Buckingham: Open University Press.

———. 2003. Resonance: Triggering a dormant dimension of the public sphere. In *European social theory: Sources and challenges*, ed. V. Kultygin and V. Zhukov, 83–99. Moscow: European Sociological Association.

———. 2005. Risk. In *Science, technology, and society: An encyclopedia*, ed. S. Restivo, 474–6. New York: Oxford University Press.

———. 2006a. Contemporary European cognitive social theory. In *Handbook of contemporary European social theory*, ed. G. Delanty, 218–29. London: Routledge.

———. 2006b. Intersubjectivity – interactionist or discursive? Reflections on Habermas' critique of Brandom. *Philosophy and Social Criticism* 32, no. 2: 155–72.

Thévenot, L. 1998. Pragmatiques de la connaissance. In *Sociologie et connaissance: Nouvelles approches cognitives*, ed. A. Borzeix, A. Bouvier, and P. Pharo, 101–39. Paris: CNRS.

———. 2001. Pragmatic regimes governing the engagement with the world. In *The practice turn in contemporary theory*, ed. T.R. Schatzki, K. Knorr Cetina, and E. von Savigny, 56–73. London: Routledge.

———. 2007. The plurality of cognitive formats and engagements: Moving between the familiar and the public. *European Journal of Social Theory* 10, no. 3, 408–22.

Trenz, H.-J., and K. Eder. 2004. The democratizing dynamics of a European public sphere: Towards a theory of democratic functionalism. *European Journal of Social Theory* 7, no. 1: 5–25.

Turner, S. 1994. *The social theory of practices: Tradition, tacit knowledge and presuppositions*. Cambridge: Polity.

Varela, F.J., E. Thompson, and E. Rosch, eds. 1991. *The embodied mind: Cognitive science and human experience*. Cambridge, MA: MIT.

Wellmer, A. 1986. *Ethik und Dialog: Elemente des moralischen Urteils bei Kant und in der Diskursethik*. Frankfurt: Suhrkamp.

Wertsch, J.V. 1998. Mediated action. In *A companion to cognitive science*, ed. W. Bechtel and G. Graham, 518–25. Oxford: Blackwell.

Wingert, L., and K. Günther, eds. *Die Öffentlichkeit der Vernunft und die Vernunft der Öffentlichkeit: Festschrift für Jürgen Habermas*. Frankfurt: Suhrkamp.

Creating shared realities through communication: exploring the agenda-building role of the media and its sources in the *E. coli* contamination of a Canadian public drinking water supply

S. Michelle Driedger

Department of Community Health Sciences, University of Manitoba, Winnipeg, MB, USA

The media's role as an *agenda-setter* has been documented in the literature, however less attention has been given to the role of the media as an *agenda-builder*. The role of news agencies in constructing and shaping news stories serve to mediate risk messages to the public by virtue of what elements of a story they report, the types of sources they use in reporting the different 'sides', and how they package a story. The media are not the only ones who seek to set and build agendas. The sources, upon which the news media rely, can and do similarly engage in agenda-building activities. They do so by appropriating the very 'media-ted' process to use the mass media as a delivery vehicle for their own communication claims through their own packaged 'take' on the story. Drawing on a national print and televised media analysis of the Walkerton, Ontario, Canada drinking water contamination event, this study explores the agenda-building role of the media and of Walkerton residents and citizen group members (Concerned Walkerton Citizens) quoted in media stories. The implications of this for risk communication will be discussed.

Introduction

Since McCombs and Shaw's (1972) introduction of the agenda-setting framework there have been a plethora of studies examining the role of media in shaping public discourse (see Rogers and Dearing 1988; McCombs and Shaw 1993). The premise – that the public learns from the media what the important issues are – is highly intuitive. As Lippmann (1922) was among the first to contend, the majority of public perceptions of certain aspects of a civil society (political environments, social and cultural issues, global issues, etc.) are not based on direct observation and experience, but rather formed second-hand through mass media reports. A key reformulation in the agenda-setting perspective, built on earlier articulations by scholars like Cohen (1963), was a movement away from the idea that the media is persuasive in directing the public about 'what to think', to exploring the role of the media in directing what the public 'thinks about'. This shift is important in that it redirects attention away from content to salience. This has prompted some to add more conceptual depth to the agenda-setting framework to examine the media's role in agenda-building. Agenda-building is concerned with how problems or topics *become* issues in a public domain (Lang and Lang 1981).

Some clarity of terms is required. By 'public' – I refer to general citizens who do not directly play a role in setting government policy or laws. In and of itself, such a definition can be problematic in that multiple actors/groups are seen as part of the 'public' – general voters/citizens, special interest groups, mass media. Each may play a role in influencing government policy or laws – voters in terms of who they elect to govern, interest groups that may seek a preferred definition of a topic of concern and apply relevant pressure to support their position, or the mass media that may amplify or attenuate the level of concern to a given topic - but none are directly responsible for the decisions made. By 'issue' – I refer to a problematic condition that holds both public and political relevance. In this sense, my conceptualization is similar to Downs' (1972) issue-attention cycle and Newig's more theoretical extension that examines the level of public attention to a social phenomenon that may influence political action (2004). For Newig, the key distinction between a 'problem' and an 'issue' is that in a public-political action sense, the focus is less on the solutions to the problem, and more on public attention given the problem. Lastly, for this analysis, 'media' refers to the *news* media in Canadian national newspapers or televised news coverage.

The news media plays an important role in civil society by reporting on events from near and distant locations. In an environment and health risk context, when risks are novel (Anderson 1997) and experienced in a very short time frame (acute risks), media attention is typically higher compared to media attention on risks (often characterized by high scientific uncertainty) experienced over a much longer period of time (chronic risks). In this case study about the *E. coli* contamination of a public drinking well water supply in one community (Walkerton, Ontario), media attention to the event was sufficiently high to garner both political and public attention. In a Canadian context, drinking water is something that is frequently taken for granted, where it is assumed that whenever the tap is turned on, clean and microbiologically safe drinking water will flow. Historically, when there have been public complaints, the majority of public concerns have revolved around taste and odour (Jardine, Gibson, and Hrudey 1999). Occasionally there have been public concerns related to byproducts formed during the chlorine disinfection process and the associated human health effects (Driedger and Eyles 2003). The nature and length of media coverage of the Walkerton *E. coli* contamination may be a contributing factor to the shift in public discourse away from chemical concerns (Driedger and Eyles 2003), from which many taste and odour complaints emerge (Parr 2005), to bacterial concerns (Driedger 2004; Parr 2005).

This paper will explore the agenda-building role of the media in the Walkerton *E. coli* drinking water contamination by drawing on a national print and televised media analysis. To fully understand the complexities of agenda-building theory, it will be important to briefly examine some core theories of the sourcing behaviour of the news media as well as of the professional behaviour of journalists. This will be set against a backdrop of the socially constructed process of news media practices more generally before examining how a less-likely (or non-'elite') source group (i.e., the Concerned Walkerton Citizens) made efforts to appropriate the very 'media-ted' (Kellner 1995) process of the news media as a delivery vehicle for their own communication strategies to advance and keep public and political attention on the contaminated drinking water issue. Although other Walkerton residents were quoted in the media, the Concerned Walkerton Citizens (CWC) was a common source used in the construction of a media story as it represented a roughly 500-resident membership. Hence, it could be argued that the CWC was recognized as 'the voice' of Walkerton residents.

Sources and the news media

Developing shared realities through routines and practices

Although the media is not 'all powerful' as some theories about media studies have suggested (i.e., broadly categorized passive audience theories – see Tate and McConnell 2001 for an overview), the media does not exist in a social, cultural, political or economic vacuum. Media stories are created by individuals working within an institution such that the media and its people 'make choices within contexts that reflect the priorities and interests of those who own and control them' (Fleras 2003, 49). Effectively, the media, rather than create 'reality', create discourses of reality, often through a reliance on sources to help provide the media with context, important facts, as well as interpretations of different events.

The theory of primary definition is based on the idea that certain elites, based on social hierarchies, enjoy a privileged access to the media, and hence hold the ability to define topics in a preferred way. The journalists' pressure of deadlines and professional demands of objectivity leads to an over-accessing of elite sources resulting in a few sources enjoying a role as primary definer by virtue of their structural positioning in certain institutions or their representative status within society (Hall et al. 1978). Schlesinger and Tumber (1999) criticize the theory of primary definers as too essentialist in that it assumes that there is only one main definition held by such sources and that it is not sensitive to the professional behaviour and roles of the media (e.g., unofficial information flows, off the record conversations, unequal media access by 'primary' definers). Ericson and colleagues (1989) further argue that the role between the news media and sources is one of negotiation; it is problematic to assume that journalists only serve to reproduce a source's 'propaganda', or that the media is a passive actor in the process.

Donsbach (2004) forwards the idea of how journalists undergo a process of 'shared reality' (based on Hardin and Higgins 1996) as a mechanism to socially validate interpretive judgements made on a daily basis in the news business. By sharing one's experiences and beliefs with others, through communication, it becomes possible to assess how reliable or valid are those judgements. The theory of shared reality posits that 'consensually validated social roles and relationships are required for the mutual creation, monitoring, and maintenance of the individual experience of reality' (Hardin and Higgins 1996, 38). Through a process of group orientation, journalists are in a constant comparative evaluation with other journalists to determine what is newsworthy (Donsbach 2004). Social interaction on the job (Patterson and Donsbach 1996), professional norms (Tuchman 1978), routinized practices including social and professional constraints of working in the news media (e.g., relationships and expectations with sources, editors, owners of news media outlets) (Croteau and Hoynes 2003; Lorimer and Gasher 2004), as well as observing the development and use of 'commonly shared scripts' by other news media (Kerbel and Ross 1999 in Donsbach 2004) can play an important role in deciding a media's shared reality. In terms of risk reporting, it is often only those stories that serve as 'triggers' that will capture media attention: questions of blame, conspiracy stories (alleged secrets or cover-ups), there are identifiable victims, links to other high-interest stories or issues, is embroiled in conflict, has a signal value (boding future ills), many people are exposed, has a strong visual impact, and can be linked to sex or crime (Bennett 1999). Key to this process is how journalists are exposed to statements and events, how they process such information at a cognitive level, and how they then activate these beliefs through reproduction (or story writing).

The agenda-building process

Agenda-building is defined as a collective process that involves the mutual collective influence of the media, government and citizens on one another (Lang and Lang 1983). In this way, agenda-building adds a conceptual depth to the traditional agenda-setting framework in that it includes the role of the public and policy-makers in shaping this process alongside media agents. It removes some of the limitations of earlier media *content* studies to focus on how issues become *salient* in a public and political context. Despite this, with some exceptions (Mathes and Pfetsch 1991; Tanner 2004; Denham 2004), few have used the analytical power of the agenda-building perspective.

Lang and Lang (1983) outline six steps needed in the agenda-building process and are paraphrased here:

(1) An event or activity needs to be covered in the media; the very act of drawing attention to some topics can affect how much people may think or talk about it.

(2) Simply being in the news is a necessary, but insufficient, condition needed to guarantee attention. Attention at a public or political level will vary based on how that issue affects people's lives.

(3) The media gives an event or activity meaning by how it is described, defined and framed.

(4) Language, through framing, conveys meaning to a broader audience (see Gamson et al. 1992; Entman 1993; Menashe and Siegel 1998; Mander 1999).

(5) The focus or attention to an activity or event can be amplified if it can be linked to other easily identified symbols (e.g., 'truth', 'public right to know'). (In a policy context, private misfortunes need to be defined as a public problem to generate large scale attention – see Stone 1997.)

(6) Using high-profile or credible sources (spokespersons or organizations) further promotes the concerns made in communication acts about an activity or event.

In addition to these propositions above, there are three important feedback loops needed for the continuous agenda-building process. Both political figures and sources need to see themselves and their own activities mirrored in the press. Information sharing within media pools (e.g., press releases, wire services, covering the same press conferences) will foster similarities in news presentations, and hence serve to create shared discourses of reality. Last, the media and political figures rely on public opinion polls or other indicators of public response (e.g., call-ins, letters to editors) as a validation of their agenda-building activities (Lang and Lang 1983).

Whether or not a condition becomes an issue in such circumstances often depends on how it is defined/presented in the media – based on how a journalist chooses to piece together the story, editorial decisions made by what is deemed 'important', and/or how sources make use of the media-ted process for their own communication acts. Thus, in the Walkerton example, for a non-elite source like the CWC to build an agenda, they need to be recognized by the media as a credible organization in order to gain media coverage, emphasis needs to be placed on how the event touched their lives (so that it can, in turn, emotionally affect others further away), and they need to turn their private misfortune (a problem affecting Walkerton) into a public tragedy (an issue affecting all Canadians).

Background to the risk event

In May 2000, the groundwater drinking water supplies of Walkerton, a small town in southwestern Ontario, Canada (approximately 180 km from Toronto), became contaminated with *E. coli*. Due to very heavy rains for a period of five days manure from a nearby non-factory commercial farm was washed into one of the town's three wells. The backflow valve, designed to protect the town's primary well from surface water run-off, failed. The bacterial and other contaminant levels overwhelmed the amount of chlorine being added to treat the well water. Shortly thereafter, the local manager turned off the primary well to a secondary one that did not have a functioning chlorinator. For a five day period, the town was consuming unchlorinated (untreated) water. The local water managers received a laboratory result showing elevated levels of *E. coli* in the drinking water supplies, but failed to advise the area Public Health Unit. Rather local managers opted to flush the system with chlorine. Within days residents were presenting with bloody diarrhea at the hospital. Throughout this time, local managers maintained that the drinking water was safe and secure. As a precautionary measure the Public Health Unit issued a boil water advisory that remained in effect for six months after *E. coli* was confirmed in the water. May 24, 2000 marks the first media story of the event where several hundred residents were reported as ill with four residents dead. The final death toll attributed to the contamination was seven dead and more than 2300 residents ill, of a town of approximately 5000 people. Due to a series of early investigations, and considerable public pressure initiated primarily by the Concerned Walkerton Citizens, the Province of Ontario announced that a public inquiry would be held.[1] The Walkerton Inquiry produced two reports (O'Connor 2002a, 2002b), one of which assessed what happened in Walkerton, and the other provided recommendations for government action. The introduction of the *Safe Drinking Water Act* in 2002 (S.O. 2002, c 32) was one of the ways in which the provincial government responded to the recommendations of the Walkerton Inquiry. One of the regulations for this Act is the *Ontario Drinking Water Standards* (O-Reg 169/03), establishing the parameters that Ontario drinking water supplies must meet.[2]

Methods: story selection and analysis

The data sources used here are a subset of a larger database of national media stories on the Walkerton case study covering a 2.5 year period (from 24 May 2000 (the first media story on the event) to 31 December 2002). Canada has only two national newspapers[3] and two[4] national broadcast stations (where most Canadian audience members can view these broadcasts without paying for cable).The compiling of that larger database consisted of using *FACTIVA* and the *Canadian Business and Current Affairs* (CBCA) electronic searchable databases to collect and compare print media coverage from *The Globe & Mail* and *The National Post* and televised news coverage from *The National – CBC News* and *CTV National News*. The broadest keyword possible was used to search these sources: 'Walkerton'. The story selection inclusion/exclusion criteria are documented elsewhere (see Driedger 2007). A total of 1436 stories were reported in national newspapers and national news broadcasts, of which 841 stories met the inclusion criteria for analysis.

The subset used in the analysis presented here (n=84) was compiled by searching the larger database for all stories that explicitly included commentary from the Concerned Walkerton Citizens (CWC) committee, or named representatives of that committee. These stories were analyzed for source placement (at the beginning, middle, or end) in the story, and the actual content included by the reporter. This was done to see whose voices were being privileged in a news story. Regardless of what a source may say, it is ultimately the reporter and editor/producer that make the decision of what to include, exclude, and order the text in the construction and presentation of the news story mediated by the shared reality and common scripts covered by other news media reporters and editors/producers. It could be argued that using the CWC as a source at the beginning or at the end of the story carries with it greater weight. This is based on a series of assumptions. First, people reading the newspaper are likely to read at least the opening paragraphs of a story that follow an interesting headline, even if they do not finish reading the story. Similarly, people watching the news may listen to the top stories even if they do not watch the entire broadcast. Second, people who are interested in the opening paragraphs of a story (or televised minutes) may be more likely to read the closing paragraphs (or watch the full segment for that story) to see how things 'end', even if they do not read the entire story (or watch the full broadcast). Third, readers/watchers in general may be more interested in quotes by Walkerton residents (as news sources) than other sources, since they were the victims of the contamination event (i.e., it happened 'to them'); particularly if such quotes are used as the story's first or last words. Hence the placement of those statements at the beginning or end of the story (as opposed to the middle) increases the likelihood that it is read/watched.

Operationalizing what counts as a story's 'first' or 'last' words is needed. In a televised story, the first or last word is captured if a source is given air-time immediately following a reporter's introduction or just prior to the on-the-scene reporter quickly summarizing the story before turning things back to the studio newscaster (which effectively ends that segment of the news). In a print media story, the first/last word is operationalized as the first/final words of the overall story. In some instances, there may be a first/final sentence by the reporter introducing/summarizing the story. If the sentence was brief, only providing broad context to the story, the overall source statement was still coded as the story's first/last words. However, if there were two or more non-source based sentences in the first/final paragraph, then the source statements were not coded as the first/last words.

Sources are not powerless in the source-reporter relationship. By choosing their words (and hence providing a 'perfect' sound-bite), they can articulate their claims in a manner that is more likely to be included in a news story than if their language is overly cautious. Hence, a thematic analysis of the language used by members of the CWC committee was also conducted to see what properties of the drinking water contamination topic was used to guide the communication strategies of CWC members quoted in the media in their efforts to exert public and political pressure to turn the topic into an 'issue' of their own construction. Recognizing the interactive and dynamic process involved in agenda-building by multiple actors, I also draw on how other sources and the media turned the Walkerton event into a national drinking water issue through agenda-building activities based on earlier analyses of the larger database.

Results

Agenda-building by the media

Tuchman (1973) coined the term 'what-a-story' to signify a news event so dramatic and novel that all of the news organizations cover the story *en-masse*; a process that facilitates the development of discourses of shared reality. The Walkerton risk event became *the* Canadian national, provincial, and local 'what-a-story'. Reporters were busy writing stories that had all of the traditional characteristics of great story-telling, where deceit, possible cover-ups and tragedy was rampant. There were *the heroes*: the initial pediatrician who first suspected that the cause of the diarrhea in two children was from *E. coli* and contacted the local health unit; the staff at the local health unit that began its investigation into possible food-related *E. coli* contamination; the local Medical Officer of Health that ordered a boil water advisory as a precaution even before receiving laboratory confirmation there was *E. coli* in the water despite assurances from the local water managers that the water was 'okay'; and eventually the Walkerton Inquiry judge who was collecting and evaluating the evidence before passing pronouncement on 'what really happened'. There were *the suffering victims*: the seven people who died (including one child), the more than 2300 who became ill, and the residents in Walkerton and surrounding communities touched and stigmatized by the contamination event. There were the *future risk others*: Canadians in general as media stories were written to hypothesize who or where could be next, as well as writing 'general interest' stories on the safety of bottled water supplies comparative to tap water,[5] casting aspersions about the safety of all drinking water sources. There were also *the villains*: the provincial government who made a series of budgetary cutbacks resulting in the elimination of the independent water inspectors who conducted spot checks on different water systems to ensure quality and safety; and the local water managers who frequently mislabelled water sample bottles to reflect multiple sources along the distribution system even though all were collected from the same location, and who also frequently falsified data on the chlorine residual test results since they were rarely undertaken.

In terms of the event's newsworthiness and important media triggers, playing the 'blame-game' was a recurrent story theme in media stories: opposition parties blaming the government; the government blaming the previous government; the provincial government blaming the local government and local managers; and the local government blaming provincial government cutbacks (Driedger 2007; Mistry 2006). However, local residents were not quick to blame. Despite inflammatory headlines suggesting otherwise, residents did not blame the local water managers for the tragedy. The local managers, two brothers (one the general manager, the other the foreman), were long-standing residents of Walkerton: they attended church, they participated in community events (e.g., volunteer firefighting), and they were friends to many in the community (Parr 2005). Many residents quoted in news stories blamed the tragedy on a systemic breakdown rather than on individuals. It was not until the Walkerton Inquiry when the local managers testified to falsifying data, and (in the case of one brother) frequently being drunk on the job, that the community tide began to shift (Parr 2005; Mistry and Driedger 2005). While there was some shift in sentiment by residents, some, including the chair of the CWC, were not prepared to abandon a longtime friend: 'I don't think there was any malicious intent there at all. Most people in town are that way, too. They know he [the local manager] made a mistake, but he didn't intentionally go out to harm anybody' (Bourette, *Globe & Mail (G&M)*, 18 December 2000, A3).

Throughout the news coverage of the event, representatives of the CWC, representing roughly 500 residents were more frequently solicited for comment on the unfolding of the what-a-story than were Walkerton residents in general (Mistry 2006). Figure 1 identifies the frequency with which the CWC was consulted as a news source, and how their voices were used based on where they were placed in a given story. In the vast majority the CWC as a news source in the middle half or the end of the news story. Upon closer examination the placement of a CWC quote in the middle of the news story was primarily used to provide comment (e.g., reaction to inquiry testimony) on the first part of the story before switching to a new story element. While perhaps not the explicit motivation of reporters, placing CWC source comments at the end of a story often served to build the CWC agenda.

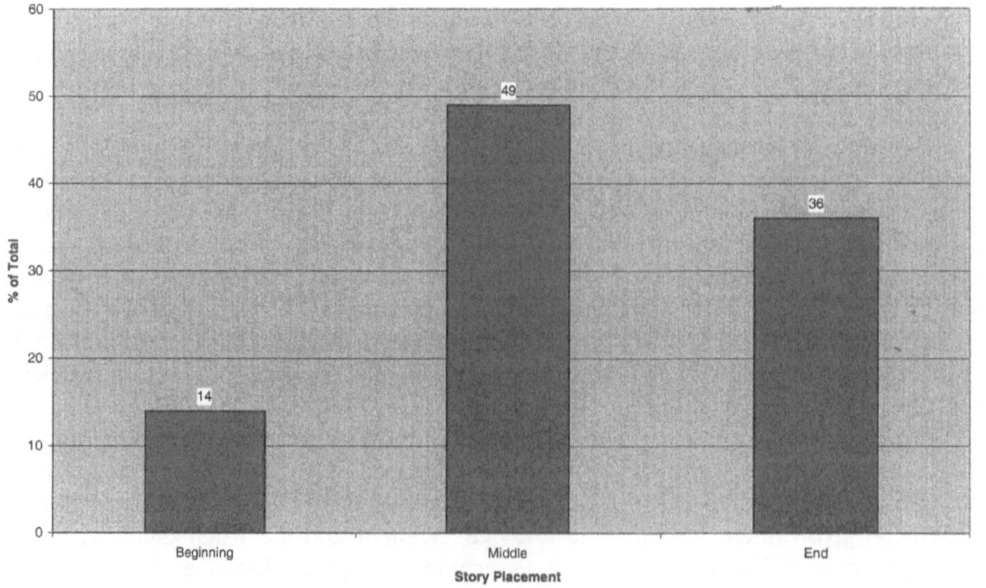

Figure 1. Placement of CWC as news source in story.

Agenda-building by the Concerned Walkerton Citizens

Through a much broader analysis of all media stories (n=841 stories), media stories frequently commented on the number of people who died or who were made ill, issues of blame, covering the various investigations and testimony given during the public inquiry, trying to piece together the sequence of events leading up to the contamination, and issues of public (mis)trust (Driedger 2007; Mistry 2006). However, while there were similarities in how the media and the CWC built their agenda around the contamination event, the matters that concerned the CWC the most had a slightly different focus. The CWC appeared to be primarily interested in the following: *finding answers* to how such a tragedy occurred (namely being the first group to demand a public inquiry, through various investigations into the matter, and trying to ensure that residents themselves had a voice in the process); identifying instances where responsible officials (primarily at the provincial level) *failed to act* in the public's best interest (where there was a less explicit emphasis on 'blame' and a greater emphasis on trying to prevent future instances); the pursuit of a more equitable *provincial compensation package* for the suffering of Walkerton residents directly affected by the contamination; continued appeals for the fact that while others were seeking blame, Walkerton residents were *still without clean water* (Walkerton as a private problem); and much broader statements calling for the need for *better national and provincial water management strategies* so that another 'Walkerton' would not happen again (Walkerton as a public issue).

Table 1 outlines some of the more powerful agenda-building quotes from the CWC and used in the media construction of the event representing the 'last words' of the story. While being quoted in the top part of the story may hold a greater possibility of being read/watched than being quoted at the end, in this analysis, the CWC was only used as the 'first words' in 11% of the stories. By contrast, the news media quoted the CWC as a source with the 'last words' in 32% of the stories.

What is notable in the extracts in Table 1 is how quickly articulate the CWC becomes (recall that the first news story was 24 May 2000). The lead members of the CWC do not have formal public relations training. Lead members consisted of a business owner of a printing company and a massage therapist. In one journalist's narrative reflection on the days, weeks and months following the initial outbreak describes the CWC, and its vice-chair, as:

> The CWC would not be denied. The politicians in Toronto were starting to listen, and Bruce Davidson, whose biggest public speaking event had been as valedictorian at massage school graduation, became an articulate, persistent voice of Walkerton's residents, perhaps *the* voice. Slightly balding with curly hair, five-foot-six-feet tall, he possessed the perfect turn of phrase, an intuitive understanding of what reporters needed, and first-hand knowledge of the town. Most importantly, he was willing to share what he knew, ask the questions few others in the community dared ask, and make himself available for a dizzying array of media interviews. Papers ranging from the *New York Times* and the *Manchester Guardian* to the weekly *Walkerton Herald-Times* talked to him. TV crews paraded through his home, sometimes lining up outside as they waited their turn. A CBC National TV crew spent three days doing "a day in the life of," following the family as they brushed their teeth with bottled water or drove the kids to a nearby town for a bath. Sociology students called asking if they could study the group as an example of an effective grassroots organization. Someone from the British Columbia government asked to use its resources. And through it all, Davidson marveled at just how much attention he and the group continued to get. (Perkel 2002, 174–5, emphasis in original)

Table 1. CWC agenda building by being granted the 'last word'.

Source	Date	Story Context[a]/Analysis theme	CWC Quote
CTV National (TV)	2 June 2000	Engineers search for the source of *E. coli* contamination in the wells Theme: just want clean water	Bruce Davidson (concerned parent)[b]: She can't dunk her head under the water and swim around the way she normally does and that cuts to the very core of who we are. *I know of people who have elderly relatives and children who've sent them away* and are very uncomfortable with the fact that what if that *something isn't taken care of yet.*
National Post (print)	15 Aug 2000, p.A6	Residents learn of new *E. coli* cases from leaked media report, not from public health unit Themes: 1) just want clean water; 2) failure to act; 3) compensation package to victims	But Bruce Davidson of the ad hoc lobby group Concerned Walkerton Citizens said the medical officer's *words are little reassurance to a town still drinking bottled water.* (Theme 1) *'Once again, the sidewalk was rolled up on Friday evening and the information store closed,'* he said. *'When you have a situation where people are having recurrences of E. coli, it's of major concern to all of us. We are more than put off by that kind of approach to our health and welfare.'* (Theme 2) Six people died and thousands more became ill in the outbreak. Mr. Davidson said many people in Walkerton are suggesting every citizen in the small town undergo blood and stool testing, an idea he fully supports. *'They're now suggesting we've been exposed to E. coli, camphyllobacter, salmonella and probably about six different parasites. At the same time, the government is putting out its compensation package,'* he said. *'It would seem rather premature to settle the car insurance claim when you're still skidding down the highway with the risk of flipping into the ditch.'* (Theme 3)

Table 1. (*Continued.*)

Source	Date	Story Context[a]/Analysis theme	CWC Quote
Globe & Mail (print)	13 Sept 2000, p. A12	CWC given standing to appear at public inquiry Theme: need national/provincial water protection	Mr. Davidson said 'I will want to see some changes in policy and procedure on how our drinking water is handled.' *'If Canada, with its vast amount of water, doesn't have safe water, then the rest of the world is in a terrible state.'*
National Post (print)	11 Oct 2000, p. A1	Provincial organization conducting tests of Ontario water finds 131 of 630 municipal systems in ill-repair Theme: Need national/provincial water protection	Bruce Davidson, head of a group called Concerned Walkerton Citizens, was equally alarmed. *'This should sound alarm bells all over the province, if not the country,'* he said after attending the health unit's press conference. *'This is a snapshot of what can happen, and I don't see why other communities' water systems should be any different.'*
CTV National (TV)	26 June 2001	Former cabinet minister questioned about budget cuts to water inspectors Theme: Failure to act	Bruce Davidson (Walkerton Citizen's Group): And quite obviously *we, here in Walkerton, paid the price for this gamble they took.*
CBC National (TV)	29 June 2001, Top story with in-depth interview with CWC and its lawyer	Ontario Premier testifies at inquiry Theme: Need national/provincial water protection	Davidson: I think so. The Premier referred to this as a wake-up call in Walkerton and I think if we look at the Premier's performance and probably the performance of many people in positions of power, *they've been hitting the snooze button on the environmental alarm clock one time too many and we've paid the price here.* And I think it's time for a fundamental re-evaluation of how we conduct ourselves and where our priorities are. And *if this inquiry and if the tragedy here leads to a successful re-evaluation of these events, then I think, not that it will have been worth it, it can never be worth it for the people here, but at least the tragedy will not be completely in vain*

[a]While the extract is for one news media source only, with respect to every story context mentioned, a similar story was present in at least one other news source.
[b]Note: CWC not yet formed.

Returning to Table 1, and specifically the bolded and italicized statements of CWC representatives, agenda-building tactics are present in the language used. In the first two quote extracts, the focus is still on the private misfortune of Walkerton residents, with references to people sending their elderly relative or children away from Walkerton for fear of their safety. Similarly, throughout this time period, the Ontario government was seeking public acceptance of a compensation package before the inquiry began and before residents understood the full health impacts of the contamination. (To this day, the full human health impacts are still undetermined, although the seven-year Walkerton Health Study is expected to be complete sometime in 2007.) In this 15 August 2000 story (see Table 1), a number of elements are touched on. Town residents who are still reliant on bottled water feel betrayed by the local public health office that failed to keep residents informed of new *E. coli* cases being diagnosed (several weeks after the initial contamination period). This risk communication gaffe does little to rebuild public trust. This also causes further worries over the health and safety of residents leading the CWC to argue that the government is being premature in offering a compensation package to residents. The CWC argues that to accept a compensation package at this juncture would be akin to settling a car insurance claim while you are having the accident.

However, once the CWC obtains official standing (in September 2000) to provide testimony of resident concerns at the Walkerton Inquiry, and hence given legitimacy in the process, the agenda-building tactics begins to shift the attention away from Walkerton as a private misfortune, to 'Walkerton' as a public tragedy that could have affected any Canadian community. References to Canada's substantial fresh water reserves and questioning its overall 'safety', suggesting that Walkerton is a 'snapshot of what can happen' elsewhere, and arguing that government officials (regardless of level) have been 'hitting the snooze button on the environmental alarm clock one time too many' all signal creative ways of negotiating CWC interests by providing perfect 'sound-bites' to capture media attention. These efforts are underscored by bringing the possible public tragedy of future risk others (Canadians) back to the private misfortune of Walkerton residents by indicating that 'if this inquiry…and tragedy here leads to a successful re-evaluation…at least the tragedy will not be completely in vain' (see Table 1). This type of messaging is carried throughout the CWC agenda-building communication acts.

Discussion

The Walkerton contamination event had many of the 'trigger' elements (based on Bennett 1999, and noted in italics below) needed to generate and capture national media attention to turn it from a passing 'what-a-story' to a bigger issue demanding public and policy attention. The media used these standard triggers to control the 'properties' (Newig 2004) of the Walkerton issue. Although residents were unprepared to play the *blame*-game, elected officials were only too happy to participate. There was a possible *cover-up* in the making through the alleged connection behind the provincial government's budgetary cutbacks eliminating independent water investigators and the failure to catch the repeated acts of falsifying treatment data by the local managers, which also had the potential to lead to *criminal activity* (charges were laid against local managers). There were *clear, identifiable victims*, including one child that died. The contamination event was *embroiled in conflict* with its search for 'truth' in what happened, and how to keep such an event from happening again. It was a story that carried with it a *high signal value* as 'The Walkerton Tragedy'. For the residents of Walkerton, everyone was affected in some way, so media attention made it a low threshold event; that is, greater media attention was unlikely to change the level of concern these residents had of their water supply. However, for the rest of the country, media attention served to create a high threshold event; put another way, the news media served to bring the 'near risk' issue in one community to the homes and lives of a far-away public, calling into question the overall safety of tap water everywhere.

As a source, and possible definer of the issue, the CWC played an equally important role as visible spokespeople for community residents to control the properties of the Walkerton event (a private misfortune) by turning it into a larger issue (a public problem): peopled died and thousands were made ill from a contamination event that could have happened anywhere in Canada. Governments of various levels failed to act in the public's best interest in the name of cost-savings. Due to the source of the contamination (manure from a non-factory farm), a number of media stories and media concern made reference to the larger issue of factory farms and environmental contamination (improperly treated and disposed animal waste entering public water ways). While the CWC never directly raised addressed the issue of farming practices, it did keep the broader issue of improper environmental management, particularly with respect to source water protection, at the fore. To build its strategy in this manner, the CWC made itself available to the news media, by providing countless interviews, allowing reporters to follow one CWC member's family for 'a day in the life of', and becoming a veritable 'quote-machine' (see Perkel 2002).

As Ericson and colleagues (1989, 377) write: 'News is a product of transactions between journalists and their sources. The primary source of reality for news is not what is displayed or what happens in the real world. The reality of news is embedded in the nature and type of social and cultural relations that develop between journalists and their sources'. These realities become 'objective' as opposed to 'subjective' through the process of sharing and communication. This study is not without its limitations. Only publicly available documents were used in this analysis. Interviews with the CWC or with reporters were not conducted to determine their 'true' motivations in building their respective agendas. Yet, agendas were built. Regulations have been put in which have standardized treated water quality in the province of Ontario. In addition, recommendations from the Walkerton Inquiry also led to the establishment of the *Nutrient Management Act* which sets binding standards on the spread of manure on farmland, the quantity allowed per unit area of land, including setbacks from public water supply systems in efforts to prevent a similar event elsewhere in Ontario (see Prudham 2004). Project next steps include exploring if and how public perceptions of drinking water risk in Ontario have changed since the government implemented the *Safe Drinking Water Act*, and whether or not this new legislation has helped rebuild (Ontario) public trust in drinking water supplies.

Conclusion

This paper has sought to explore the agenda-building role of the media and its sources concerning a drinking water contamination event in one Canadian community. In so doing, this research carries with it a number of practical implications for risk communication researchers. First, media studies need to de-emphasize its essentialist approach to power – either bestowing too much power in the hands of the sources (theories of primary definers), the media (passive-audience models), or audience groups (active-audience models). Rather, by adopting a combined theoretical approach of understanding shared realities and agenda-building, it is possible to begin unravelling the complexities used in news story constructions. Too many media analyses have unproblematically assumed the agenda-setting role of the media, often ignoring how the media, government officials, the public and interest groups build agendas that serve their respective interests.

Second, non-elite source groups need to be given more attention in media studies. Some may argue that studying the role of a source group (n=84) used in 10% of all the possible stories compiled in the larger study sample (n=841) is of little consequence. It could be said that the CWC agenda-building efforts failed because they were not given the same 'air' time as more official sources (i.e., government officials, representatives, witnesses at the Inquiry, etc.). However, to say the CWC (or other non-elite groups like them) failed would miss the point. Considerable risk research has concentrated on the accessible actors (officials, media, interest groups, etc.), often ignoring how the issue under study concerns the public, except through the constructed lens of 'the other'. Part of the reason for this is the high cost (not to mention time) involved in doing primary data collection with members of the public who are touched by a risk event. Using the media is a poor, but often necessary, surrogate measure to understand how the public may view a risk. Provided that the emphasis of such studies is not on the content (i.e., the media leads the public to think in a certain way) but rather on the salience (i.e., the media and others build agendas to add pertinence to what the public may think about) of a risk event, it may be possible to use the media as a window to how the public may be responding to an issue. Nonetheless, media attention alone is insufficient, and public concern may continue long after the media has moved onto other issues. Consequently, it is important to balance such studies by using other data collection techniques (e.g., focus groups, surveys) with the public to expand an otherwise narrow view.

Third, and related to the second point, risk communication failures in areas of public and environmental health are typically more problematic when it is characterized by scientific uncertainty. While in this case study, microbial or bacterial contamination is not characterized by scientific uncertainty, unlike the case of chlorinated disinfection byproducts in drinking water, there is still insufficient understanding of the long-term health impacts of those exposed to microbial pathogens in drinking water. At a contextual level, the Walkerton drinking water issue now serves as one of the greatest risk communication 'symbolic secondary events' (in agenda-building terms) for other drinking water events elsewhere in Canada. (The Walkerton event has also been compared to other drinking water outbreaks in the developed world – see Hrudey et al. 2002, 2003). Walkerton residents, and Canadians in general, were surprised that just over one year after the Walkerton risk event, a new community was experiencing a microbiological contamination event of its own (the town of North Battleford, Saskatchewan, experienced a *cryptosporidium* outbreak that affected more than 1900 residents). Other communities (e.g., Erickson, British Columbia) have also been described as a 'Walkerton waiting to happen' (Pynn 2001).

While neither the CWC nor the media are directly responsible for this shift in public concern, their role is not inconsequential. As a society, we are constantly exposed to 'media-ted' messages – in the news media, popular literature, televised programmes, advertisements, and so forth. As an agenda-builder, the media create and sustain a contextual society that capitalizes on the production and consumption of media images, messages and symbols as the basis for content. Within the context of environmental risks, the role of news agencies in constructing and shaping news stories serves to mediate risk messages to the public by virtue of what elements of a story they report, the types of sources they use in reporting the different 'sides', and how they package a story. However, the media are not the only ones who seek to set and build agendas. The sources, upon which the news media rely, can and do similarly engage in agenda-building activities. They do so by appropriating the very 'media-ted' process by using the mass media as a delivery vehicle for their own communication claims regarding their own packaged 'take' on the story. It is only by studying the interactions between the different groups involved internally and externally that we can begin to unravel the complex strategies used to communicate risks in the media.

Acknowledgements

This research was funded through a research grant from the Social Sciences and Humanities Research Council (410-2004-2025), and, in part, through funding from the Canada Research Chairs program.

Notes

1. This overview of events is pieced together from media reports, a journalist's narrative account of the events based on interviews with residents and officials of Walkerton (Perkel 2002), and Report 1 of the O'Connor public inquiry into the event (O'Connor 2002a).
2. In Canada, it is the provinces that have constitutional jurisdiction over setting regulations for drinking water supplies, with the exception of water supplies that fall on federal lands or on First Nations Indian reserves. There is a federal body, the Canadian Federal-Provincial-Territorial Committee on Drinking Water that sets the public health guidelines with respect to Canadian drinking water supplies. However it is up to each province and territory to determine if it will adopt the guidelines set by this Committee, or if it will adopt more stringent or more relaxed guidelines. Typically, the province devolves the responsibility for implementing these guidelines to the municipalities. Hence, it is the Government of Ontario that is primarily responsible for setting regulations for Ontario drinking water supplies. Until the Walkerton *E. coli* outbreak, Ontario was governed by *Ontario Drinking Water Objectives* (last revised 1994) that were not legally enforceable.
3. The reality of the Canadian market is that the *Globe and Mail* and *The National Post* are only widely circulated within Ontario. Most Canadian provinces read the local provincial papers. In Canada, there is no widely distributed national equivalent to newspapers in the United States, like, for example, *USA Today*.
4. More recently, a third broadcaster has entered the market in Canada: *Global News*. However the database used to collect news stories did not catalogue transcripts from this broadcaster.
5. There are different regulations and testing requirements in Canada for tap water compared to bottled water. Municipal tap water systems are tested more often and held to a higher standard in Canada than bottled water supplies which is only governed under the rules of the *Food and Drug Act*.

References

Anderson, A. 1997. *Media, culture and the environment*. London: UCL Press.

Bennett, P. 1999. Understanding responses to risk: Some basic findings. In *Risk communication and public health*, ed. P. Bennett and K. Calman, 3–19. Oxford: Oxford University Press.

Bourette, S. 2000. Walkerton water chief to speak up: Neighbours, friends, reporters await Koebel's account of E. coli tragedy. *The Globe and Mail*, December 18, A3.

CBC - The National. June 29, 2001. Host: Brian Stewart.

CTV National News. June 2, 2000. Host: Tom Clark.

———. June 26, 2001. Host: Lloyd Robertson.

Cohen, B.C. 1963. *The press, the public and foreign policy*. Princeton: Princeton University Press.

Croteau, D., and W. Hoynes. 2003. *Media society: Industries, images, and audiences*. 3rd edn. Thousand Oaks, CA: Pine Forge Press.

Denham, B.E. 2004. *Sports Illustrated*, the mainstream press and the enactment of drug policy in Major League Baseball. *Journalism* 5: 51–68.

Donsbach, W. 2004. Psychology of news decisions: Factors behind journalists' professional behaviors. *Journalism* 5: 131–57.

Downs, A. 1972. Up and down with ecology: The 'issue attention cycle'. *Public Interest*, 38–50.

Driedger, S.M. 2004. *Perception of equivocal health risk: Drinking water in a post-Walkerton climate*. Research Grant funded by the Social Sciences and Humanities Research Council, 2004–2007.

———. 2007. Risk, the media, and public perceptions: A comparison of print and televised news stories of Walkerton, Ontario's *E. coli* O157:H7 outbreak in drinking water. *Risk Analysis* 27, no. 3: 775–89.

Driedger, S.M., and J. Eyles. 2003. Different frames, different fears: Communicating about chlorinated drinking water and cancer in the Canadian media. *Social Science & Medicine* 56: 1279–93.

Entman, R. 1993. Framing: Toward clarification of a fractured paradigm. *Journal of Communication* 43: 51–8.

Ericson, R.V., P.M. Baranek, and J.B.L. Chan. 1989. *Negotiating control: A study of news sources*. Toronto: Toronto University Press.

Fleras, A. 2003. *Mass media communication in Canada*. Scarborough, ON: Thomson-Nelson.

Gamson, W.A., D. Croteau, W. Hoynes, and T. Sasson. 1992. Media images and the social construction of reality. *Annual Review of Sociology* 18: 373–93.

Gillis, C. 2000. Walkerton outbreak traced to its roots: Manure from cattle farm: Medical officer fears it may happen again. *National Post*, October 11, A1.

Hall, S., C. Critcher, T. Jefferson, J. Clarke, and B. Roberts. 1978. *Policing the crisis: Mugging, the state, and law and order*. London: Methuen.

Hardin, C.D., and E.T. Higgins. 1996. Shared reality: How social verification makes the subjective objective. In *Handbook of motivation and cognition*, ed. R.M. Sorrentino and E.T. Higgins, vol. 3, 28–84. New York: Guilford.

Hrudey, S.E., P. Payment, P.M. Huck, R.W. Gillham, and E.J. Hrudey. 2002. Walkerton: Lessons learned in comparison with waterborne outbreaks in the developed world. *Journal of Environmental Engineering Science* 1: 397–407.

———. 2003. A fatal waterborne disease epidemic in Walkerton, Ontario: Comparison with other waterborne outbreaks in the developed world. *Water Science and Technology* 47: 7–14.

Jardine, C.G., N. Gibson, and S.E. Hrudey. 1999. Detection of odour and health risk perception of drinking water. *Water Science & Technology* 40: 91–8.

Kellner, D. 1995. *Media culture: Cultural studies, identity, and politics between the modern and the postmodern*. London: Routledge.

Lang, G.E., and K. Lang. 1981. Watergate: An exploration of the agenda-building process. In *Mass communication review yearbook*, ed. G.C. Wilhoit and H. de Bock, vol. 2, 447–68. London: Sage.

———. 1983. *The battle for public opinion: The president, the press, and the polls during Watergate*. New York: Columbia University Press.

Lippmann, W. 1922. *Public opinion*. New York: Free Press.

Lorimer, R., and M. Gasher. 2004. *Mass communication in Canada*. 5th edn. Oxford: Oxford University Press.

Mander, M.S., ed. 1999. *Framing friction: Media and social conflict*. Chicago, IL: University of Illinois Press.

Mathes, R., and B. Pfetsch. 1991. The role of alternative press in the agenda-building process: Spill-over effects and media opinion leadership, *European Journal of Communication* 6: 33–62.

McCombs, M.E., and D.L. Shaw. 1972. The agenda-setting function of mass media. *Public Opinion Quarterly* 36: 176–87.

———. 1993. The evolution of agenda-setting research: Twenty-five years in the marketplace of ideas. *Journal of Communication* 43: 58–67.

McLaughlin, T. 2000. NDP denied standing at inquiry Walkerton judge wants to avoid politicizing tainted-water hearing. *Globe and Mail*, September 13, A12.

Menashe, C.L., and M. Siegel. 1998. The power of a frame: An analysis of newspaper coverage of tobacco issues – United States, 1985–1996. *Journal of Health Communication* 3: 307–25.

Mistry, B. 2006. Risk and drinking water: A media analysis of drinking water risk perception in Ontario, post-Walkerton. MA thesis, Department of Geography, University of Ottawa.

Mistry, B., and S.M. Driedger. 2005. Canadian media presentations of Walkerton, Ontario's E. coli O157:H7 contamination of drinking water supplies: An examination of language and discourse of televised news coverage. Paper presented at Canadian Association of Geographers Annual Meeting, June, in London, Ontario.

Newig, J. 2004. Public attention, political action: The example of environmental regulation. *Rationality and Society* 16: 149–90.

O'Connor, D.R. 2002a. Part one report of the Walkerton commission of inquiry. Ontario Ministry of the Attorney General. Queen's Printer for Ontario. http://www.attorneygeneral.jus.gov.on.ca/english/about/pubs/walkerton/.

———. 2002b. Part two report of the Walkerton commission of inquiry. Ontario Ministry of the Attorney General. Queen's Printer for Ontario. http://www.attorneygeneral.jus.gov.on.ca/english/about/pubs/walkerton/.

Parr, J. 2005. Local water diversely known: Walkerton, Ontario, 2000 and after. *Environment and Planning D: Society and Space* 23: 251–71.

Patterson, T.E., and W. Donsbach. 1996. News decisions: Journalists as partisan actors. *Political Communication* 13: 455–68.

Perkel, C. 2002. *Well of lies: The Walkerton water tragedy*. Toronto: McClelland and Stewart.

Prudham, S. 2004. Poisoning the well: Neoliberalism and contamination of municipal water in Walkerton, Ontario. *Geoforum* 35: 343–59.

Pynn, L. 2001. B.C. health officer fears 'shades of Walkerton': Kootenay town refuses to chlorinate its drinking water. *National Post*, January 3, A4.

Rogers, E.M., and J.W. Dearing. 1988. Agenda-setting research: Where has it been? Where is it going? In *Communication yearbook 11 (555–594)*, ed. J.A. Anderson. Newbury Park, CA: Sage.

Schlesinger, P., and H. Tumber. 1999. Reporting crime: The media politics of criminal justice. Oxford: Clarendon Press [excerpted in H. Tumber, ed. *News: A reader*, 257–66. Oxford: Oxford University Press].

Stone, D. 1997. *Policy paradox: The art of political decision making*. New York: W. W. Norton.

Tanner, A.H. 2004. Agenda building, source selection and health news at local television stations. *Science Communication* 25: 350–63.

Tate, E.D., and K. McConnell. 2001. The mass media and violence. In *Communications in Canadian society*, ed. C. McKie and B. Singer. 5th edn. Toronto: Thompson Educational Publishing Inc.

Tuchman, G. 1973. Making news by doing work: Routinizing the unexpected. *American Journal of Sociology* 79: 110–31.

———. 1978. *Making news: a study in the construction of reality*. New York: Free Press.

Vallis, M. 2000. *E. coli* under control in Walkerton: health officer: Spate of new cases: Certain number of infections expected in summer. *National Post*, August 15, A6.

Narratives of risk

Gaspar Mairal

Antropología Social, Universidad de Zaragoza, Spain

Risk is a probabilistic notion which has been used by scientists since the sixteenth century. It is not a perception but a calculation or a narration. This paper proposes an historical perspective to show how risk jumped from mathematical calculation to the narrative. A book by Daniel Defoe, *A Journal of the Plague Year*, which was published in 1720, is the best example of an early narrative of risk. Together with the example of Defoe's book, some ethnographic episodes from Spain are described and analysed: the protest against the Almaraz nuclear power station, the sinking of the Prestige and the subsequent oil slick, the flooding of a campsite in Biescas where 86 people died. This paper suggests an exploration into the narratives of evil, harm and suffering to classify them in terms of uncertainty, risk, fear, panic and terror. *The Diaries* of Viktor Klemperer (1996) are used as the narrative material to be interpreted in order to identify different kinds of narrative. Also the literature written by some of the Holocaust survivors, Primo Levi for example, becomes a very relevant source to understand how the identity of victims comes from a narrative challenge: to tell the truth.

> ...between the action of narrating a story and the temporal character of human existence there is a correlation which is not purely accidental, but it represents the shape of a transcultural need. In other words: time becomes human time if it is articulated in a narrative way, and the narration gets its full meaning when it turns into a condition of time existence.[1]
>
> (Ricouer 1990)

Introduction

Between 2000 and 2001 I participated in a project[1] which brought together several European researchers with some experience in risk research in order to exchange their knowledge and work. Two workshops took place in Birmingham (2000) and Madrid (2001) where other colleagues, from different fields, were invited. Our discussions and debates produced two documents, *Situated Risk* and *Representing Risk*. We took as our starting point a description of risk in situations which were familiar to us thanks to our past experience and at all times aiming to sustain any conceptualization with an empirical foundation. We agreed that the 'representation' of risk was a common feature on a variety of situations.

> What became apparent in the course of these discussions was that, although the types of hazards involved and the history of the various situations varied enormously, clear similarities could frequently be identified in terms of the socio-cultural processes and dynamics shaping these risk situations. A subsequent review by the members of the Network's co-ordinating group concluded that central to the dynamics described by

participants in the Birmingham workshop, and cutting across the diverse configurations of the different risk situations described, was the process of *representing risk*.[2]

Our view was also oriented against the reification of risk. Risk was not an object, as I had found in my previous research on dams construction, but there were 'objects of risk' and 'objects at risk'. So a dam was an object able to produce a notion of risk and the land or the houses were objects which were at risk. Risk is then the relationship that a human group can establish between both objects. As seen below we tried to shape a relational definition of risk:

> Risk can be considered as a frame that creates contexts which bring together an 'object of risk' (a hazard or source of potential harm), an 'object at risk' (a target of potential harm) and an evaluation (implicit or explicit) of human consequences. Risk is therefore a relational order through which connections between people and 'things' are constituted and which guides their interpretation.[3]

If we do not rectify risk, the objective condition is not a part of risk but of those objects which contextualize[4] risk through a relationship. From this point of view the problem of risk, if it is objective or subjective, must be considered in another way. Risk is something which tends to the future, it does not happen, but could happen, it does not exist, but it could exist. What, then, is 'real'?

Risk is probabilistic

My proposition is that risk is not a fact, it is rather an artefact, a tool that is used to bring together objects, facts, events or any other entities which can produce harm and others which can be harmed. This linkage derives from the notion of probability which had its first development in the sixteenth century when Pacioli,[5] Cardano[6] and Tartaglia,[7] three courtiers and algebra experts, applied their mathematical knowledge to gambling (cards, dice, coins). The basis of probability calculation would have been found around 1650 when it was debated if it was possible to determine the probability of winning a game. In 1654 Fermat and Pascal gave the first definition of probability. This said that the probability of something happening was equal to the quotient between the number of favourable and possible cases. In 1657 Huygens[8] wrote the first book about the probability to be published. Laplace completed this scientific development in 1810 with his work *Analytical Theory of Probabilities*. The probability is, as we can see, a modern idea, belonging to the world of science or expert knowledge.

This is exactly what happens with regard to the siting of the nuclear power station of Almaraz in Spain, where some local people claimed that the number of congenital malformations was much higher than the Spanish average and they attributed their origin to radioactive activity. No other situations like this one have been reported in Spain, where a dozen nuclear power stations are operating. A very famous Spanish TV journalist, Mercedes Milá, presented on 30 November 2004 on Tele5, the most popular Spanish TV channel, a story called *Diario de una amenaza: centrales nucleares (A Diary of a Threat: Nuclear Power Stations)* entirely devoted to the Almaraz case. It is important to notice that local people who were interviewed used a similar notion of risk. They told stories of other people who had some contact to the nuclear power station or lived in its surroundings and had suffered some disease or harm which was precisely described. From this connection, which had been established thanks to a story, they mentioned the probability of a radioactive causality. It is important to point out that they were using stories more than data and that in a certain way they told the stories to demand data. If data did not fit to their notion of risk they reactivated the stories, because as a last resort the risk was based upon stories. This was the narrative framework of this TV report, first we heard the local people's stories which introduced some vague references to expert procedure as statistical estimates and then other narrations, increasingly expert, were introduced. Little by little people who had been leading this local opinion, doctors, epidemiologists and the director of this nuclear power station, were giving this narrative expert substance using the concept of epidemiology. Everybody, except for those not recognizing any risk, asked for an epidemiologic study which has never been carried out. For several years this problematic situation has been a narrative of risk with a great national interest when it was followed by an enormous audience of millions of people.

The most basic definition of probability links two facts, the number of favourable chances with the number of possible cases and this is its fundamental framework. If we assume this kind of relationship, not in terms of the resulting quotient, but in terms of the resulting consequences, we have a narrative frame in terms of 'When A and B connect C could happen' (Figure 1). This is a transformation of probability in narration because the dimension of time has been put into the proposition. The following step should take us to estimating how the two objects are brought together and here I wish to introduce 'narrative linkage' as a concept to identify this representational process. Inasmuch as two objects, things, events, etc. not sharing a common attribute, form a new entity, the risk, is essentially through the narration. Narration should have a structure of time and the risk narratives have all the same subject: probability. This kind of narration tells us that there is a probability that harm is produced when A and B connect. A nuclear station power, for instance, is not a risk but the probability that a nuclear power station might affect the health of some people is really a risk. If we remember the case of the Almaraz nuclear power station, this was the basic framework of a reiterative story:

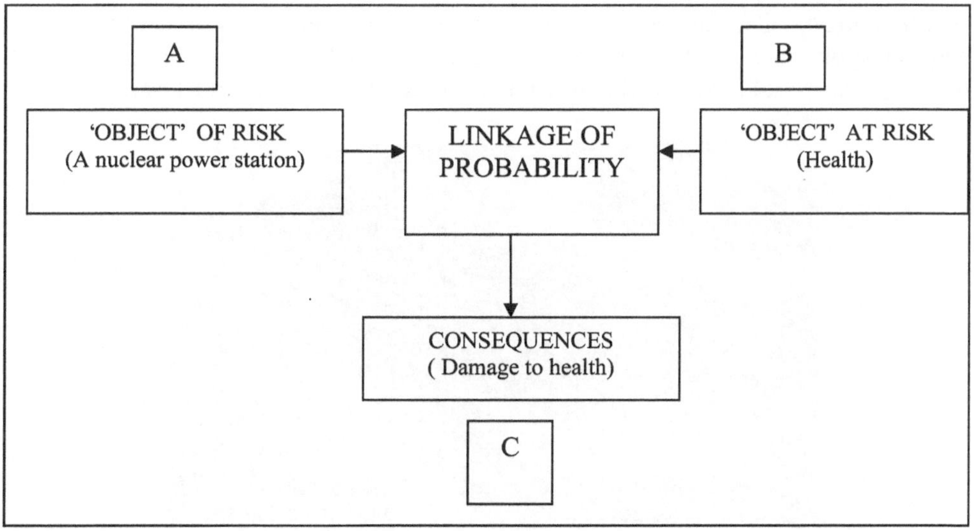

Figure 1. A narrative frame of risk.

the radioactivity (A) connects to the health of people (B) and illness could be its consequence (C). This story was enough to fill one hour of TV as its most important topic.

Danger is an attribute of things, facts or other entities while risk is a relationship that we establish between objects, facts or entities. I can say that a «tsunami» is very dangerous but I never say that a «tsunami» is risky. Nevertheless I can describe my proximity to the «tsunami» as risky. So the risk is not derived from the object but from my relationship to the object. Beyond this simple example I would like to mention a short description of a disaster which happened in the Spanish Pyrenees on 7 August 1996. That afternoon a vast downpour[9] struck a small valley and the Arás stream flooded a campsite where 86 people died (Figure 2). After this tragedy both criminal and administrative proceedings were opened and while in 1999 the administration was absolved in the criminal court which established that this flooding was «exceptional and unpredictable», in 2005 the autonomous Government of Aragon and the Spanish Ministry of Environment were sentenced in the administrative court to pay 11.2 million Euros in compensation. A very long sentence was made public in December 2005 and I have used the words of an expert witness and translated one of his statements: «Danger is not predictable, but risk certainly is». It is important to define a certain context for this statement.

This stream had a cone of debris through which any vast volume of water and solid materials was able to run after one of the highly intense storms which are quite frequent in the summertime in the Pyrenees. In fact, two disastrous floods had taken place in 1929 and 1959, although in the absence of any buildings, what was really affected was a road. Then major engineering works were carried out to control any new flooding. The final stretch of this stream was channelled and several small dams were built upstream to avoid new damage to the road (Figure 3).

Nevertheless, in 1986 a campsite was located on this cone of debris with the knowledge and permission of the administration which ignored a technical report in which this location was assessed as being 'high risk'. In 1996 this new flood produced a terrible wave which inundated the campsite and swept away the people who where there (Figure 4). The immediate statements given to journalists by political and technical spokesmen mentioned that such a flooding could never have been predicted and only 'fate' could explain this terrible event, as if it were an old Greek tragedy.

Figure 2. The 'Barranco de Arás'.

Figure 3. Works carried out before the catastrophe.

This argument remained for years as the official cause of this disaster. A subsequent investigation confirmed, however, that an official report[10] had advised against the location of this camping area on this cone of debris and this report had a significant influence in the final verdict issued.

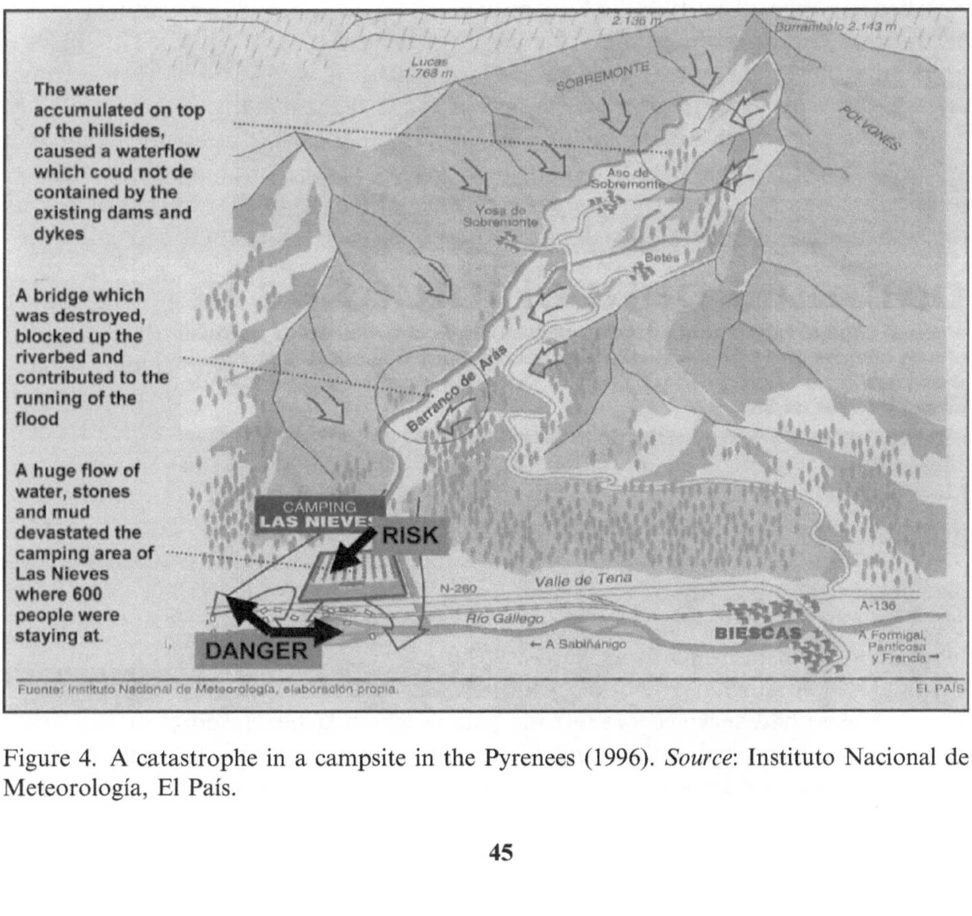

Figure 4. A catastrophe in a campsite in the Pyrenees (1996). *Source*: Instituto Nacional de Meteorología, El País.

This brief account of events helps us to understand the expert witness's words.[11] The Arás stream was dangerous because it could potentially break its banks in spite of the preventive works. We can say then that danger was an attribute of this torrent, but the probability of this danger materializing with consequences was the risk. If the advice of not locating a camping area on the cone had been taken into account, the risk would have not existed. In the Arás stream there was a danger and in the location of the campsite a risk. The former is an attribute but the latter is a relationship. From the same point of view the first criminal sentence evaluated the danger to establish that there was no responsibility attached to the government authorities, but what the second sentence evaluated was the risk, establishing that the authorities were responsible. I would like to think about this comparison. It shows the way in which expert knowledge has conceived danger and the risk. Danger is the attribute of a fact but risk is the product of our actions. It is a construction, but not invented from nothing but derived from something which can be recognized as belonging to the reality, i.e., danger.

Risk is historical

Risk is, from the beginning, a scientific and then modern concept linked to the idea of probability which, as seen before, was first developed in the sixteenth century from algebraic calculation. It was a scientific development which helped to overcome the annoyance felt by modern and scientific minds when they faced uncertainty. In the sixteenth century people like Pacioli, Cardano and Tartaglia wished, of course, to win when they played cards in the European courts instead of suffering from a vulgar uncertainty. They introduced a good amount of scientific knowledge and thanks to probability they transformed uncertainty into risk. I do not know if they became wealthy, but they certainly created the new concept of probability.

Could there be a historical theory of risk? If so, it would mean then that the notion of risk was created and developed according to historical conditions, those which led to the emergence of a scientific society. The basis of the concept of risk is scientific more than cultural if we understand that science is an expert and culture to be common knowledge. Nowadays we live in a permanent crossroads between science and culture and their hybridization is a feature of modern society. The notion of risk jumped from science to culture and it would be useful to study the historical process in which this jump took place. This is going to be my contribution in this paper.

I am sure that 1722 was a landmark because that was the year in which Daniel Defoe published a book, *A Journal of the Plague Year*. I wish to use this book to show how Defoe was aware that he, by then a well-known journalist, was contributing, as an expert, to preventing a plague that was probably going to affect the city of London. Though he introduced some data, he did not use mathematical calculations of probability but a narration to tell the citizens of London that a new plague could affect them and to achieve this goal he used a story and a kind of highly elaborate journalism, which lead Anthony Burgess to state that Defoe was truly the first English journalist.[12] The book is written as if it were the memoirs of a man, a merchant, who had survived the terrible plague which lashed London in 1655 and then tells what happened at that time. Defoe had been indeed one of the survivors of the 1655 plague and he was using his own memories constructing a sophisticated narration in which the artistic quality of the writing mingled with a first-hand knowledge of the facts and a good amount of information which Defoe collected in libraries and archives. Burgess[13] again writes in the introduction to the Spanish edition that this book has a double truth, that of the conscientious historian and a deeper one, his creative imagination. This final appreciation indicates very well how science and culture are already acting together in this wonderful book.

By about 1720 following the outbreak of 1665, the plague had practically disappeared in Europe. But it reappeared in Marseille, and the news, arriving from that city to London, spread uncertainty. Defoe worried about future events if the plague were to reach London again and wished to protect the population. This is the way in which Defoe constructed his narration, with a description of events already happened and he took them as an example to introduce the idea of the probability of these events happening again. This is a 'narrative matrix'.[14] In my opinion the narrations of risk tend to be a consequence of a matrix or narrative model.

At the beginning of his book which is written in first person, the narrator remembers a past time, 1665, the year of the plague:

> We had no such thing as printed News Papers in those Days, to spread Rumours and Reports of Things; and to improve them by the Invention of Men, as I have liv'd to see practis'd since. But such things as these were gather'd from the Letters of Merchants, and others, who correspond abroad, and from them was handed about by Word of Mouth only; so that things did not spread instantly over the whole Nation, as they do now. (Defoe 2001, 3)

We can look at an historical change because the story that Defoe tells us testifies it. The time of the 'matrix narrative' had no journalism, but the narrative itself is a masterly lesson of journalism. We can see the effect of the newspapers and the replacement of rumour by the written new. This is a very significant fact, the birth of journalism, for understanding the historical context for the emergence of the narratives of risk. Defoe's book is also important for the use of expert knowledge together with literary imagination. The book is written in relation to a background of facts and data. As Godwin writes in his introduction to the English edition:

> Backed by careful research, his unfussy prose triumphantly carries off the mild literary fraud of *A Journal*, which at one level is the most striking historical document to come out of the plague. But it is really a novel, revelling in Defoe's infallible ear for the cadences of real speech, and his revolutionary desire and ability to set them down. The narrator's own voice is a masterpiece of understated realism, adapting its very structure to reflect uncertainty, shock, and the faltering linkages of memory. (Defoe 2001, xiii)

A Journal of the Plague Year is a novel but also an historical narration with at least two features to be emphasized. In the first place we find the expert knowledge which permits Defoe to reconstruct some events, dated 55 years before, as a tale full of credibility and verisimilitude. From my point of view, this is a condition to recognize a narrative of risk, also there is always a part of expert knowledge in it. In second place we have the use of a narrative and realistic device to reinforce this tale and produce the effect which was pretended, to protect the people of London against a new plague. The probability is something which becomes real thanks to the narration of what happened before. Here the narration, instead of being a mathematical calculation, materializes the probability. The nature of this narrative action which, on the other hand, takes risk from science to culture, resides on time. The calculation is substituted by a time framework, just as Defoe put together his narration, linking the past to the present and the future. In this sense I consider that the best definition for risk should be historical[15] and probabilistic and then either mathematical or narrative.

Risk is not a perception but a narration

The next example serves to see how a narrative of risk is the result of previous narrations which, like an earthquake, produce after-tremors. On 19 November 2002 an oil tanker, the *Prestige*, sank in the Atlantic Ocean about 250 km from the northwest coast of Spain (Figure 5). The Spanish authorities had ordered this tanker to be towed out to sea in an attempt to prevent an oil slick reaching the coast of Galicia, the northwest region of Spain, where traditional fisheries are very important. In spite of this effort there was highly destructive oil slick affecting the coasts of Galicia, Asturias, Cantabria, the Basque country and even the south Atlantic coast of France.

There has been long debate and extremely intense political controversy over this decision because there was another option which consisted in towing the damaged oil tanker to a secure port and then extracting the oil from the tanks. This option had to face the problems of a damaged oil tanker reaching port and it was also possible that the tanker might spill the oil inside the estuary or catch fire.

On 3 December 1992 the *Aegean Sea*, also an oil tanker, ran aground at the entrance of Corunna, the most important city in Galicia, spilling out about 80,000 tonnes of oil. There was a huge oil slick and the oil tanker itself broke and burnt

Figure 5. The sinking of Prestige (2002).

(Figure 6). A column of black smoke rose up to one kilometre and engulfed the city. It was a nightmarish sight which produced panic. In 1976 another oil tanker, the *Urquiola*, had also sunk at the entrance to the Corunna estuary and produced an oil slick. Considering the recent history of the oil slicks in the Corunna estuary and the effects on the population which felt intensively affected, we might estimate the presence of a shared narrative of risk[16] which told of the burning of the *Aegean Sea* and previous oils sticks. The option of bringing the *Prestige* to the port of Corunna to extract its oil, was immediately rejected out of hand. The recorded tapes[17] of all the communications show that the order to move the *Prestige* away from the coast was given two hours after the first information about the difficulties of navigation caused by a leak. It was to all intents and purposes a slogan: '¡Sink the tanker as far as possible from the coast!' My description of this case seeks to emphasize the narrativity of risk, the 'matrix' effect and how in this case the narration prevailed over calculation. So the response to the risk was more cultural than expert.

Let us consider the hypothesis that when we find a public notion of risk, there are people who tell stories to other people and these stories refer to something that happened before or could have happened. The stories tend to adopt a standardized form whose basic construction includes a narrative frame of risk: the relationship between A and B can probably produce C. In this respect I am not very fond of perception as the main phenomenon to conceptualize a shared notion of risk. Instead I prefer to estimate that more than perception there is narration. The narratives of risk circulate through a diversity of stations, though in general they have expert sources.

A narrative model

Risk is a part of a broader human experience: our relationship with evil, harm and suffering. If the risk becomes probability, uncertainty becomes possibility, fear the proximity, panic the action and terror the experience. To a certain extent, these are the variations of a relationship or linkage with evil, harm and suffering. I have tried to formulate a model in which different features help to characterize every variation.

Figure 6. The city of Corunna (1992).

I do not think that risk should be researched as an isolated phenomenon, but as part of a world of human experiences.

The reading of Viktor Klemperer's diaries[18] showed me how a narration, a long narration, could reveal the relationship between a man and evil. Klemperer, a professor of modern languages at the University of Dresden, was a Jew married to a gentile. His diary covers, almost day by day, a period (1933–1945) which started with the coming of the Nazi party to the power and ended when he and his wife returned to their home after the liberation. That is why this diary is an extraordinary document and a first-hand source of information on everyday life in Germany during the Third Reich. Klemperer, who died in East Germany in 1960, had no interest in the distribution of his diaries and so they were only discovered in 1995 by one of his old students, Walter Nowojski, and published in 1998 for the first time. One main theme covers the whole book, the racial and criminal policy applied by the Nazis to the Jews all the time they governed Germany. We know that this policy underwent a process of development from the Nuremberg laws until the final solution. The diary of Klemperer follows this process in everyday life. First we can see how rumours began to spread and afterwards how a chain of events occurred. Klemperer lost students at university, he could not publish any of his writings, he was expelled from his academic position, he and his wife were taken out of their own house which was given to 'Aryan' occupants, he could not take books out of libraries, he had to sell his car, he had to wear a Star of David on his coat, they have to move to houses only for Jews, he was sent to work in a factory or forced to work on the streets, he was sent to prison for one month, he could see how their neighbours were called for a journey and taken to Theresiensdat as a first stage to going to one of the death camps in Poland. They knew what Auschwitz was and were expecting at any moment to be taken there. But they had an opportunity to survive and remain among the final small group of Jews in Dresden for being a mixed couple. In fact, they were expecting to be called at any moment and had prepared their luggage and minds for the end, but the allied Air Force bombed Dresden and practically destroyed the city. In this nightmare of terror, the Klemperers experienced the proximity to death and acted consequently. Nevertheless the subsequent chaos helped them to leave the city and wander about the country to live as refugees for the last few months of the war.

The reading of this diary inspired me to identify a process which went from uncertainty to terror and was probably experienced by many European Jews. Klemperer's diary is an exceptional testimony. Uncertainty fills the first part of this narration, when rumours arrive at the Klemperer's house. At the same time there is a certain acceptance of this situation which is generally evaluated as temporal or exaggerated. 'Everything will soon return to what it was before' is repeated very often. All conversations end up at the same feeling of uncertainty. Then narratives are mainly oral, rumours or talking in a whisper.

The enactment of the Nuremberg laws on racial policy introduced knowledge in the lives of people like Viktor Klemperer and these new rules, widely disseminated came to represent an 'object of risk'. The new situation forced a re-evaluation of events derived from the newspapers and the Nazi propaganda. Many Jewish families, in the Klemperer's circle tried to emigrate, but it was not easy to obtain permission. Klemperer himself was torn between his deeply-rooted German feelings[19] and a notion of risk which was becoming more precise. When he tried to get out of Germany it was too late: the Nazi government had stopped any Jewish emigration.

Sometimes he felt fear and then any notion of risk disappeared with the proximity of a terrible outcome – arrest and deportation to the death camps. At other times he was insulted or assaulted by Nazi fanatics when they saw him wearing the Star of David. Nevertheless there was one episode of fear which reveals what a situation such as this was like. The war had begun and Klemperer was denounced for having forgotten to black out a window according to the instructions in case of an air raid. He spent one month in prison, feeling that he might be transported to the gas chamber at any time or just executed. Naturally he had no opportunity of writing a single word during this horrific time, but when he was released, at home he wrote an account of this imprisonment. This part of the whole narration has a special character which is more literary; one notices easily that a certain stylistic ambition was there. Fear is a feeling without knowledge because the latter vanishes when we are going through it. So there are situations when harm, suffering or evil are so close and their presence is felt so much, that any knowledge is erased. The panic is action and expresses an encounter with evil. It moves people to act and can produce knowledge though it is always a new one. Terror can only be the experience of the survivors. Terror produces victims, but they cannot tell us anything. Klemperer suffered the experience of the bombing of Dresden, lost contact with his wife when they both were escaping from the explosions and re-encounters her afterwards. Here we do not have anything but what has already happened. Thus terror can only be represented through memories because it is almost impossible that someone could write directly this kind of experience when it is being suffered. The people who survived the Holocaust and wrote about this cruel experience are the best example. The writing of Primo Levi, Jean Améry, Jorge Semprún, Imre Kerstéz or Tadeusz Borowski and some others, can teach us a lot about situations in which someone has lived or encountered the suffering or the evil.

The whole narration, just as it develops in the long and detailed diaries written by Klemperer, shows their life, his and his wife's, and the action of the Nazi's Jewish policy from the time of hate and xenophobia to the time of extermination and final solution. A basic narrative structure develops here: it is the growth of evil and its relationship with the lives of some people whose suffering and death is described with emotion and precision. This experience constitutes the *issue* which gives content to the narration and it evolves for some years showing a *frame* of *time* which could be characterized as an increasing linkage with harm, suffering and evil. This frame of time includes different relationships going from uncertainty to risk, fear, panic and terror, having each a predominant *style* of narration and rhetoric: oral, written, mass, literary, memorial, metaphoric and metonymic.

The narration

Terror displays an extensive and deep knowledge in the case of the survivors because they are aware that they have got a narrative challenge: to tell the truth. Primo Levi, for example, wrote three books[20] to narrate his terrible experience in Auschwitz and he committed suicide less than one year after the publication of the last one, *The Drowned and the Saved* (1986). In his second book *The Reawakening* (1963) he tells his long return trip after the liberation to Turin where he lived and how he, for the first time, visited Germany in a devastated Munich. Then he mentions his personal

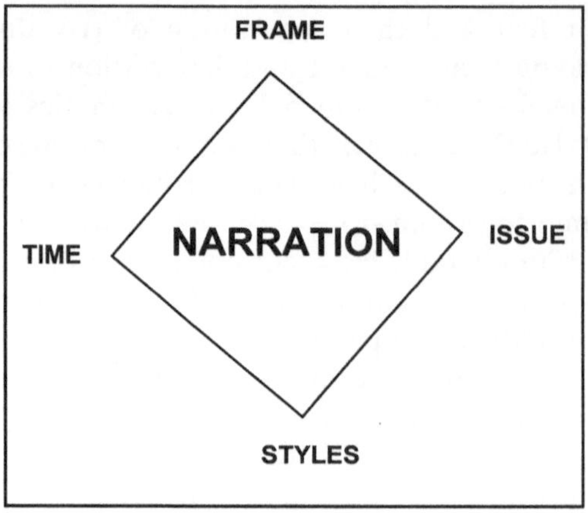

impression of the German people who did not pay attention to them, a group of the Holocaust survivors.

> It seemed to me that all of them would have to question us, reading our faces to recognize us and then listen humbly our story. But none of them look at our eyes and none of them accepted the challenge (...)[21]

This episode of Primo Levi's memories let us see clearly this narrative challenge and how it was already present just after his liberation. Being a survivor meant for Primo Levi to become a writer, though he really worked as a chemist all his life. When he thought that after three books he had won the challenge, a creative block took possession of him as he confessed to Bianca Giuddetti, one of his friends, just three days before the suicide.

> He, who had lived to testify, said that except for some details, he had to reread his books to raise his awareness to come out the year that he spent in Auschwitz.[22]

I present a table (see Figure 7) to distribute this set of relations with evil, harm and suffering in connection to a sample of diverse features and trying to make up a narrative model in which the risk is a part. According to my previous argument, the linkage between 'object of' and 'objects at', defines the character of a relationship

	LINKAGE	KNOWLEDGE	ACCEPTABILITY	EXTENT	DURATION	STRENGH	STYLES
UNCERTAINTY	Possibility	Little	High	Wide	Long	Low	Oral
RISK	Probability	Much	Medium	Situated	Medium	Medium	Mass Communication
FEAR	Proximity	Nothing	Low	Close	Short	High	Literature
PANIC	Action	Emergent	None	Closer	Short-Medium	Very High	No style
TERROR	Experience	Very much	None	Located	Very Long	Very high	Memories

Figure 7. A narrative model.

which can be in terms of possibility to narrate uncertainty, probability to risk, closeness to fear, action to panic and experience to terror. The amount of available knowledge makes also a significant difference because uncertainty shows a scarce amount of knowledge whereas risk needs some expert and massive knowledge. Fear is so acute that it blocks any knowledge and even dissolves the existing one. The pass from risk to fear is due to the awareness of vicinity to an object of risk which becomes then an object of fear breaking down the narrative of risk. There is not a previous story but a later one, which has a predominant literary form. In a certain way the description of fear has been a creation of literary and visual arts. Panic has no knowledge but can force the emergence of a new one, not being expert or institutional, bur spontaneous and self-governing. Terror promotes a deep knowledge as the result of long and intense remembering.

Other features, included in this table, are related to a degree of acceptability, extent, duration and strength. Finally different kinds of narrative are estimated. From the oral narration which is more frequent when we have a situation of uncertainty to the communication of risk through the mass media, the literary or visual narration of fear and the action of remembering the past to articulate memories when the narration pretends to tell the experience of terror. Different narratives and a variety of rhetorical expressions provide a wide field to be researched. This should be a promising research line if we focus on the classification of these different kinds of narratives.

Notes

1. This project, *Network for the Research into the Construction of Environmental Risk* was funded by The European Science Foundation (ESF) and included researchers from five European universities: Asa Boholm (Göteborg), Françoise Lafaye (Bordeaux 2), Gaspar Mairal (Zaragoza), Amalia Signorelli (Napoli) and Peter Simmons (East Anglia).
2. *Representing Risk*. Network for the Research into the Construction of Environmental Risk. http://www.unizar.es/risk/ (2000–2001).
3. *Conclusions*. Network or the Research into the Construction of Environmental Risk (2000–2001). See: http://www.unizar.es/risk
4. Thompson and Dean (1996) distinguished between 'probabilist' and 'contextualist' conceptions of risk.
5. Pacioli, L. (1494) *Summa de Aritmética, geometría, proportioni et proportionalità*, Venecia.
6. Cardano, J. (1663) *Opera Omnia*, Amsterdam.
7. Tartaglia, N. (1556) *General Trattato di Numeri et Misure*, Venecia.
8. Huygens, C. (1657) *Ratiociniis in alae ludo*, Amsterdam.
9. The rainfall has been calculated between 200 mm and 250 mm. There was a maximum of 500 mm for eight minutes.
10. Emilio Pérez Bujarral was the author of this technical report.
11. This expert witness was Pedro Montserrat, a well known botanist who has been working for many years in the Pyrenaic Institute of Ecology.
12. Defoe's narrative has a main subject which is very coherent with my proposal to read *A Journal of the Plague Year* as an original narrative of risk, the individual alone or together with other fellow people, confronting horrible and unexpected events. Robinson Crusoe is one of literature's archetypes, a hero of modernity, colonialism and self-made man.
13. The English original corresponds to an edition from Penguin English Library (1966)
14. I used this concept to analyse how the 'mad cows' affected Spain in 2000. The public opinion, under the influence of mass media, connected this case with a collective poisoning

caused by an olive oil adulterated with colza oil. From 1983 almost 1000 people had died. The narrative of 'colza' case acted as a 'matrix' to promote a new narrative about the mad cows. See Mairal 2003.

15. I have read a very interesting inventory of historical ideas about risk in Jackson, Allum, and Gaskell 2006. http://www.qualitative-research.net/fqs/.
16. In my visit to the Oceanographic Museum in Corunna I could see at the entrance a big photograph of this black smoke column emerging from the Hercules Tower, the most characteristic monument of Corunna. The way this picture was shown suggested something like a lesson to be learnt by the visitors.
17. See 'Las cintas secretas del Prestige' ('The secret tapes of the Prestige'), *El País*, 11-12-2005.
18. Klemperer 1998. There is a Spanish translation from 2003.
19. Viktor Klemperer, a cousin of Otto Klemperer the famous musician, was not very fond of Zionism and, though his father had been a rabbi, he had joined a Protestant church.
20. *If This is a Man* (1958), *The Reawakening* (1963) and *The Drowned and the Saved* (1986).
21. Levi, P. *Los hundidos y los salvados* (Barcelona: Muchnik Editores).
22. Anissimov, M. (2001) *Primo Levi o la tragedia de un optimista*, Editorial Complutense, p. 534.

References

Defoe, D. 2001. *A journal of the plague year*. New York: The Modern Library.

Jackson, J., N. Allum, and G. Gaskell. 2006. Bridging levels of analysis in risk perception research: The case of the fear of crime. *Forum: Qualitative Social Research* 7, no. 1: 20. http://www.qualitative-research.net/fqs/.

Klemperer, V. 1998. *Ich will Zeugnis ablegen bis zum lezten (Tagebücher 1993–1945)*. Vol. II. Berlin: Aufbau Verlag GmbH.

Mairal, G. 2003. A risk shadow in Spain. *Ethnos* 68, no. 2: 179–91.

Ricouer, P. 1990. *Time and narration*. Vol. 1. Chicago, IL: The University of Chicago Press.

Thompson, P., and W. Dean. 1996. *Competing conceptions of risk*. http://www.piercelaw.edu/risk/vol7/fall/thompson.htm.

Resilience at risk: epistemological and social construction barriers to risk communication

Richard Stoffle[a] and Jessica Minnis[b]

[a]*Bureau of Applied Research in Anthropology, University of Arizona, Tucson, AZ, USA;* [b]*Social Science, College of the Bahamas, Nassau, Bahamas*

This paper is about the persistent failure of social scientists to bring into the Environmental Impact Assessment (EIA) process socially constructed environmental concerns held by potentially impacted communities. The failure to communicate perceived risks results from a two-communities divide based on both epistemological differences and obfuscation due to vernacular communication. The analysis provides robust modeling variables that can bridge this social-environmental divide. The case involves data collected from members of traditional communities regarding their perceptions of the potential impacts of proposed Marine Protected Areas (MPAs). The study is situated in the Bahamas where the government has approved setting aside 30 No-take MPAs to protect their sea. This analysis is based on 572 interviews conducted during eight field trips with members of six traditional settlements in the Exuma Islands and Cays in the central Bahamas. Confidence in the findings is high because the sample involves 34% of the census population of these settlements and the findings have repeatedly been returned for review and approval by the members of these settlements.

Introduction

This paper is about the communication of perceived risks from potentially impacted communities to social scientists charged with documenting and weighing these impacts in an Environmental Impact Assessment (EIA). Two-Communities theory (Caplan 1979) is offered as a heuristic frame for both studying and discussing this problem. The Two-Communities theory was formulated to explain why scientific knowledge produced by academic-based researchers is not used by government policy makers. The theory assumes that these kinds of people live in separate worlds with different and often conflicting values, different reward systems, and different languages. Caplan concludes that the resultant communication difficulties are culturally based and Kerr (2007, 1415) shows that Two-Communities communication is a persistent problem. Although some have argued that the theory is more of a metaphor than a model (Dunn 1980), the Exuma analysis documents that the theory can be operationalized for risk communication research on the relationship of risk bearing communities and the scientific community.

Local communities often perceived risks to their relationships with the environment, but social scientists frequently are intellectually and procedurally blocked from bringing these impact concerns to an EIA. It is argued here that complex nature–society co-adaptations tend to be conveyed by local people with surrogate imagery that masks more than illuminates the perceived risks. Complex co-adaptations, even once translated into social science variables, are normally ignored because they are beyond most extant EIA models. So this analysis indicates that there are robust variables for modeling the interface between natural and social dimensions of the EIA process. These research findings are most useful in situations where a traditional community is co-adapted with its natural resources and the relationships between the two are potentially altered by a proposed development or conservation project.

Social-ecological modeling

Social Impact Assessment (SIA) has a responsibility to bring critical human variables into the EIA process. Even though it is analytically possible to study social impacts apart from associated natural impacts, it is clear to human ecologists and others that these two are inseparable. The EIA process through regulation and practice has ghettoized human and natural studies and thus provides few models for integrating the two. Recent EIAs have moved to resolve differences in research and analytical methods so as to produce a holistic model of potential project siting impacts that more accurately reflects what can occur to people and nature together.

Social and natural variables are used in decision models to help design, site, evaluate, and monitor Marine Protected Areas (MPAs). Often the natural variables are better understood and more robust than are the social variables. Despite the efforts of social scientists to provide guidance that would bring critical social variables into the modeling process (Bunce et al. 2000; Mascia et al. 2003), little progress is evident. Three examples illustrate this modeling problem and the social science responses: (1) Sala et al. (2002) modeled MPA placements in the Upper Gulf of California, Mexico and was critiqued by Sanchirico et al. (2003), (2) Roberts et al. (2001) modeled the famous Soufriere MPA in St. Lucia, West Indies and was critiqued by Christy et al. (2003), and (3) Mumby et al. (2006) documented the social value of increases in the natural fish stocks outside a Bahamas MPA, but the social science data are much weaker than the biology.

The present analysis argues that even though there is a fundamental divide between scientists and communities at risk from MPAs, these risks can be communicated and ultimately modeled. The analysis compares recent field data with the Bahamas national MPA modeling and thus addresses both this model and the implications of modeling choices. The Commonwealth of the Bahamas is a special case of coral reef and marine ecology protection because the government adopted into law the recommendations of a science-based report that evaluated approximately 30 No-take Marine Protected Areas (MPAs) using a model that equally weights social and natural variables (Stoner, Hickson, and Dahlgren 1999).

This model assumes that the presence of a small coastal community is a positive advantage for siting and managing the proposed MPAs. This modeling decision reflects a common assumption that establishing successful MPAs requires local support and cooperation (B. Stoffle et al. 1994; R. Stoffle et al. 1994; Keller and Recchia 1998; National Research Council 2001; Berkes 2004; Aswani and Lauer 2006). This study supports that assumption.

A second modeling question addressed by this research is what dimensions of community are most robust for predicting their response to nearby MPA conservation. The Bahamas MPA modeling uses simple social demographics; however, this study argues for the use of three more complex and robust social variables (agency, resiliency and identity) for predicting the community-level responses to MPA siting.

A third modeling assumption is that MPAs are more likely to impact communities that are economically dependent on cash from marine products, but that a shift to other economic sectors like ecotourism can off-set the loss of commercial fishing. This study suggests that subsistence-based traditional communities are very threatened by MPAs, and economic sector shifts are not an appropriate mitigation for either the natural or social impacts of MPAs.

Methods

Especially important to the siting of MPAs in the central Exumas are 20 traditional coastal communities, typically less than 200 in population, located in the Exuma islands and cays where three large No-take MPAs were proposed by the Bahamian government. Six traditional communities were selected for study, with the approval of the Bahamas Department of Fisheries and after the political and social leadership of each settlement was approached with a study description and agreed to proceed. Each community is near and thus potentially impacted by one of the three proposed MPAs (Figure 1). Before this study began, each community had publicly responded to the nearby MPA proposal – with the two northern communities being strongly against the MPA, the two central communities being neutral, and the two southern communities being strongly positive. Interviews were designed to understand what MPA impacts were perceived by these sets of communities and how potential impacts contributed to why communities responded differently to their proposed MPA. In the first round of interviews it became clear that local people were expressing themselves in vernacular language and expressions that would have to be unpacked in order to make their perspectives accessible.

This analysis derives from 572 interviews conducted over six years. Twenty-eight students and three professors participated in the study. Interviews were guided by the Sea Attachment instrument which focused on seven knowledge domains, each related to one kind of potential relationship between the people and the sea. A series of follow-up instruments (Quality of Life, Diachronic Sea/Land Mapping, Grubbing, and Tourism) were designed to provide specific data on what emerged as key issues. Extensive oral histories with community elders strengthen diachronic knowledge of natural resource and social adaptive processes.

Each interview was conducted in private and guaranteed to be held confidential. Approximately one third of all census registered members of these settlements were interviewed at least once. Oral histories were developed over multiple field sessions and involved dozens of hours of interview time plus feedback-confirmation sessions.

Interviews were conducted in iterative waves as the field teams returned eight times. These mixed methods (Tashakkori and Teddlie 1998; Beebe 2001) permitted new insights to emerge from the analysis of previous field data before returning with

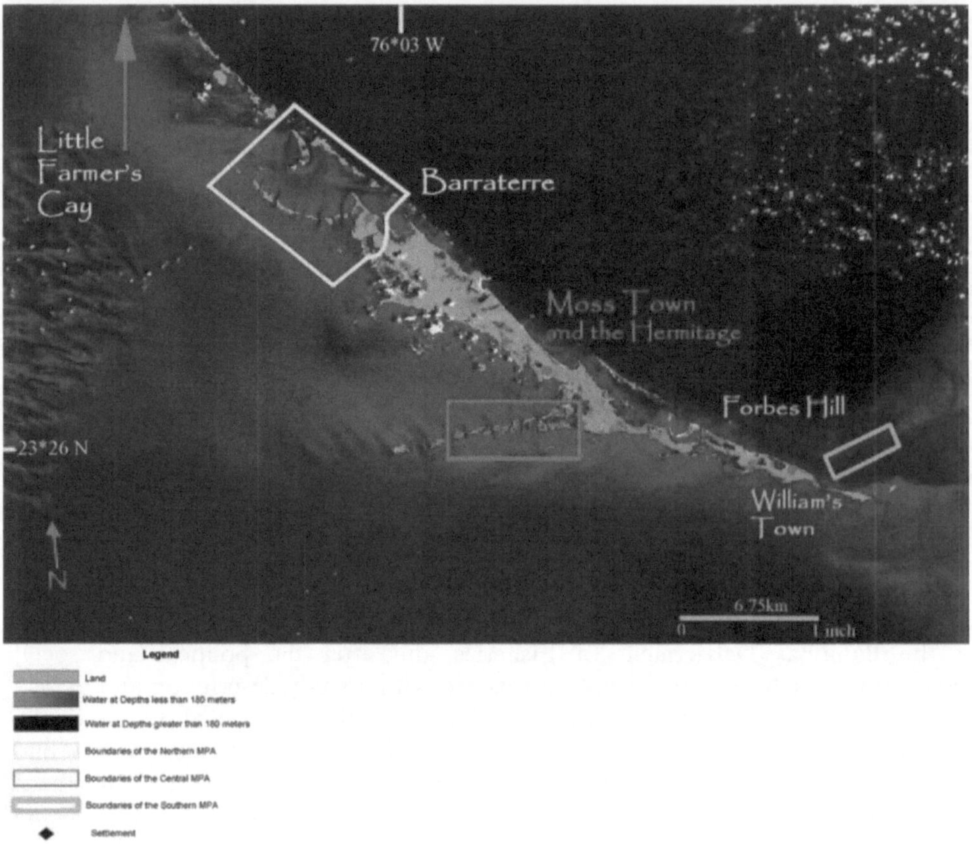

Figure 1. View of Exuma island and cays with five of six study communities marked.

expanded follow-up queries and new instruments. All major findings were triangulated (Campbell and Fiske 1959; Jick 1979) by multiple survey questions, oral history, and documents. Key findings were hand delivered to most people who had been interviewed and discussed openly in each settlement so as to receive feedback and confirmation. The findings are viewed has having a high confidence level, as this non-statistical version of the concept is understood in cultural anthropology.

Results

The research supports the following: (1) the community is the best unit of social analysis, and these are traditional communities having continuity back to slavery, (2) three social variables - agency, resilience, and identity – robustly predict community responses to MPAs, and (3) a wide-spread pattern of vernacular communication of risk perception obfuscated underlying risks which in turn had to be unpacked by systematic data collection.

Community as unit of SIA analysis

The Two-Communities theory must be operationalized to fit into specific research situations. Much has been written on the definition of the *science community*

(Goldman 1999, 221–71), so this analysis focuses on kinds of *local risk communities*. Social Impact Assessment (SIA) is the proper term of reference for studies that predict impacts from potential MPAs and assess human responses (Goldman 2000; Weiant and Aswani 2006). The SIA literature on the impacts of conservation projects variously argues for a focus on the individual resource user, the household of the resource user, and the resource dependent community. Each of these approaches has analytical utility, but in the past decade the emphasis has shifted away from individual stakeholder and household analysis to the study of the local risk community. In general, the community analysis approach works best when the community is small in scale, old in place duration, stable in membership, and composed of people who share a sense of history, have a common identity, and are culturally co-adapted with their land and sea. The members of the community should not have *ontological insecurity* (Giddens 1990), but should instead believe that their future generations will have sustainable access to natural and social resources. Indeed, Conroy et al. (2001), building on Ostrom (1992), suggests that for conservation studies and consultations the term *community* be restricted to those having shared beliefs, stability of membership and complex, multilayered, long-term interactions with each other.

Members of the six communities immediately stipulated in initial interviews that they represent a people who have always learned about and cared for their land and sea in the Exumas. They thus defined themselves as a kind of community (which is termed a *traditional community*) to be dealt with during the MPA siting, design, and management phases.

The African ancestry people of the Exumas arrived in 1784 as slaves of Loyalist citizens of the English crown. Some had been slaves in the American colonies (creoles), but others were recent arrivals from Africa. Both types of former African people arrived as unfree laborers for planters who were escaping the aftermath of the American colonial revolt. Because at this time there were no traditional American Indian people in the Exumas, both the Loyalist planters and the African slaves were confronted with learning and using a new ecosystem.

Most Loyalists planters failed at cash-crop and slave-based agriculture and left their lands and slaves alone within a decade or so. Plantation lands, adjoining marine ecosystems and the former slaves remained together, with the people often taking the name and the land of the former plantation owners. With few options for leaving and little competition for the local resources the former slaves, now self-identified as community members, stayed and converted knowledge of the marine ecosystem into sustainable environmental practices, and assumed de facto ownership (via usufruct) of their community's ecosystems.

The social-ecological systems developed by the traditional people of the Exumas persisted through time, surviving human (such as overseas labor) and natural (such as hurricanes) perturbations. It is argued here that through environmental learning, including acquiring knowledge of marine biodiversity, the traditional people of the Exumas made a resilient way of life which is valued today.

These six communities from the islands and cays of the Exumas meet Conroy's criteria for a local natural resource community. For various reasons these Exumian communities tend to act in unison on issues such as outside development and conservation. Community members, for example, were uniform in their support for establishing a nearby MPA. Despite overwhelming support for establishing a nearby

MPA, there were strong negative public responses by members of the two northern communities and many other respondents qualified their support.

SIA variables: agency, resilience, and identity

Community MPA responses included almost unanimous support for the idea of establishing a nearby MPA while at the same time resisting establishing certain kinds of MPAs. These apparently conflicting responses can best be understood in terms of three social variables: agency, resilience, and identity.

Agency: 'When you come into my house you show respect'

Members of local communities consistently said that they should be in charge of decisions regarding their traditional lands and seas, so outsiders should always respect this stipulation. The vernacular phrase 'When you come into my house you show respect' implies that outsiders must know about and acknowledge local history and resource ownership through usufruct. Failures to properly respond to this stipulation caused negative responses to the MPA proposals and proponents. Because the Bahamian national government does not recognize usufruct ownership of either land or sea, scientists have not seen the utility of understanding what risks MPAs present to 'my house'.

Agency is the contemporary term used to describe the sense that what is desired by the community will in fact occur. The QofL instrument asked about the person's perceived ability to influence the outcomes of events affecting them and their settlement and how this influence compares with that of the people in the community 30 years ago, and will compare with the people of the community 30 years in the future. When asked 'In comparison to members of my family and people living in this settlement 30 years ago, I have (1) much more influence, (2) a little more, (3) about the same amount, (4) a little less influence, (5) much less?', of the 128 respondents, the majority (57.6%) perceived themselves to have more influence, while 14.4% perceived themselves to have the same and 28% perceived less influence. When asked 'How satisfied are you with your ability to influence the outcome of events affecting your settlement these days?', 50% are satisfied, 25.7% are comfortable, and 24.2% are dissatisfied. Perceived positive trends in agency are viewed as causing the next generation to have more (70.3%) or the same (20.3%) ability to influence events. The QofL responses indicate that these communities have a strong sense of agency, which they say derives from working together as a community, being self sufficient, and having a local government system that gives them a direct voice in the Bahamas Parliament.

If an MPA proposal assaults agency, what will result? The northern communities were presented with an MPA proposal that they soundly rejected by personal, community, and political responses. When this study began, the northern MPA proposal had been taken out of consideration by the Department of Fisheries, so in one sense this was an agency victory for the northern communities. When, however, the QofL responses are presented by region of settlement, the northern were less satisfied (44%) with their ability to influence events than central (54%) and southern (48%) communities. People from the north report that confidence in their ability to influence the outcome of events was shaken by the northern MPA proposal and the proposal process which by-passed the communities.

Having been through an unsuccessful MPA siting process, even when the final decision supported the local community position, appears to have had implications for future MPA proposals. When, during the QofL interviews, people were asked 'Who should design and control a local MPA (local people, national government representatives, or mixed local and national)?' 64.1% favored a partnership with the national government, rather than a local model (18.8%) or a national model (17.1%). Support for the national MPA model is lowest in the north (2.6%), and slightly higher in the central (6.8%) and the southern (8.8%) communities. Interviews document that after the northern MPA proposal failed, many northern people simply did not trust in an MPA proposal exclusively designed and managed by the national government and their science advisors.

Resilience: 'How we gonna live if you take our sea?'

The MPAs were designed by scientists to protect the sea and increase its biodiversity, but local people expressed the concern 'How we gonna live if you take our sea?' To the outside scientist, taking some of the sea as an MPA would protect the sea elsewhere, but local people neither use nor own through usufruct all of the sea. Also, science-based recommendations that people shift their production to other economic sectors fail to recognize the social and cultural implications of traditional natural resource uses. 'Gonna live' implies social support from community as well as food for an individual.

The people of these communities speak with pride about their common history as African ancestry people forcibly brought as slaves to this area in the late 1780s and then abandoned within a generation. By remaining on the lands of their former plantation settlement, these people co-adapted (some would call co-evolved) with this socially and physically isolated portion of an enormous archipelago island system. Together as a community, less as individuals and households, they built a resilient way of life involving redundancies in natural resource use strategies and webs of interdependent social relationships.

Resilience means the magnitude of disturbance that can be absorbed or buffered without the system undergoing fundamental change in its functional characteristics. When people become traditional, learn about their ecosystems, adjust their adaptive strategies to protect them from natural and social perturbations, they then can be said to have developed a resilient way of life (Stoffle, Toupal, and Zedeno 2003). The term *environmental multiplicity* builds on the narrower but established term *occupational multiplicity* (Comitas 1964), to describe their system of resilient adaptations. Conceptually these terms describe a range of multi-stranded and redundant connections among the members of a traditional community and between them and their primary natural use areas (Stoffle 1986).

There are too many adaptations to be fully discussed here, so these are organized by where they occur in natural and social space. There are six natural use spaces and two social spaces (Table 1) that constitute the core of environmental multiplicity. This table points to where people interact with each other and the natural resources (under community control through usufruct). Because of redundant natural use areas and social relations, there are alternative ways of socially and naturally achieving community goals. Everyone, for example, has the ability to fish, but at any given time or period the fishing can occur with a team of women who do hand

Table 1. Environmental multiplicity in natural and social spaces.

Natural use spaces		Social spaces	
		Reciprocity from subsistence	Wage labor
Marine	Littoral	Grubbing, sea weed collecting	
	Near shore coral reefs and cay cuts proximal deep water	Subsistence fishing alternate crew fishing	Bone fishing guides Lobster fishing
Terrestrial	Forests	Gathering medicine and spiritual plants for family and community healing	Collecting and weaving palm top
	Agricultural lands	Home gardens Clearing of lands	Cash crops
	Pastures	Domestic use animals	

fishing called grubbing or a small male crew in a boat. When all the males are available to fish, only a few crews will fish at a time and then share the catch with each 'cooking pot' in the settlement. Thus, crews who fish only catch sufficient fish for the evening meal for all the community. Different crews have different fishing areas, thus spreading fishing pressure over a wide area by taking turns fishing. Men who are not fishing often work with others on community projects that also benefit the families of the crew who is currently fishing. Natural resources can derive from both land and sea, thus assuring that crop loss due to drought can be offset by the marine resources, which in turn can be replaced with land products when hurricanes disrupt the marine ecosystem. The environmental multiplicity system established by these traditional communities assures that people do not starve or live in fear, but instead live well and secure being always supported by their nature and society.

Community relationships are facilitated by symmetrical reciprocity; which is an implicit, non-legal contractual obligation, unenforceable by any authority apart from one's sense of honor and shame. Social rules specify that natural resources be gathered and shared with others, who should in turn offer labor or other resources. Symmetrical relationships are designed not to be extinguished, thus potentially sustaining relationships between individuals and social groups over generations. Given the subsistence economy of these settlements over the past 222 years, most community support obligations do not involve cash exchanges. The main exception to this is the need for cash to purchase rum which is exchanged in reciprocity. Thus social structure of the community relies on materials from direct production, personal labor, and some cash.

A shift to a cash-base economy potentially threatens social reciprocity. In the QofL interview *community support* was defined as the ability and willingness of other members of the settlement to come to your aid as needed. Such support could come from group organizations such as local churches, family members, or neighbors. When asked 'In comparison to members of my family and people living in this settlement 30 years ago, is there (1) much more support, (2) a little more, (3) about the same amount, (4) a little less support, (5) much less support?' of the 133 respondents, most said there was less support (57.1%), some said it was about the same (20.3%), and some said it was better than 30 years ago (22.5%). Follow-up interviews identified the increase in wage labor and a growing standard of living (consumption) as the reasons for declining community support. Still, when asked about trends most respondents (54.3%) believe that community members recognize this as a social problem that will be solved in the future by returning to past levels of direct production and community support.

Resilience is impacted in a positive way when an MPA restores the adaptive strategies formerly lost by a traditional community. These losses may have occurred because a resource declined below a sustainable level, outside fishers eliminated spawning aggregations, or foreign tourists purchased community usufruct lands. The MPA can restore fishing stock, restrict outside fisher access, and officially empower local communities as owners and managers of traditional resource areas.

Resilience is weakened to the extent that the MPA eliminates extant adaptive strategies. This can occur if the MPA mitigates the loss of a reciprocity-based subsistence fishery with cash and wage labor opportunities or if the MPA fails to protect traditionally used cays and mangroves from tourism and other developments. Traditional communities throughout the Caribbean use symmetrical reciprocal exchange based on direct production (Burpee, Morgan, and Dragon 1986); so, if an MPA reduces or eliminates access to natural resources needed in such exchanges it can weaken community social structure. Often MPA planners attempt to mitigate this impact by stimulating cash-based production, but this analysis shows that cash itself can weaken community support. A weakened community may lack the ability or will to support its members in times of crisis.

Identity: 'This project will kill us!'

'This project will kill us!' makes no sense to scientists who see no possibility that people will die just because they cannot use a portion of the sea near their community. Such vernacular expressions are easily dismissed as unfounded emotional outbursts (Wardman 2006). Humans, however, are social beings whose identity is tied in real ways to relationships with others and social organizations. Thus a threat to 'kill' a person can involve a risk to identity as well as to body.

Identity as it occurs at the individual, community, ethnic group, and national levels is an important cultural dimension for all people, but it is especially critical in the post-slavery former European colonies of the Caribbean (Munasinghe 2001). Caribbean people, especially those of African ancestry, live in societies that were not made for them (Alleyne 2002). Once independent, they were free to establish an identity not tied to the former colonial powers (Olwig 2002). This identity formation process is perceived by some as the key social process in the Caribbean today.

Community identity is recognized as a key variable in US Federal government guidelines, prepared in compliance with the Magnuson Stevens Fishery Management

Act (MSA) National Standard #8, which specifies that fishery dependent communities can be identified by data documenting that residents perceive fishing to be important to the continuity and self-identification of their community.

Identity is a complex human phenomenon (Gupta and Ferguson 1997a) and should be studied with a clear operational definition that makes sense in a specific research setting (Gupta and Ferguson 1997b). One approach to defining the identity of coastal communities is to measure the *cultural centrality* of the sea. The 14-page Sea Attachment instrument, which was administered in all six settlements, asked questions about seven knowledge and use domains including: Material Arts, Sea Biology, Underwater Landscapes, Land Biology, Expressive Arts, Identity Symbols, and Settlement Stories. The resulting 69 interviews document that these community members were both knowledgeable of and emotionally connected with the sea. They write, and read in public, poetry about the sea, bring its products into their arts, and sustainably use it on a daily basis. The main church of each settlement is situated on the highest ridge where members can view the sea from its many windows, and the most popular hymns have a marine theme. They say without their uses of sea they would not exist.

Although in most cases an MPA will only modify some local access to the sea, a large No-take MPA can impact community identity by preserving all customary access to the sea or eliminating all such access. For example, the proposed northern MPA encompassed almost all of the traditional marine and terrestrial resource use areas for the community of Barraterre (Figure 2). It should be noted that even though the Bahamian No-take MPA does not limit terrestrial uses on cays within its boundary, boaters cannot legally have marine products while in the MPA, so if they caught fish elsewhere they could not stop in the MPA on the way home to collect plants in the cays. A strict interpretation of no-marine-products-in-a-boat-in-the-MPA rule would preclude all Barraterre fishing because they could not return to their home which was completely surrounded by the MPA.

The leeward cuts involved in this MPA create unique underwater places that attract many species of marine animals. Due to their leeward position and proximity of these cays to the community of Barraterre, these cays and cuts traditionally are primary subsistence fishing grounds. The cuts between the windward cays have extremely strong currents and are difficult to fish. The leeward cuts with the slower currents have coral reefs on either ends with sea grass beds in the middle; there, fishers target snappers, groupers, grunts, and jacks. These cuts have been traditionally fished for food by both men and women; the latter traditionally frequented the leeward locations by sculling (single oar propulsion) their boats to a desired spot where they fish by line, dive for or hook conch, and even catch sea turtle. Women also use these leeward cays to gather medicinal plants and palm top for basket weaving material.

The northern MPA was so large and totally positioned over traditional marine use areas that it basically eliminated fishing and cay uses for the people of Barraterre. Their strong negative response to the MPA was (the issue of agency was also a factor) because they believed it would 'destroy them' – that is, by eliminating all subsistence fishing activities and associated reciprocity so they could no longer be the community they had become over the past 222 years. The community believed its identity was at stake.

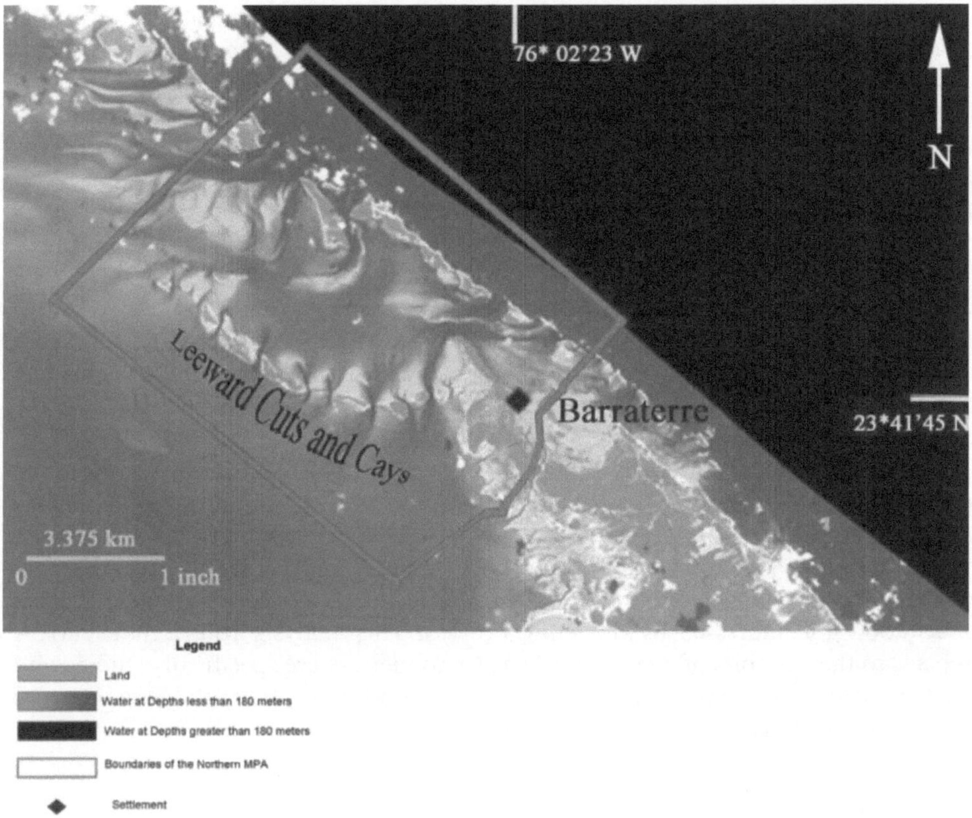

Figure 2. Northern No-take MPA around the community of Barraterre.

Discussion

This analysis argues for the utility of studying risk communication as a process occurring between two communities: scientists and local people. Epistemological and social constructions obfuscate the communication and application of risk knowledge between these two communities. Exumian risk communities communicated their concerns about the potential impacts of proposed No-take MPAs, but these concerns were framed in vernacular terms that had be translated into social science language and ultimately EIA models. Expressions like 'this MPA will kill us' are easier to dismiss than to model; however, with follow-up data collection all initial vernacular risk communications were made transparent and worked into variables that can be measured and modeled.

The marine environment was a lifesaver and a community builder in the past and it remains a *safety zone* in the minds of Exumian people today. Many of the marine and terrestrial use patterns developed after slavery have persisted simply because they are the foundation for a high quality way of life that has stood the test of time. As a *region of security* the marine environment has become a part of how people view themselves and where they teach the lessons of life to new generations. This region is viewed as belonging to the community it has given life.

The study did not have a research instrument that specifically studied the meaning of vernacular expressions of perceived risk from No-take MPAs. All survey instruments provided some useful interpretative insights, but critical to understanding what was being said was the presence of Bahamian students and faculty on the study team. The QofL interviews proved most useful for profiling the communities and understanding specific responses to proposed No-take MPAs. The field of QofL studies has become well established around the world, after having been developed at the Institute for Social Research at the University of Michigan (Campbell, Converse, and Rodgers 1976). Bandura (1999) documents that QofL is a strong measure of community agency. Still, other instruments are needed to contextualize the QofL responses, learn the history of the community, and measure the cultural centrality of the sea. Small samples and a few elder oral histories will suffice for profiling ideal communities; but more research is needed the less the community under study matches the Conroy et al. (2001) criteria.

The people in this study recognized that portions of their traditional way of life were being placed at risk by No-take MPA proposals and communicated these risks through vernacular expressions (only some of which are discussed here). Because marine scientists do not share this traditional culture and vernacular language (as explained in the Two-Communities theory) locally perceived risks were neither understood nor addressed by the siting process. This analysis argues that if MPA impacts to the variables of agency, resilience, and identity are specifically studied and if all vernacular risk expressions are fully understood, then common ground can be found and MPAs can be successfully sited.

Acknowledgements

We thank the Bahama Department of Fisheries for approving this research, and the people of the Exumas Islands and Cays for their participation. Funding was provided by the NSF Biocomplexity in the Environment Program (Grant OCE-0119976) to the Center for Biodiversity and Conservation at the American Museum of Natural History. This study is the joint effort by the University of Arizona and the College of the Bahamas. The COB team included Kendra Arnett, Chervain Dean, Tarah McDonald, Ward Minnis, Tavarrie Smith, and Yasmin Skinner. The UofA team included Alex Carroll, Clinton Carroll, Fletcher Chmara-Huff, Jill Dumbauld, Heather Fauland, Richard Gilmour, Arin Haverland, Cory Jones, Shawn Kelley, Noreen Lyell, Aja Martinez, Amanda Murphy, Nathaniel O'Meara, Kathryn Payne, Terra Pierce, Peter Poer, Daniel Post, and Kathleen Van Vlack.

References

Alleyne, M. 2002. *The construction and representation of race and ethnicity in the Caribbean and the world.* Kingston, Jamaica: University of the West Indies Press.

Aswani, S., and M. Lauer. 2006. Incorporating fishermen's local knowledge and behavior into geographical information systems (GIS) for designing marine protected areas in Oceania. *Human Organization* 65: 81–102.

Bandura, J. 1999. Social cognitive theory: An agentic perspective. *Asian Journal of Social Psychology* 2: 21–41.

Beebe, J. 2001. *Rapid assessment process.* Walnut Creek, CA: Altamira Press.

Berkes, F. 2004. Rethinking community-based conservation. *Conservation Biology* 18: 621–30.

Bunce, L., P. Townsley, R. Pomeroy, and R. Pollnac. 2000. *Socioeconomic manual for coral reef management*. Maryland, MD: National Ocean Service, NOAA.

Burpee, G., J. Morgan, and A. Dragon. 1986. *Household survey of Grenada: Findings and documentation of procedures*. Ann Arbor, MI: Institute for Social Research, University of Michigan.

Campbell, A., P. Converse, and W. Rodgers. 1976. *The quality of American life*. New York: Russell Sage Foundation.

Campbell, A., and D. Fiske. 1959. Convergent and discriminate validation by the multi-trail-multimethod matrix. *Psychological Bulletin* 56: 81–105.

Caplan, N. 1979. Two-communities theory and knowledge utilization. *American Behavioral Scientist* 22, no 3: 459–70.

Christy, P., B. McCay, M. Miller, C. Lowe, A. White, R. Stoffle, D. Fluharty, L. McManus, R. Chuenpagdee, C. Pomeroy, D. Suman, B. Blount, D. Huppert, R. Eisma, E. Oracion, K. Lowry, and R. Pollnac. 2003. Towards developing a complete understanding: A social science research agenda for marine protected areas. *Fisheries* 28: 22–6.

Comitas, L. 1964. Occupational multiplicity in rural Jamaica. *Proceedings of the American Ethnological Society*, 41–50.

Conroy, C., A. Mishra, A. Rai, N. Singh, and M. Chan. 2001. Conflicts affecting participatory forest management. In *Analytical issues in participatory natural resource management*, ed. B. Vira and R. Jeffery, 165–84. New York: Palgrave Publishers.

Dunn, W.N. 1980. The two-communities metaphor and models of knowledge use. *Science Communication* 1: 515–36.

Giddens, A. 1990. *The consequences of modernity*. Palo Alto, CA: Stanford University.

Goldman, A. 1999. *Knowledge in a social world*. London: Oxford University Press.

Goldman, L., ed. 2000. *Social impact analysis: An applied anthropology manual*. New York: Oxford – Berg.

Gupta, A., and J. Ferguson, eds. 1997a. *Culture, power, place*. North Carolina: Duke University Press.

———, eds. 1997b. *Anthropological locations: Boundaries and grounds of a field science*. California: University of California Press.

Jick, T.J. 1979. Mixing qualitative and quantitative methods: Triangulation in action. *Administrative Science Quarterly* 24: 601–11.

Keller, G., and C. Recchia. 1998. Editorial - lessons from marine protected areas around the world. *Parks* 8: 1–4.

Kerr, R. 2007. Pushing the scary side of global warming. *Science* 316, no. 8: 1412–15.

Mascia, M., P. Brosius, T. Dobson, B. Forbes, L. Horowitz, M. McKean, and T. Turner. 2003. Editorial: Conservation and the social sciences. *Conservation Biology* 17: 649–50.

Mumby, P., C. Dahlgren, A. Harborne, C. Kappel, F. Micheli, D. Brumbaugh, K. Holmes, J. Mendes, K. Broad, J. Sanchirico, K. Buch, S. Box, R. Stoffle, and A. Gill. 2006. Fishing, trophic cascades, and the process of grazing on coral reefs. *Science* 311: 98–101.

Munasinghe, V. 2001. *Callaloo or tossed salad? East Indians and the cultural politics of identity in Trinidad*. New York: Cornell University Press.

National Research Council. 2001. *Marine protected areas: Tools for sustaining ocean ecosystems*. Washington, DC: National Academy Press.

Olwig, K. 2002. *Global culture, island identity*. London: Routledge.

Ostrom, E. 1992. Community and the endogenous solution of commons problems. *Journal of Theoretical Politics* 3: 343–51.

Roberts, C., J. Bohnsack, F. Gell, J. Hawkins, and R. Goodridge. 2001. Effects of marine reserves on adjacent fisheries. *Science* 294: 1920–23.

Sala, E., O. Aburto-Oropeza, G. Paredes, I. Parra, J. Barrerra, and P. Dayton. 2002. A general model for designing networks of marine reserves. *Science* 298: 1991–3.

Sanchirico, J., R. Stoffle, K. Broad, and L. Talaue-McManus. 2003. Modeling marine protected areas. *Science* 301: 47–8.

Stoffle, B., R. Stoffle, D. Halmo, and C. Burpee. 1994. Folk management and conservation ethnics among small-scale fishermen of Buen Hombre, Dominican Republic. In *Folk management in the world's fisheries*, ed. C. Dyer and J. McGoodwin, 115–38. Boulder, CO: University of Colorado Press.

Stoffle, R. 1986. *Caribbean fishermen farmers*. Ann Arbor, MI: Institute for Social Research, University of Michigan.

Stoffle, R., D. Halmo, T. Wagner, and J. Luczkovich. 1994. Reefs from space: Satellite imagery, marine ecology, and ethnography in the Dominican Republic. *Human Ecology* 22: 355–78.

Stoffle, R., R. Toupal, and N. Zedeno. 2003. Landscape, nature, and culture: A diachronic model of human-nature adaptations. In *Nature across cultures*, ed. H. Selin, 97–114. London: Kluwer Academic Publishers.

Stoner, A., M. Hickson, and C. Dahlgren. 1999. *Scientific review of marine reserve network proposed for the commonwealth of the Bahamas by the Bahamas Department of Fisheries*. Nassau, Bahamas: Bahamas Department of Fisheries.

Tashakkori, A., and C. Teddlie. 1998. *Mixed methodology: Combining qualitative and quantitative approaches*. Thousand Oaks, CA: Sage Publications.

Wardman, J. 2006. Towards a critical discourse on affect and risk perception. *Journal of Risk Research* 9, no. 2: 109–24.

Weiant, P., and S. Aswani. 2006. Early effects of a community-based marine protected area on the food security of participating households. *SPC Traditional Marine Resource Management and Knowledge Information Bulletin #19*, 16–31.

Scientised citizens and democratised science. Re-assessing the expert-lay divide

Rolf Lidskog

Centre for Urban and Regional Studies and Man-Technology-Environment Research Centre, Örebro University, Sweden

During the last decade there have been growing calls for increased public inclusion in risk regulation. This paper investigates three of these proposals for a new relationship between science and the public, namely New Production of Knowledge, Postnormal Science, and Scientific Citizenship. These all concern how science can be democratised and how new relations between expertise and citizens can be negotiated and designed. By critically discussing the similarities and differences between these proposals, this paper examines the implications of the call for public inclusion in risk regulation. By way of conclusion, some warnings are raised concerning the belief in public inclusion as a cure-all for making knowledge production and risk regulation more publicly credible and socially robust. The space created for public inclusion may work as means for legitimating decisions, diluting accountability and persuading the public, with the consequence that the expert-lay divide may be reproduced rather than transformed.

Knowledge beyond our senses?

In June 2003, Reuters reported that plunderers sold barrels for two dollars each to the inhabitants of an Iraqi community in the district of Madaenna, south of Baghdad. The people used the barrels to wash clothes and store food. After a while their children became sick, and one of them, a three-month-old baby, got burns on her back. Later it was revealed that the barrels originated from an Iraqi nuclear research centre and its waste plant; yellow cake – uraniferous waste – had been stored in the barrels. The plunderers had poured out the contents, rinsed the barrels, and then sold them to the local people. They had no information that the barrels were contaminated, and when children became sick it was impossible for them to make a connection between cause and effect, between eating food that had been stored in the barrels, or wearing clothes washed in them, and the illness of the children.

Even if this story may be a 'radioactive rumour', a possible media fiction, this narrative conveys the truth that scientific findings and recommendations often appear meaningless and irrelevant from the perspective of lay people. Another apposite example of this can be found in the Belorussian journalist Aleksijevitj's interview with people involved in the Chernobyl catastrophe. A village woman tells about her meeting with experts:

> We've always lived on bread and potatoes, but now they said we mustn't! You didn't know whether to laugh or cry.... They told us to wear gauze masks and rubber gloves when we were working in the vegetable garden.... Then along came some pompous idiot of a scientist and gave a talk to the village club and said we had to wash the firewood.... You couldn't believe your ears. (Aleksijevitj 1997, 46, my translation)

Both these stories are examples of how difficult it can be for lay people to understand expert advice aiming to protect them from certain risks. The latter story also shows that scientific recommendations can clash with their everyday lives and practical understanding. For instance, the mothers of Chernobyl schoolchildren were advised to wash their clothes every day, to wash out the radioactive dust. However: 'Dirt – that meant ink-, earth- and grease-stains to them, not some kind of short-lived isotopes' (Aleksijevitj 1997, 133, my translation).

We are, as Ulrich Beck (1992) and Anthony Giddens (1990) strongly emphasise, dependent on different expert systems and expert knowledge. As laypersons we are unaware of these risks and must rely on science, the social institution with the competence and authority to establish their nature and extent. This applies not only to technological disasters such as that in Chernobyl, but also to present-day creeping and diffuse environmental problems, for example acidification, ozone depletion, and climate change. Thus we live in a society where many risks are fundamentally research-dependent, and where experts not only have a central role in guiding policy-making, but also in guiding people's choices in their everyday lives. In these cases, citizens seem to have no choice but to trust experts in order to achieve a healthy and safe life, and governments may wonder why lay people do not always follow their recommendations; why do they ignore some risks when there seems to be strong scientific evidence that these risks present a real threat to health and the environment? In this case, science seems to have the responsibility to tell the truth to people, and risk communication is an important instrument for informing people about these risks and relevant coping strategies.

However, this picture of science-dependent risk as being impossible for lay people to grasp can be questioned, at least if it is taken too far. There are plentiful examples of scientific findings and recommendations that have been shown to be irrelevant or incorrect, and of lay people's ways of understanding and managing a situation having been more adequate than recommendations from experts. Among the most well-known cases is Brian Wynne's (1989, 1992, 1996) study of Chernobyl's radioactive fall-out in West Cumbria. Sheep-farmers complained to the authorities that the appointed scientific experts misinterpreted the consequences of the fallout because they ignored important knowledge, not least the farmers' knowledge about variations in types of soil in the area. Hence, those representing the local community showed a greater reflexive capacity than those representing the scientific community. In this case, research dissemination is not the solution, but instead a better-developed relationship between science and the public.

Another example is the case of popular epidemiology where lay people experience health effects and start to trace their causes, despite the fact that governmental bodies, private companies, and expert communities persistently argue that there is no reason to suspect that the environment is contaminated (Brown

1997).[1] There are many examples where lay people's findings have been proven to be valid, and experts have had to change their original standpoints. Hence, there also seems to be reason for lay people to mistrust science, and science itself fosters this mistrust by failing to recognise the public and other stakeholders as knowledgeable actors who can deal with complexities and uncertainties.

We are living in a paradoxical situation where lay people have become increasingly conscious of various risks at the same time as these threats in many cases seem to be further away from – in many cases beyond – their perception as well as control. They often have to trust science for guidance in how to act at the same time as science has often proven to be incorrect. This has led to the frequent emergence of conflicting claims, interests, and viewpoints about risks, and we face a competitive situation where different actors struggle to be seen as legitimate providers of knowledge on environmental risks.

During the last decade a number of proposals have been developed aiming to tackle this situation of conflicting and contested knowledge claims. They all concern how science can be democratised and how new relations between expertise and citizens can be negotiated and designed. The proposed solutions differ, but all advocate inclusive strategies within knowledge production: that citizens should not only be seen as passive receivers of scientific knowledge, but also as active partners in the production of knowledge; and that science needs to become more reflexive and transparent, making its own assumptions and values visible. Earlier demarcations between science and lay people should be transgressed through a democratisation of science and a scientisation of the citizenry.

But to what extent and in what ways is this relevant when approaching risks that are heavily dependent on science? And if citizens are welcome to deliberate and negotiate on these kinds of risks, what can they contribute and what implications will it have? By introducing and assessing some recent proposals for a democratisation of science, this paper aims to investigate the role of experts and lay people in the production of knowledge about risks. It will do so by introducing and evaluating three proposals for a new relation between science and citizens.

My discussion primarily concerns risks to health and the environment, and special attention is given to risks that are remote from ordinary people's perceptions, and which therefore are heavily dependent on expert systems. The reason for this focus is that whereas both the scientific and political communities today emphasise the need for inclusive strategies to be able to deliver credible knowledge and legitimate political decisions, at the same time it is claimed that the public has a limited capacity to understand certain kind of risks and should therefore not have any substantial influence over the management of these kind of risk. This view is not least to be found within the nuclear sector, where many governmental agencies as well as nuclear waste companies claim that citizens have to be taken into consideration only because they are a part of a political reality, and that they have nothing to add from a knowledge perspective (Lidskog and Sundqvist 2004). Thus, a decisive question is to what degree and in what sense citizens can contribute to the knowledge production with regard to these kinds of expert-dependent risks.

The paper comprises five sections, this introduction being the first. In the next section, the problem is specified in greater detail by discussing invisible risks, risk management, and public inclusion. Risk management is currently in the seemingly paradoxical situation of asking for more science as well as more public involvement. The third part analyses three recent proposals for a new relation between science and citizens: 'New Production of Knowledge', 'Postnormal Science', and 'Scientific Citizenship'. All these proposals share a number of similarities, but also display some differences. In the fourth part I discuss the relevance of these approaches to the discussion of invisible risks. In the fifth and concluding section, I assert that science and risk management need to be opened up to these kinds of risks as well. This requires the creation of deliberative spaces where citizens and scientists can interact, negotiate, and deliberate. However, at the same time the warning must be raised that these spaces can provide uneven opportunities for persuasion, and there is a risk that the dichotomy expert-lay will be reproduced if scientists' pre-defined views of risks are taken for granted.

Invisible risk and public inclusion

It is often said that a feature of today's threats to the environment is that they are increasingly diffuse and beyond the direct perception of lay people (Beck 1992). In some cases the consequences are delayed in time; the results of today's action will first be perceptible in the future – my own or that of my children. In other cases the consequences are distributed in space; the regional distribution of, for example, airborne pollutants may make it hard for lay people to establish a relationship between cause and effect. Thus the spatial and temporal character of many current environmental problems makes them invisible in the sense that they are beyond the layperson's perception. In practice many invisible risks seem to consist of a combination of these two features – their effects are both spatially spread and temporally delayed. When consequences perceptible to laypeople do appear it may be too late to act (such as in the case of the depletion of the ozone later, which is first noticed by human beings when bodily symptoms such as skin cancer appear). Science is therefore crucial for gaining knowledge about these environmental risks, and is consequently central to regulatory agencies' risk management as well as for the actions of individuals.

To a large degree risk management has been – and still is – a science-centred activity. It is traditionally defined as a process where experts establish the probability and magnitude of the hazards, other experts evaluate the benefits and costs of various options, and then political priorities are invoked to choose the best option (Amendola 2001). However, risk regulators and others responsible for the management of risk gradually recognised that the public's perception of risk was very different from the view held by the experts. This was interpreted as being due to public ignorance or misunderstanding of science, and the proposed solution was risk communication. By informing – and sometimes even educating – it about the 'real' risk, it was believed that the public would correct its judgement and accept risks that experts and regulators had found to be acceptable. However, when companies and agencies started to communicate about risks, it became clear that the public was anything but a homogenous category and that it had divergent views on risks. As a result, risk communication had to contend with the complexities of risk perception. Based on these findings, governments and companies formulate risk communication strategies for certain target groups, thereby hoping to influence their understanding and behaviour. In so doing, science's pivotal role was strengthened. It is not only relevant in exploring the nature of certain risks, but also how people understand risks, and in proposing ways of effectively communicating them to the public. Thus, according to this view, risk assessment concerns objective analysis aiming to produce factual knowledge about specific risks, while risk communication is about the distribution/transmission of this factual knowledge to the public.

However, the capacity of science to deliver trustworthy knowledge is currently contested. As Latour (1998, 208) puts it, we have moved from a culture of science to a culture of research, which is characterised by uncertainty, controversy, involvement, and values. Opportunities arise for contesting and negotiating specific risks when the public sees science as having no privileged access to truth and risk assessment as resting on many questionable assumptions. This situation – which is extensively discussed elsewhere – results in new conditions for risk management and risk communication (see e.g., Höijer, Lidskog, and Uggla 2006; Jasanoff 2005; Löfstedt 2005; Renn 1998; Todt 2003). In the absence of uncontested knowledge which can be distributed to the wider society, risk communication has to be transformed from a means of distributing information to a vehicle for mutual learning and deliberation where all involved parties influence each other, leading to a changed and broader understanding.

Within social-scientific risk research there seems to be a near consensus that greater involvement by non-experts is needed in risk assessment, management, and communication processes (Belzer 2001; Freudenberg 2001; Owens 2000; Renn 2001; Sjöberg 2000; Spash 2001). These views are not confined to the realm of theory. There are a number of political declarations at the national and international levels emphasising the importance that citizens get more involved in the policy process. One example is the Aarhus convention (UN 1998), which states that citizens should be entitled to access to information, public participation in decision-making, and access to justice in environmental matters. Another is Environmental Impact Assessment (EIA) which is a legislative tool for both democratising environmental decision-making and also broadening the knowledge base for a decision by including not only expert knowledge but also that of citizens. Further examples are the Seveso II directive (EU 1997), the European Union White Paper on Governance (EC 2001) and the UK Royal Commission on Environmental Pollution (RCEP 1998).

Thus, there seems to be a trend within science and regulation – at least rhetorically – towards moving beyond the old dichotomies of objective vs. subjective risk, and expert vs. lay knowledge in risk management (Lidskog, Soneryd, and Uggla 2005). Instead, more open and integrated approaches to risk assessment, risk management, and risk communication are suggested, and new relationships between experts and lay people are required. In the search for new relations between science and the public, inclusion, participation, deliberation, and transparency are key words (Gouldson, Lidskog, and Wester-Herber 2007). However, the plea for public inclusion is easy to understand when dealing with risks easily conceivable by the public, such as noises, odours and traffic accidents. But when approaching more invisible risks this strategy could be questioned. What knowledge can the public contribute when risks are not perceptible to them, such as in the case of radioactivity? To what extent should experts listen to the public in cases of invisible risks? And if they do listen to the public, are not the grounds of justification rather democratic than epistemic? In the case of invisible risks, a governmental agency may claim that citizens' right to influence the regulation of a specific risk should be restricted for reasons of safety and human health. Thus even if democratised science and inclusive risk management seem to be relevant goals for some environmental problems, traditional science-centred regulation may be judged to be the most suitable in cases such as radiation protection.

However this idea is based on the viewpoint that there are certain kinds of risk about which lay people have nothing to contribute, and that they therefore have no other alternative than to trust science. As will be discussed in the next section, this view is disputed, and instead lay people are seen as knowledgeable and reflexive actors with the capacity to deliberate even on invisible risks. Furthermore, the debate on who should be seen as legitimate knowledge producers not only takes place between science and citizens, but within science as well. There are plenty of situations where scientists vie with each other to disseminate their specific understandings of an environmental problem, struggles that take place within as well as between disciplines. Also, when facing problems in society, there is a need for a plurality of expertise where no single one can legitimately assume the role of expert over all experts (Giddens 1994, 186).

Thus, science does not produce an uncontested view on invisible risks, and it is difficult to draw a clear-cut demarcation between scientists and lay people. Instead it paves the way for boundary work where demarcations are created, negotiated, and transformed. The next section will discuss this by presenting three proposals for redesigning the relationship between science and citizens.

Towards a democratised science

During the last decade a number of proposals have been developed claiming that science must become more inclusive, that new relations between expertise and citizens have to be negotiated and designed. Three of the most well-known proposals are 'New Production of Knowledge', 'Postnormal Science' and 'Citizen Science'. All these proposals claim that the territories and populations of science and society are undergoing reconfiguration. However, they differ in their evaluation of the structural causes of this change, how far-reaching these processes are, and which normative conclusions should be drawn from this development.

New Production of Knowledge

In 1994, Gibbons and his colleagues presented their study *The New Production of Knowledge. The Dynamics of Science and Research in Contemporary Society*. In this book, they claim that a growing contextualisation and socialisation of knowledge has implied that we are now facing a new form of knowledge production. The traditional form – Mode 1 science – was characterised by hierarchical structures, internal control systems, and disciplinary borders within which research problems were formulated and solved. The new kind of knowledge production – Mode 2 knowledge – is non-hierarchic, heterogeneous, and open to other forms of knowledge than that of science. Thus, the knowledge production in society has moved from the closed context of disciplinary science to a broader transdisciplinary, social, and economic context.

In their follow-up study *Re-thinking Science. Knowledge and the Public in an Age of Uncertainty* (Nowotny, Scott, and Gibbons 2001) they radicalise their standpoint, claiming that there has not only been a shift in knowledge production, but in the wider society as well. We are facing a more profound social transformation that not only creates Mode 2 knowledge, but a Mode 2 society, a society transcending the categorisations of modernity. Thus, a theory of de-differentiation is presented, where previously discrete categories – state, market, culture, and science – are now transgressed and brought together. The driving force in this transformation is the emergence of a knowledge society. Two generations of mass higher education have brought about a dramatic increase in the number of knowledgeable actors. These actors are not solely located in the traditional academia. Instead a number of knowledge sites have emerged in society such as research institutes, governmental agencies, industrial laboratories, and think-tanks. We have a situation of more open systems of knowledge production, where science and society are co-mingling and are subject to the same evolutionary processes, which is leading to an erosion of boundaries between different forms of rationality.

This transformation has far-reaching implications. One is the emergence of a contextualised, or at least context-sensitive science where the authority of science is more closely tied to concrete cognitive practices, their results and impacts. Thus, to become socially robust, it is not enough for science to ponder about reliability and validity. Instead it must also reflect on its implications (including unintended consequences). Science has thus moved from being a culture of autonomy to a culture of accountability.

Furthermore, this contextualisation not only affects the form of science, but also its content. Instead of possessing a core of cognitive values and social practices that demarcate it from other knowledge practices, science finds that its epistemological core is empty – or rather is crowded with heterogeneous practices. This means that there is no other way to create trusted knowledge than through consultations, discussions, negotiations, contestations, mediations, and deliberations in the agora. The agora is, however, not an inclusive place for all citizens, but rather a public space populated by highly articulate, well-educated people – a category of people created by the mass higher education of contemporary society.

New Production of Knowledge claims that the decline of science's normative-cultural authority has resulted in an urgent need for science to gain credibility. The way to achieve this is to create arenas for the critical discussion of science, where people – mostly well-educated ones – can participate. Furthermore, because of the complexity of the reality with which science is dealing, there is a need to re-think science, to create a science that is richly contextualised, socially robust, and epistemologically eclectic. Even if Nowotny and her colleagues are rather vague in their description of the practical implications, it is possible to ascertain a number of implications for the relationship between experts and lay people.

Firstly, expertise is no longer located in the traditional sphere of science, but is socially distributed in society (located at different knowledge sites). Secondly, this expertise has to address issues that cannot be reduced to purely scientific and technical matters. Instead it has to deal with context-dependent issues. Thirdly, this means that science is losing its earlier pretension of being of a general and universal character. Instead socially robust knowledge is characterised by its relevance to a specific context. Fourthly, by making implications central, new possibilities arise for making science accountable. Fifthly, lay people are a heterogeneous category, and within this category there are a growing number of well-educated citizens with the capacity to act in the agora (where experts from different areas meet, interact, and negotiate).

However, even if the New Production of Knowledge argues forcefully for a democratisation of science to be achieved through discussion and contestation, the agora seems to be an exclusive place, both in terms of actors and arguments. The agora is here seen as rather elitist; it is not populated with all kinds of lay people, but with highly articulate, well-educated people. Despite the emphasis on there no longer being any core cognitive values and social practices that demarcate science from other knowledge forms, it is those already trained in scientific thinking who are accorded the capacity to contest science. As in the case of the original agora in ancient Athens, it seems that many categories of people are excluded from the space of deliberation. Furthermore, by claiming that the scientific expertise is still one of the most powerful mediating authorities in the agora, it seems as if the democratisation of science is more of a spatial than a substantial kind; it is still the expertise – now spatially distributed to different knowledge sites – that has to bring together and orchestrate many highly contextualised and heterogeneous knowledges. Thus, the expertise still seems to be the key actor, and large number of lay people seems to be excluded from deliberative spaces. Despite this, the deliberations in the agora are judged to be a central mechanism for creating a socially robust and publicly trusted knowledge.

Postnormal Science

Technical catastrophes such as Seveso, Bhopal, and Chernobyl have become powerful symbols of the pathologies of our technological civilisation. *Postnormal Science* – a concept introduced by Jerome Ravetz and Silvio Funtowicz – emerged as a response to this awareness (Funtowicz and Ravetz 1990, 1993; Ravetz and Funtowicz 1998; Ravetz 1999; De Marchi and Ravetz 1999). Thus, in contrast to the New Production of Knowledge, it is not the development of a knowledge society that serves as the foundation for recent changes in science, but instead the new characteristics of industrial risks.

Obviously, decisions on risk have always contained various sorts of uncertainty and value-commitments. But today we face system uncertainties and decision stakes that are extremely high compared to those of earlier risks. Traditional science cannot provide relevant knowledge for guiding action in the face of these new hazards where extreme uncertainty is combined with the possibility of prolonged, and often even irreversible harm.

Traditional, disciplinary science is 'normal science' in the Kuhnian sense, a puzzle-solving activity within an unquestioned and implicit paradigm. However, when approaching these new kinds of hazards, the knowledge that normal science produces is insufficient. A high degree of validity and reliability of scientific findings within the laboratory context is not enough when facing the complex conditions of the real world. Thus, when science leaves the laboratory, when it makes knowledge claims about conditions outside the closed laboratory world, it tends to lose its reliability. As Ravetz and Funtowicz (1998, 45) put it: 'the more "scientific" knowledge is, the more constricted is the area of its application in relation to the real problems facing humanity'. In this case, the validity and reliability of science is achieved at the expense of practical relevance. The more complex the problems, the less relevant knowledge is gained through laboratory science. Thus, there is an increased recognition that environmental problems by their nature should not be reduced to a standardised laboratory-based analysis, but instead require elements of enriched local knowledge for their solution (Ravetz and Funtowicz 1998, 47).

This situation – high system uncertainty and high decision stakes – constitutes the basis for the development of what Funtowicz and Ravetz call 'Postnormal Science'. This involves better cooperation between disciplines, but foremost concerns an opening up towards other kinds of knowledge claims. The way to bridge the gulf between the world of the laboratory and the world of society is through public inclusion, where groups other than scientists are welcome to take part in the process of knowledge production. Extended peer communities constituted by citizens, environmental movements, and political organisations should be formed to judge the quality and relevance of research.

Local knowledge is accorded special importance. The reason for this is that contextually generated knowledge about local circumstances is often more relevant to action than traditional, universal, abstract, and 'context-independent' scientific statements. Local knowledge and personal experience can create a unique and important body of knowledge on environmental problems, something which science lacks. As De Marchi and Ravetz (1999, 756) put it:

> When stakeholders in an issue, either counter-experts or lay, are treated as potential peers, sharing the definition and management of a problem, they can mobilise resources of local knowledge and understanding which complement the generalised knowledge of scientists or official experts.

Postnormal Science makes a strong plea for the greater involvement of stakeholders in knowledge production. The reason for this is that we need to understand the contextual character of a problem if we are to solve it. This has several implications for the relationship between experts and lay people. First, Postnormal Science tries to explain the political and social dimensions of scientific practice without reducing it to politics. The reasons for including stakeholders and the public are not only expressed in terms of democracy and justice, but also knowledge and quality. Secondly, it explains why science is necessary but not sufficient for solving many problems. Thirdly, it provides a changed role for the expertise: from producing facts to providing testimonies. Fourthly, whereas New Production of Knowledge primarily opens science up for 'semi-scientific knowledges' – knowledge that is possible to gain for those who have an academic training – Postnormal Science has a much broader approach, where all kinds of categories of people and organisations are welcome to enrich the process of knowledge production. Postnormal Science welcomes 'extended facts' – e.g., housewives' epidemiologies, as in the case of Woburn (see Note 1) – in the knowledge process.

However, it should be noted that at the same time as it has a broader approach to public inclusion, it also restricts the role of the public. Whereas New Production of Knowledge invites a part of the public – the well-educated – to play a leading role alongside science, Postnormal Science welcomes broader categories of people, but only gives them a supporting role; their task is to enrich the traditional production of scientific knowledge. Science seems to still be in the driver's seat, whereas an extended peer community is given the task of evaluating and correcting it. The reason for this may be that science is seen – as in the case of normal science – as a producer of universal and generalised knowledge, whereas local knowledge is contextual and case-specific. This can be explained with reference to Postnormal Science's realistic account of risk – risks are not socially constructed, but reflect certain characteristics of a world external to people.

Citizen Science

In *Citizen Science* (1995), Alan Irwin addresses the question of the proper relationship between science and the citizen within the field of environmental risks. Traditionally, a science-centred view of society has dominated, where all kinds of conflicts or tension between science and citizens are explained with reference to public ignorance or irrationality. There is a clash between science's universal and 'de-contextualised' character and lay people's local understandings. From this science-centred perspective, there is a need to educate the citizens, which is believed to lead to greater acceptance for a (scientifically guided) policy. Thus, science is placed at the centre of policy-making, whereas the public are seen as passive spectators, as witnesses rather than participants.

Irwin is critical of this understanding, at the same time as he tries to avoid the opposite standpoint, that people alone – without any reference to scientific findings – should decide about risk issues. Referring to the Sociology of Scientific Knowledge (SSK), he claims that science is anything but a homogenous practice, and that the diversity of the public is an important characteristic of contemporary society. Science's ambition to create abstract, universal and formalised knowledge, and its ignorance of the context-specific, contextually generated understanding of lay people is one reason for the clash between science and the public. A way forward, Irwin suggests, is to adopt a symmetrical understanding of different kinds of knowledge, and to start from the public understanding of a certain problem rather than a science-centred view that often alienates the public.

Irwin searches for a way to relate these different understandings to each other, privileging neither the public nor science. *Citizen Science* is the term coined by him for the arena where different knowledge claims can meet and cross-fertilize. In this arena science has engage with situations that give rise to citizen concerns and engage with non-scientifically generated understandings and expertise. However, citizen science is not only a challenge for science, but also for the wider society. It is a challenge for all people to establish a framework for expanded dialogue between different categories of people and to scrutinise different knowledge claims. To find a balance between telling and listening, giving and receiving should be a goal of everyone involved in the knowledge production. The way forward, Irwin suggests, is not for science to invite citizens to discuss problems pre-formulated by scientists, or for scientists to only be ready to answer questions raised by the public. Instead more 'public-centred' methods are needed, such as science shops and other citizen-guided activities.

In *Citizen Science* local knowledge is seen as closely related to living; people live in specific places, a fact which largely influences their understanding of environmental problems. Their contextually generated knowledge is highly practical, case-specific, instrumental in orientation and has no claim to universal status.

The perspective elaborated in this book is radicalised and deepened in a more recently published book. In *Science, Social Theory and Public Knowledge* (2003) Irwin and his colleague Mike Michael emphasise that 'the public' is no longer a local phenomenon. The public is no lay community, but a highly knowledgeable community embroiled in networks of alliances and forming hybrid groups. Furthermore, the risks currently challenging the public are in no way local, but to a large degree of global character.

The expert-lay divide is therefore obsolete, and Irwin and Michael replace it with their concept of 'ethno-epistemic assemblages'. This emphasises that all knowledge production and bodies of knowledge – irrespective of how far they have travelled – are situated; knowledge is always produced in, and part of the context of local cultural conditions. It also emphasises the dynamic character of these kinds of networks with their various practices and heterogeneous actors. The implication of their perspective is that 'the public' and 'science' cease to exist as distinct entities, and are replaced by assemblages, coalitions, and hybrid groups which deliver statements that weave in and out of the expert and lay domains.

The proposal has several implications for the expert-lay divide. Firstly, by adopting a symmetrical approach it avoids constructing a firewall between science and the public. Secondly, and taking after the SSK-approach with its emphasis on context rather than cognition, it offers a way to distinguish between the self-image of science and its practices showing that its claim to universal knowledge is an ambition rather than reality. Thirdly, it avoids seeing science as a homogenous entity, but sees it instead as a heterogeneous social practice. Fourthly, it emphasises the local dimension of environmental risks, without ignoring its global dimensions; global risks are always embedded in local contexts. Fifthly, risks and knowledge are not seen as context-free and free-floating, but instead as part of people's everyday lives and not separated from their social contexts.

Citizen science makes a number of arguments for public inclusion and how this inclusion should be designed to enable a cross-fertilisation between different knowledges. However, it is rather unclear how the quality of knowledge (irrespective of whether it is produced by scientists or citizens/ethno-epistemic assemblages) should be evaluated; Irwin warns for populism (letting the people decide) as well as scientism (letting scientists decide) but gives few clues as to how to mediate in situations with conflicting knowledge claims.

How far should science and the public swim?

According to the proposals presented, we today face a situation where the public is swimming upstream, taking part in the defining research problem, whereas science is swimming downstream, taking part in the discussion of practical implications of its research.[2] The scientisation of society and the democratisation of science have resulted in the transgression of old boundaries between science and society, experts and lay-people, and facts and values. At least on the level of rhetoric, there is a strong emphasis on science and the public needing to interact more and come closer to each other.

The three proposals all agree on the need for greater involvement of stakeholders in knowledge production and assessment. The reason for public inclusion is not primarily one of democracy or efficiency, but of knowledge. Because human beings are knowledgeable actors, public inclusion provides them with new possibilities for creating relevant and trusted knowledge. To do that it is necessary to create spaces for interaction between citizens and science. Furthermore, they all claim that science should take greater accountability for its implications. In short, science has to humanise itself by bringing people back into science and by making explicit its basic assumptions, value-laden decisions, and practical implications. The public, in turn, has to scientise itself by using its reflexive capacity to critically evaluate science and produce knowledge of relevance to the understanding of environmental risks. Thus science and the public should not be severed from each other, with risk communication serving to bridge the gulf between them. Instead a new social contract between science and citizens should be established, one which does not build on traditional risk management with its linear model of science-policy-public.

As we have seen above, the proposals also have a number of differences. Above all they differ as to how far the public and the science have swum – and should swim – up or downstream respectively, and to what extent it is only a part of the public and a part of the science community that have left their original positions and begun to approach each other. Their different standpoints are partly based on their different views of the relation between science and lay people, on whether integration or separation is seen as the hallmark of the proposed new relationship. The New Production of Knowledge forcefully argues that a general process of de-differentiation has resulted in the epistemological core of science becoming crowded with heterogeneous practices. Thus, earlier demarcations between different forms of rationalities are eroded, and a 'democratised science' is emerging. In contrast to this, Postnormal Science advocates a division of labour, where local people and extended peer reviews constitute a more context-sensitive form of rationality than science's more abstract and standardised knowledge. In managing environmental risks, both kinds of rationalities are needed to create relevant knowledge. Citizen Science has an ambiguous view, simultaneously adopting a symmetrical understanding of all forms of rationality and advocating a demarcation between scientific and lay knowledge.[3] However, later developments within Citizen Science emphasise an erosion of the boundaries between the expert and lay spheres, and argue that expertise is widely distributed in society causing a hybridisation of the previously discrete categories of expert and public.

From predefined demarcations to deliberative negotiations

What can be negotiated, contested, and deliberated by experts and lay people? To what extent and in what way is the proposal presented above relevant when discussing risks which seem to have no direct perceptible consequences and hence no immediate local implications? Postnormal science and citizen science claim that persons directly affected by an environmental phenomenon are better aware of its symptoms and are in a more pressing situation that others, whereas in the New Production of Knowledge it is rather experts – scientists as well as well-educated professionals – who have the capacity to grasp the implications of environmental risks.

When dealing with local implications it seems relevant to include affected lay people in risk management and knowledge production. Moving towards imperceptible risks, the question emerges of what lay people can contribute to the process of knowledge production. However, Postnormal Science aspires to provide a way to manage this kind of risk (De Marchi and Ravetz 1999). These risks are still hypothetical (there being no palpable or demonstrable damage), system uncertainties are high, and the decision stakes are high (the consequences of unintended effects may be severe and irreversible). Despite these characteristics, lay people are accorded an important role. The reason for this is that in this case science alone cannot give guidance for the regulation, and the regulatory agencies have no choice but to widely consult with stakeholders and to create extended peer communities. Thus, there seems to be a belief in the reflexive capacity of the public, where their perspective is not solely connected to 'living' but also to 'knowing'. They are thereby accorded an evaluative capacity for deliberating over invisible risks.

However, in this case the public seems to be invoked primarily as a means to complement already produced scientific knowledge by evaluating its relevance and enriching it with local knowledge. In its proposal for extended peer review (broader knowledge assessment) science seems to still be in the centre, and the task for the extended peer community is only to try to correct it occasionally. Thus, there is a risk that the hierarchy between scientific and local understandings is reproduced, where local knowledge only serves as a secondary input. This seems to also be the case for the New Production of Knowledge. Its new public space for production and transformation of science – the agora – is primarily populated by well-educated people. Thus, it is not a place that welcomes the public in all its diversity, and where all kinds of arguments can flourish. In that sense, there still seems to be a hierarchy of knowledges where only those close to a scientific understanding can take part in knowledge production, which leads to specific configurations of experts and the public, and which in turn results in reproducing earlier hierarchical demarcations between science and the public.

Instead of a hierarchy of different kinds of knowledge – wherein science is seen as superior – it is more appropriate to advocate a view of different knowledge fields all containing specific expertise. In the case of the radioactive fallout in West Cumbria, it was the sheep-farmers who provided expert knowledge about the variation in soil types, whereas in other cases it could be academic professionals such as lawyers or engineers who provide relevant knowledge. Thus, instead of arguing whether it is well-educated people (New Production of Knowledge) or lay people (Postnormal Science, Citizen Science) who can – and should – contribute to the production of knowledge, it may be better to see it as a question open for discussion and deliberation. Instead of suggesting a predefined idea of the proper role of experts and citizens, it is better to create spaces for deliberation and negotiation (cf. Jasanoff 2005). Different knowledge fields may to various degrees be seen as relevant when approaching a specific issue; both locally derived knowledge and knowledge acquired through higher education could be relevant when approaching risks. At the same time, it should be noted that this proposal, with its emphasis on argumentation and deliberation, runs the risk of having unequal effects. The reason for this is that more powerful actors have better possibilities than others to articulate their standpoints and influence the discussion, and it is an institutional challenge to create arenas and forms for deliberation that are responsive to this.

Conclusion: a cautious welcome to inclusion

It could be claimed that the examples given in the introductory section of this paper concern deficits in science communication of no relevance to the issue of knowledge production. The people in the Iraqi community were not informed that the barrels were contaminated, and the scientists who visited the village club in the Chernobyl area lacked insight into risk communication and the importance of understanding the local people to make their scientific advice understandable and relevant for those affected.

However, to reduce these stories to only being a problem of risk communication would mean adopting the 'deficit model' of public understanding of science (Irwin and Wynne 1996; Wynne 1995). According to this model, knowledge is first produced in a closed circle of scientists, and it should then be disseminated to the public (who in many cases are unable to understand science properly). The assumptions behind this model are heavily criticised, both for reasons of democracy as well as of knowledge (Yearley 2005, ch. 8). Often expert claims are ignored by citizens not because they are incomprehensible, but because they are deliberately rejected. To take another example from the Chernobyl case, a resident provided the following reflection over the behaviour of the Soviet experts who studied the consequences of the catastrophe (Aleksijevitj 1997, 139, my translation):

> Their route: by car from the hotel to the offices of the regional committee of the party, then back again also by car. They study the situation from local newspaper reports. Suitcases with sandwiches from Minsk. Tea they make with mineral water. Brought that with them too. It was the receptionist at their hotel who told me. People don't believe what's in the papers or on the radio or on TV, they watch what the higher-ups do. That's more trustworthy.

This quotation shows that the public does not restrict its evaluation to scientific findings but may include the issue of trustworthiness (Wynne 2001). Thus, it is not primarily knowledge in isolation that is evaluated by the public, but rather to what extent expertise and institutions are trustworthy; can we trust these experts and institutions who are supposed to be managing these risks? Thus, even if citizens lack the resources to discover and measure radioactivity, they are not totally at the mercy of scientists' advice and recommendations. Even if they take no part in the production of knowledge, they have a reflexive capacity to critically evaluate different knowledge claims. Thus, just because a risk appears to be imperceptible does not necessarily imply that it is incomprehensible for the public.

At the same time warnings should be raised against attaching unlimited evaluative and deliberative capacity to citizens. The solution to conflicting interpretations over risk is not to replace scientism with populism. Instead, as the three reviewed proposals forcefully argue, the challenge is to develop new relations between science and the public, and to create deliberative spaces – spaces for exchange across institutionalised boundaries – for the public evaluation and contestation of science.

At the same time, there is a need to raise some warnings against an uncritical adoption of the plea for public inclusion. The reason for this it that while a scientisation of society certainly is taking place, doubts can be raised over whether there really is a democratisation of science. Citizens are rarely included in the knowledge production of science, but instead are invited to take part in different kinds of public consultation activities such as environmental impact assessments and public panels on controversial issues (such as in the case of GMOs and nuclear waste disposal). In these settings they are given the opportunity to evaluate proposals and articulate standpoints. But when citizens speak back to science, does it have any effects on research agendas, knowledge production, and decision-making? I will therefore conclude by raising some critical points, stating that the welcoming of public inclusion should be of a cautious kind.

Firstly, the move from a deficit theory towards a dialogue theory should be viewed with some suspicion. The reasons for this is that there are ample examples from public participation programmes and risk communication projects where inclusive strategies have been used as a means for legitimating decisions, diluting accountability, and persuading the public. Thus, the plea for public inclusion may often reinforce a modernistic, hierarchical understanding of science and society, where public inclusion primarily serves as a means for educating them and obtaining their consent.

Secondly, there is a need to question predefined meanings provided by science, and its construction of the expert-lay divide. Even if citizens have no access to expert systems that produce knowledge on certain aspects of a phenomenon, this does not mean that they cannot critically evaluate it, or that they must subordinate their evaluation and understanding to that of science. Citizens have the ability to contest and re-negotiate the boundaries of expert knowledge, and to create alternative meanings – such in the above quotation about citizens' interpretation of the Russian experts' visit to Chernobyl. At the same time, it should be acknowledged that neither the citizens nor the scientists have an unlimited deliberative capacity, and that there often are limited resources with which to perform these deliberations.

Thirdly, to label citizens as lay people is still relevant in a very important sense: in contrast to others involved in knowledge production and risk management, citizens are not professionally active. It is not their professional task to take part in public meetings, contest knowledge claims, evaluate different proposals, and elaborate standpoints. Instead they have to reallocate leisure time to take part in knowledge production. Simply put, public inclusion is a time-consuming activity, and even in cases where the citizens are economically compensated for their involvement, they still participate at the expense of their leisure time or time at their workplace.

Fourthly, public inclusion runs the risk of fostering the goal of consensus-making at the expense of revealing power relations. The reason for this is that the quest to create socially robust knowledge may involve a wish to reconcile different interests and knowledges, and result in a loss of adversaries and critical perspectives (cf. Mouffe 1999). However, there is no way eliminate power, and therefore it is an important task for citizens to bring power relations to light and thereby make it possible to evaluate their implications. Associated with the issue of power is the role of politics and policy-making. An exclusive focus on the need for a new contract between science and citizens runs the risk of concealing the political dimension of controversial issues. A call for public inclusion in knowledge production and risk management may be seen as an easy solution to controversial issues. Thus, there is a risk that public inclusion is used as a general cure for a multitude of problems – e.g., lack of trustworthiness of science, lack of political legitimacy, and regulatory failures – and the political character of many risks may be concealed.

Fifthly, there is a need to organise deliberative spaces that transcend the national territory. Today, many deliberative spaces are subordinated to the nation-state, and issues are thereby nationally framed. At the same time processes of globalisation and de-regulation imply that risks are hardly confined to national territories. Thus, in the face of globalisation, there is an increased need to create deliberative spaces across political borders (Archibugi, Held, and Köhler 1998; Beck 2000; Held 2004; Held and Koenig-Archibugi 2005; Lidskog 2005; Lidskog and Elander 2007). People are exposed to transboundary risks and to expert systems that are spread over the world. The development of science-public relations should therefore not be restricted by the territorial borders of the nation-state. Hence, it is of great importance to create deliberative spaces not only across the institutional borders between science and citizens, but also across political borders between nation-states. The current debate over genetically modified foods is an example of an issue which should not be restricted to science-citizen discussion within separate nation-states. Thus, there is a need for political experimentation, creating – and testing – different institutional forms for the deliberate handling of competing discourses, framings and understandings of risks, also invisible ones.

To sum up, there is a need to develop new relationships between science and citizens which do not reproduce the dichotomy between experts and lay people. The spaces created for negotiation and deliberation should not be determined by scientifically produced facts. Instead science and risk management should be made more accessible, even in those cases where risks are claimed to be invisible to citizens.

Acknowledgement

This article is written as a part of the project 'Risk regulation in late modernity', financed by the Swedish Research Council. Earlier versions of it have been presented at the conferences *Environment, Knowledge and Democracy*, Marseille 6–7 July 2005 and *New Perspectives on Risk Communication: Uncertainty in a Complex Society*, Göteborg University 31 August–2 September 2006. I wish to thank Göran Sundqvist, Science & Technology Studies, Gothenburg University, for constructive criticism of an earlier version of this paper.

Notes

1. One of the most well-known cases of popular epidemiology is that of Woburn, Massachusetts. A child was diagnosed with leukaemia, and his mother started to investigate the possible sources, suspecting that it was caused by contaminated water. In 1974 she asked the health department to investigate the fresh water in the area, but they refused to do so. She then conducted a survey among the local inhabitants and found that there was an increased level of leukaemia concentrated to the eastern part of Woburn. The health department dismissed the investigation on the grounds that it had not been conducted by an expert. The municipal board contacted a research laboratory, but no causal relation was found. However, lay people continued to conduct investigations, and with the help of a biostatistician they succeeded in showing that there was an overrepresentation of leukaemia in the area. Later it was found that two fresh-water wells were contaminated by chemicals (see Brown and Mikkelsen 1990).
2. The metaphor is somewhat misleading, because it seems to be based on a science-centred model – a diffusion model in the words of Latour (1987, 142) – where science constructs and delivers knowledge, and people are only included when it comes to the implementation phase. Even if the swimming metaphor aims to problematise this division of labour, it still conveys the image of a technocratic top-down approach, where science is at the top and citizens at the bottom.
3. In *Citizen Science*, it is unclear how Irwin reconciles a symmetrical approach to rationality with his separation between expert and lay understandings (stating that living in the environment provides specific ways of knowing it, which constitutes the foundation for lay epistemologies). In his and Michael's later book *Science, Social Theory and Public Knowledge*, they claim that it has become exceedingly difficult to talk of the public at all, and that there instead exist hybrid groups and assemblages (Irwin and Michael 2003, 85). Instead of there being a divide between expert and lay, they argue that actors now embody

both lay and expert knowledge. Thus, even if expert and lay no longer serve as empirical categories, they still appear to be relevant analytical concepts for Irwin and Michael. An alternative interpretation is that the expert-lay divide is not constituted by different rationalities – both categories are crowded with heterogeneous practices – but rather by context; they consist of the same kind of rationality (or rationalities), but depending on which social contexts they appear in they are categorised as expert knowledge or lay knowledge.

References

Aleksijevitj, S.A. 1997. *Bön för Tjernobyl* [*Prayer for Chernobyl*]. Stockholm: Ordfront.

Amendola, A. 2001. Recent paradigms for risk informed decision making. *Safety Science* 40: 17–30.

Archibugi, D., D. Held, and M. Köhler, eds. 1998. *Re-imagining political community: Studies in cosmopolitan democracy*. London: Polity.

Beck, U. 1992. *Risk society. Towards a new modernity*. London: Sage.

———. 2000. *What is globalization?* London: Polity Press.

Belzer, R.B. 2001. Getting beyond 'grin and bear it' in the practice of risk management. *Reliability Engineering and Systems Safety* 72: 137–148.

Brown, P. 1997. Popular epidemiology revisited. *Current Sociology* 45: 137–56.

Brown, P., and E.J. Mikkelsen. 1990. *No safe place. Toxic waste, leukaemia, and community action*. Berkeley: University of California Press.

De Marchi, B., and J. Ravetz. 1999. Risk management and governance: A post-normal science approach. *Futures* 31: 743–57.

EC. 2001. *White Paper on Governance: Report of the Working Group on Democratising Expertise and establishing Scientific Reference Systems*. May. Brussels: European Commission.

Freudenburg, W. 2001. Risky thinking: Facts, values and blind spots in societal decisions about risks. *Reliability Engineering and Systems Safety* 72: 125–30.

Funtowicz, S., and J.R. Ravetz. 1990. *Uncertainty and quality in science for policy*. Dordrecht: Kluwer.

———. 1993. Science for the post-normal age. *Futures* 25: 739–55.

Gibbons, M., C. Limogenes, H. Nowotny, S. Schwarzman, P. Scott, and M. Trow. 1994. *The new production of knowledge. The dynamics of science and research in contemporary societies*. London: Sage.

Giddens, A. 1990. *The consequences of modernity*. Cambridge: Polity Press.

———. 1994. Risk, trust and reflexivity. In *Reflexive modernization. Politics, tradition and aesthetics in the modern social order*, ed. U. Beck, A. Giddens, and S. Lash, 184–97. Cambridge: Polity Press.

Gouldson, A., R. Lidskog, and M. Wester-Herber. 2007. The battle for hearts and minds. Evolutions in organisational approaches to environmental risk communication. *Environmental & Planning C* 27: 56–72.

Held, D. 2004. Democratic accountability and political effectiveness from a cosmopolitan perspective. *Government and Opposition* 39: 364–91.

Held, D., and M. Koenig-Archibugi, eds. 2005. *Global governance and public accountability*. Malden, MA: Blackwell.

Höijer, B., R. Lidskog, and Y. Uggla. 2006. Facing dilemmas. Sense-making and decision-making in late modernity. *Futures* 38: 350–66.

Irwin, A. 1995. *Citizen science. A study of people, expertise and sustainable development*. London: Routledge.

Irwin, A., and M. Michael. 2003. *Science, social theory and public knowledge*. Maidenhead: Open University Press.

Irwin, A., and B. Wynne, eds. 1996. *Misunderstanding science? The public reconstruction of science and technology*. Cambridge: Cambridge University Press.

Jasanoff, S. 2005. *Designs of nature. Science and democracy in Europe and the United States*. Princeton, NJ: Princeton University Press.

Latour, B. 1987. *Science in action: How to follow scientists and engineers through society*. Cambridge, MA: Harvard University Press.

———. 1998. From the world of science to the world of research? *Science* 280: 208–9.

Lidskog, R. 2005. Siting conflicts – democratic perspectives and political implications. *Journal of Risk Research* 8: 187–207.

Lidskog, R., and I. Elander. 2007. Representation, participation or deliberation? Democratic responses to the environmental challenge. *Space and Polity* 11: 75–94.

Lidskog, R., L. Soneryd, and Y. Uggla. 2005. Knowledge, power and control. Studying environmental regulation in late modernity. *Journal of Environmental Policy & Planning* 7: 89–106.

Lidskog, R., and G. Sundqvist. 2004. On the right track? Technology, geology and society in Swedish nuclear waste management. *Journal of Risk Research* 7: 251–68.

Löfstedt, R. 2005. *Risk management in post-trust societies*. Basingstoke: Palgrave Macmillan.

Mouffe, C. 1999. Deliberative democracy or agonistic pluralism? *Social Research* 66: 745–58.

Nowotny, H., P. Scott, and M. Gibbons. 2001. *Re-thinking science. Knowledge and the public in an age of uncertainty*. Cambridge: Polity Press.

Owens, S. 2000. 'Engaging the public': Information and deliberation in environmental policy. *Environment and Planning A* 32: 1141–8.

Ravetz, J.R. 1999. What is post-normal science? *Futures* 31: 647–53.

Ravetz, J.R., and S.O. Funtowicz. 1998. Commentary. *Journal of Risk Research* 1: 45–8.

RCEP. 1998. *18th report on environmental standard setting*. London: Royal Commission on Environmental Pollution.

Renn, O. 1998. The role of Risk communication and public dialogue for improving risk management. *Risk decision* 1: 5–30.

———. 2001. The need for integration: Risk policies require the input from experts, stakeholders and the public at large. *Reliability Engineering and Systems Safety* 72: 131–5.

Sjöberg, L. 2000. Perceived risk and tampering with nature. *Journal of Risk Research* 3: 353–67.

Spash, C.L. 2001. Broadening democracy in environmental policy processes. *Environment and Planning C: Government and Policy* 19: 475–81.

Todt, O. 2003. Designing trust. *Futures* 35: 239–51.

UN. 1998. *Convention on Access to Information, Public Participation in Decision Making and Access to Justice in Environmental Matters, 25th June 1998, Aarhus*. Geneva: United Nations Economic Commission for Europe.

Wynne, B. 1989. Sheep farming after Chernobyl. A case study in communicating scientific information. *Environment* 31(10–15): 33–9.

———. 1992. Misunderstood misunderstanding: Social identities and public uptake of science. *Public Understanding of Science* 1: 281–304.

———. 1995. Public understanding of science. In *Handbook of science and technology studies*, ed. S. Jasanoff, G.E. Markle, J. Peterson, and T. Pinch, 361–88. London: Sage.

———. 1996. May the sheep safely graze? A reflexive view of the expert-lay knowledge divide. In *Risk, environment & modernity. Towards a new ecology*, ed. S. Lash, B. Szerszynski, and B. Wynne, 44–83. London: Sage.

———. 2001. Creating public alienation: Expert cultures of risk and ethics on GMOs. *Science as Culture* 10: 445–81.

Yearley, S. 2005. *Making sense of science. Understanding the social study of science*. London: Sage.

Risk communication, prenatal screening, and prenatal diagnosis: the illusion of informed decision-making

Michael Siegrist, Marie-Eve Cousin and Carmen Keller

ETH Zurich, Institute for Environmental Decisions (IED), Zurich, Switzerland

Pregnant women's knowledge of prenatal tests was assessed in face-to-face interviews. The focus of the study was on how physicians communicate test results to their clients and whether women possess enough information to make informed decisions. Thirty-five women located in the German-speaking part of Switzerland participated in the study, which was based on half-standardized interviews lasting about an hour. Standardized questions were used to measure women's knowledge of prenatal tests. In selecting women to participate in the study, we ensured that they were counseled by different physicians. Results showed that most participants overestimated the risk of having a baby with Down syndrome. Most women had difficulties in interpreting a positive test result associated with blood tests (i.e., PAPP-A), but almost all women knew that this test provides a probabilistic outcome. Most women had very accurate knowledge of the risk of a miscarriage associated with invasive tests (i.e., amniocentesis, chorionic villus sampling). A substantial percentage of the women, however, did not realize that the invasive tests are also probabilistic tests. Overall, results of our study suggest that approaches to risk communication vary across physicians. There seems to be no standardization in counseling. In addition, our results suggest that many pregnant women may lack the knowledge necessary to make informed decisions regarding prenatal tests.

Introduction

Down syndrome (Trisomy 21) is the most frequent clinically significant chromosome aberration at birth. Infants with Down syndrome are at high risk of various physical abnormalities and diseases. In addition, Down syndrome children can show mental retardation (Smith-Bindman et al. 2001). Many parents, therefore, are interested in the prenatal detection of affected fetuses. Based on this information, they can adapt to circumstances in which their child will have a disability, or they can decide to terminate the pregnancy.

Several screening methods for detecting fetuses with Down syndrome are available. Current research aims to further improve prenatal diagnosis. The European Union project called Special Advances in Fetal Evaluation (SAFE) aims to implement routine non-invasive prenatal diagnosis and cost effective neonatal screening methods (www.safenoe.org). While large amounts of resources are being invested to improve prenatal diagnosis, far less is being invested to examine how test results should be communicated to parents. Pregnant women need information to decide whether to undergo testing or not, and they must know the implications of positive and negative test results (Smith, Shaw, and Marteau 1994). It should be emphasized, however, that selecting the best manner of advising patients so that they are able to make good informed decisions is a widespread problem across all medical fields (Bogardus, Holmboe, and Jekel 1999; Thomson, Edwards, and Grey 2006).

Physicians must communicate screening results to their patients. This is a complex task, and difficulties associated with it must not be neglected. In their definition of risk communication, Edwards, Elwyn, and Mulley (2002) emphasize that it is a two way exchange of information and opinions about risks, and that the ultimate goal of risk communication is to improve decisions about clinical management. It is relatively easy to agree upon a definition of risk communication. There is much less agreement on the question of how to achieve this goal. It is safe to conclude that risk communication is still more an art than science.

In the present research we examined pregnant women's knowledge about and understanding of prenatal screening and diagnostic tests. We were interested in learning whether women are able to make an informed decision or not. Furthermore, we were interested whether additional information would have an impact on women's decisions. We focused mainly on Down syndrome, because this is the most frequent chromosome aberration.

Down syndrome

The risk of delivering a Down syndrome child is positively correlated with the age of the mother. If the mother is between 15 and 29 years old, then the odds are 1:1500 of delivering a Down syndrome child; the odds are 1:100 for mothers between 40 and 44 years old. Despite increasing use of prenatal tests, the number of liveborn Down syndrome children has remained constant in the eastern part of Switzerland (Binkert, Mutter, and Schinzel 2002). The reason for this is not only that the use of prenatal tests has increased, but so has the maternal age.

Prenatal screening for Down syndrome

Non-invasive tests are used for prenatal screening. The first-trimester test combines information from maternal blood test, ultrasound and maternal age. It is often used for Down syndrome screening. The first-trimester test involves determining the levels of pregnancy-associated plasma protein-A (PAPP-A) and free beta human chorionic gonadotropin. Ultrasound can be used because Down syndrome fetuses have excess skin on the neck. The false positive rate of such a screening procedure is 5%, and the test sensitivity is 90% (Miny 2001). The triple test, or AFPplus Test, has a lower sensitivity (Miny 2001). This test can be used to detect Down syndrome or Spina bifida. It can be conducted between the sixteenth and eighteenth week of the pregnancy, whereas the first-trimester test can be done between the eleventh and fourteenth week (Müller, Imhasly, and Leuthold 2005).

Women for whom the non-invasive test indicates a higher risk may undergo a more invasive diagnostic test, such as amniocentesis or chorionic villus sampling (CVS). For the amniocentesis, amniotic fluid in the womb is collected. This fluid contains fetal cells, which are examined. In the CVS procedure, tissue is taken from the placenta, which contains fetal cells that can be tested for Down syndrome. Both tests, amniocentesis and CVS, entail a risk of miscarriage. In a brochure for pregnant women, the University Hospital of Zurich states the risk as 1% (Klinik für

Geburtshilfe 2004). In a brochure by an NGO that counsels pregnant women, the risk for CVS is given as 0.5–2%, and for amniocentesis as 0.5–1% (Neuhaus n.d.).

Parents' understanding of screening and diagnostic tests

The decision-making process of whether to conduct prenatal tests is difficult, because women must make several decisions within a short time period (Santalahti et al. 1998). First, women must determine whether to have a screening test. Given a positive test result, women must decide whether to have an invasive diagnostic test. Finally, if test results indicate a fetal abnormality, women must decide whether to terminate the pregnancy or not.

Al-Jader, Parry-Langdon, and Smith (2000) examined whether women had sufficient understanding of the antenatal screening process to make informed decisions. They interviewed 34 women who were 20 weeks pregnant. Results showed that a majority of the women were not aware that the screening tests were voluntary. About two thirds of the women did not know that the blood test was a screening tool, and that subsequent tests might be necessary. Results of this study suggest that most women may not have the knowledge and understanding of prenatal tests necessary for making informed decisions. In another study, a validated multi-dimensional measure of informed choice was used to examine decision making in women who underwent a first-trimester screening for Down syndrome (Jaques, Sheffield, and Halliday 2005). Based on these standardized measurements, the authors concluded that only two thirds of the respondents could be categorized as having made an informed choice. Hunfeld et al. (1999) examined the physician-patient interaction following a fetal anomaly scan. A few days later the pregnant women filled out a questionnaire concerning the results of the scan. Results suggest that the majority of the women correctly recalled the information given about the unborn baby's anomaly. However, some women were not able to reproduce some of the information that is essential for making an informed decision.

Communicating screening results may influence women's well-being. The screening tests have a high false positive rate. In other words, a high proportion of women are informed that they have a high risk of delivering a Down syndrome child when in fact the fetus does not have the Down syndrome. Such false positive test results seem to have a negative impact on women's well-being (Georgsson Öhman et al. 2006). Being informed of an increased risk of having a Down syndrome child caused strong reactions of anxiety and worry.

Communication of test results

Screening and diagnostic tests yield probabilistic rather than deterministic information. Communicating probabilistic information to people who are not trained in statistics or the concepts of probability is a challenge. One idea for overcoming these difficulties is to communicate probabilities in words rather than numbers (e.g., it is rather likely that you will have a Down syndrome child). Results of various studies suggest that verbal labels for probabilities may have different meanings for different persons (Beyth-Marom 1982). In the context of presenting negative screening results to pregnant women, it has been shown that it is better to communicate risks using numbers rather than verbal descriptors (Marteau et al. 2000).

There are several types of information and different numerical formats that can be used to convey statistical information to the patient (Gigerenzer and Edwards 2003). Single event probabilities (e.g., 30% chance of having a Down syndrome child), conditional probabilities (e.g., sensitivity), or relative risks (e.g., probability of having a Down syndrome child compared to the base rate) can be communicated. Past research indicates that people understand numerical information better when it is expressed as frequencies as opposed to probabilities (Cosmides and Tooby 1996; Gigerenzer and Hoffrage 1995).

Participants distinguished between low and high risks better when the information was presented as frequencies rather than as probabilities (Siegrist 1997). People found risk expressed as '1 in x' more worrying than when the same risk was expressed as a percentage (Abramsky and Fletcher 2002). In this study, participants compared the two numerical expressions. Therefore it is not known whether perceived risks would differ. Despite the research comparing different risk communication formats, it is still an open question of how statistical information should be presented to patients so that they would be able to make informed decisions.

As outlined here, there are many different approaches to informing women about the outcomes of prenatal tests. Little is known, however, about how physicians inform their patients of the outcomes of such tests.

Rationale of the present research

In the present research, we were interested in learning more about how physicians inform women about prenatal tests. We examined the knowledge of pregnant women about screening and diagnostic tests. We were interested in learning whether pregnant women possess the knowledge that is needed for making informed decisions. We utilized an exploratory and qualitative approach.

Method

Participants

Thirty-five women from the German-speaking part of Switzerland participated in the study. In selecting women to participate in the study, we did our best to ensure that they were counseled by physicians from different institutions. Considering the whole sample, participants were attended medically by 17 male and 15 female gynecologists. Only participants with experience with prenatal diagnosis were selected (either invasive or non-invasive tests). Table 1 shows the prenatal tests the women of our sample had experience with. The women's age ranged between 21 and 42 years (M=33.9). Seventeen of the women were pregnant, 10 had recently delivered a baby, and eight had a baby delivered more than one year ago. The education level of the participants was above average.

Procedure

The interviews consisted of four parts. In the first part, we asked open questions, and participants characterized their pregnancy. In the second part, participants answered structured questions on a portable computer. Questions were related to the evaluation of the physician, their information-seeking behavior, various risk

Table 1. The types of screening (with outcome) and diagnostic tests participants had experience with.

Form of prenatal test	Number of Participants
Blood test (positive)	1
Blood test (positive), CVS	8
Blood test (negative)	11
Blood test (negative), amniocentesis	1
Blood test (negative), CVS	1
Amniocentesis	1
CVS	10
Ultra sound	2

Note: CVS = Chorionic villus sampling.

estimates, and their attitudes toward abortion. In the third part, half-standardized questions were posed. These questions were related to the support of the physician and the information provided by the physician. Knowledge related to invasive and non-invasive tests was also assessed. In the third part, participants received information about the age-specific risk of having a Down syndrome child and the sensitivity and false positive rate of blood tests (i.e., PAPP-A). Participants reported their hypothetical decisions based on this information. The interviews lasted between 60 and 90 minutes. All interviews were conducted by the second author.

Some participants had delivered more than one child. Therefore, questions were asked in regard to the sole or the most recent pregnancy.

Results

Risk of having a child with Down syndrome

The risk of having a child with Down syndrome is strongly influenced by the woman's age. Sixty percent (n=21) of the participants reported that they were informed by their gynecologist about their age-risk of giving birth to a Down syndrome child. Seventy-two percent (n=13) of the women who were 35 years and older, and 35% (n=6) of the women who were younger than 35 years old were informed by their physician about their age-risk of giving birth to a Down syndrome child. Results suggest that older women were more often informed about their age-risk than younger women. The risk information was conveyed as a number (n=6) or as a verbal description (n=15).

Participants were asked to state their age-related risk. Women who had already delivered their baby were asked to provide risk estimates for their recent pregnancy. Past research suggests that most people understand frequencies better than probabilities (Hoffrage et al. 2000). Therefore, we asked participants to provide frequency estimates: 'Please provide the following estimate: Out of 1000 pregnant women of your age, how many are pregnant with a child with Down Syndrome? _____ women'. Participants' estimates expressed as probabilities and base rates are shown in Figure 1. Results suggest that most women overestimated the risk of having a Down syndrome child. In other words, most women provided values that were higher than their actual age-risk of giving birth to a Down syndrome child. We also

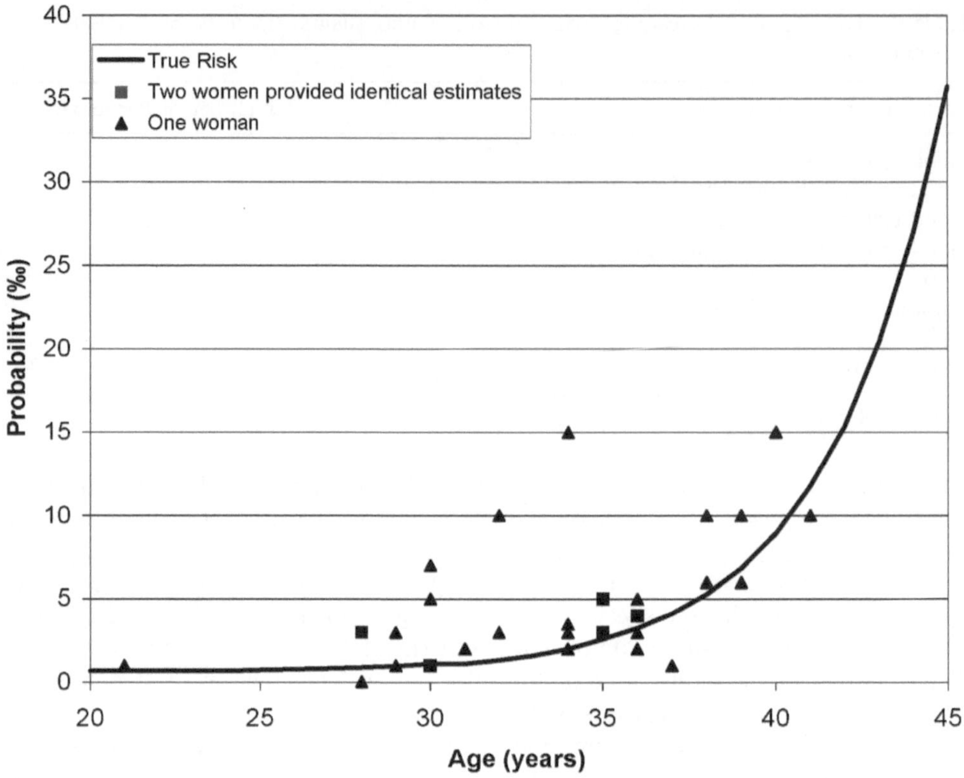

Figure 1. Women's risk estimates and age-related base rates. Estimates of two women with estimates over .70 are not shown.

analyzed only the data of women who reported that they had been informed by their gynecologist about the age-related risk of having a child with Down syndrome. However, virtually the same results were observed, indicating that these women also overestimated the risks.

Knowledge about prenatal tests

Most women knew that screening tests are probabilistic and not deterministic. Twenty-eight participants correctly answered yes to the question, 'If you received a positive test result, is there a possibility that the child is healthy?' Three women responded no, and four women did not answer this question. Results of further questions suggest, however, that the women did not have a clear idea about the size of the probability of having a child with Down syndrome given that the blood test was positive.

Participants' answered questions that tested whether they understood that false positive and false negative tests results are also possible for invasive tests. Responses to this question are shown in Table 2. Results suggest that most women were not aware that invasive tests are also probabilistic tests like blood tests. We observed similar results for women who had and for women who did not have invasive tests. A majority of the women who answered these questions believed that a positive test

Table 2. Participants' responses to questions testing whether they understood that false positive and false negative tests results are possible also for invasive tests. Numbers in brackets refer to women who had invasive tests.

	Yes	No	No answer
If you have a positive test result, is there a chance that you will deliver a healthy child?	11 (6)	18 (13)	6 (2)
If you have a negative test result, is there a chance that you will deliver a child having the Down Syndrome?	11 (7)	18 (12)	6 (2)

result implies that there is no chance that the child will be healthy, and the women believed that with a negative test result it was certain that the child did not have Down syndrome. Results suggest, therefore, that most women perceived invasive tests as deterministic and not as probabilistic tests.

Invasive tests are associated with some risks. The probability of a miscarriage due to an invasive prenatal test varies between 0.5% and 2.0% (Neuhaus n.d.). All of the participants knew that invasive prenatal tests are associated with some risks. Their estimations are shown in Figure 2. Results suggest that most participants accurately knew the risk of a miscarriage. Only few women over- or underestimated the risk.

Information about the blood test

During the last part of the interview, participants received information about their age-risk of having a Down syndrome child, the false positive rate and the test sensitivity. Utilizing a decision tree (Sedlmeier and Gigerenzer 2001), we explained to participants the meaning of a positive and of a negative result in a blood test. As the base rate, we used the specific age risk of the women. In addition to the decision tree,

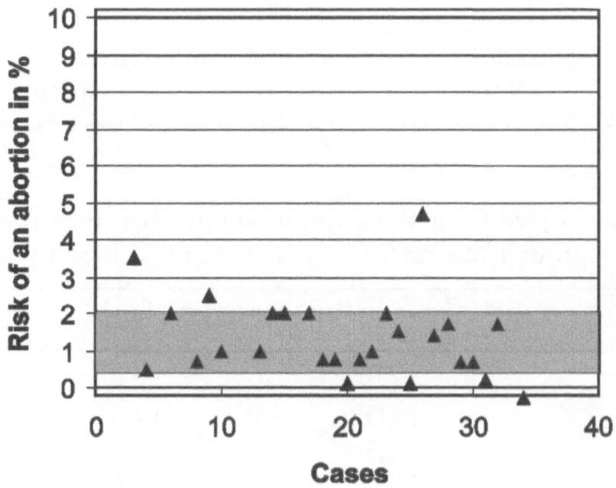

Figure 2. Estimated risks of a miscarriage due to an invasive prenatal test. The grey area shows expert' estimations.

results of the Bayesian reasoning were illustrated utilizing a graphical display. We explained to participants how many of 1000 women of their age would have positive test results even though they had a child without Down Syndrome. Participants were also informed about how many women would have negative test results even though they had a Down syndrome child.

Participants were asked how they perceived the accuracy of the blood test, given the information presented to them. They could indicate whether they perceived the blood test as more accurate or as less accurate prior to the information we presented to them. Responses are shown in Table 3. Results show that a majority had perceived the blood test as more accurate prior to our explanation of what a positive result means. The second most frequent response was that participants perceived the blood test as less accurate before receiving more information from us. For only a few participants, our explanation of the meaning of a negative or a positive test result was a reassurance of their prior beliefs.

We further asked participants the following hypothetical question: would you have opted for a blood test, if you had had the present knowledge about the blood test? Results in Table 4 show that a substantial portion of the women who had a blood test would not opt for a blood test if they were fully informed about the interpretation of a positive and a negative test result.

Discussion

Communication of screening and diagnostic results is an important part of prenatal testing. Parents must make an informed decision based on test results and their values. A non-invasive blood test is often used for prenatal screening. The false

Table 3. How was the perception of the accuracy of the blood tests prior to the new information? Numbers in brackets refer to women who had blood tests.

	Less accurate	No change	More accurate	Missing
29 years and younger	0	0	3 (3)	3
30–34 years old	3 (3)	1 (1)	6 (4)	1
35–38 years old	3 (2)	3 (2)	4 (3)	2
39–42 years old	3 (1)	0	3 (0)	0
Total	9 (6)	4 (3)	16 (19)	6

Table 4. Would you have opted for a blood test if you had had the present knowledge about the blood test? Numbers in brackets refer to women who had blood tests.

	Yes	No	Missing
29 years and younger	4 (4)	0	2
30–34 years old	2 (1)	4 (3)	5
35–38 years old	3 (2)	5 (2)	4
39–42 years old	1 (0)	3 (0)	2
Total	10 (7)	12 (5)	13

positive rate of this screening procedure is relatively high. Therefore, women with a positive test result undergo a more invasive diagnostic test like amniocentesis or CVS. These tests have a risk of miscarriage. Should the invasive tests yield positive results, then parents must decide whether to terminate the pregnancy or not. Under time constraints, parents face a number of decisions. It is crucial that parents understand the screening and diagnostic procedure and the interpretation of a positive and a negative test result. Otherwise they cannot make an informed decision. Results of past research suggest that many women may not be in a position to make an informed decision (Al-Jader, Parry-Langdon, and Smith 2000; Jaques, Sheffield, and Halliday 2005). In the present study we examined how women in the German speaking part of Switzerland are informed about prenatal tests and what they know about screening and diagnostic tests.

We conducted face-to-face interviews with 35 women. All of them had some experiences with prenatal diagnosis. Fourteen women had experience only with screening tests. Twenty-one women had experience with invasive tests (amniocentesis or CVS). Women were well educated and interested in the topic of prenatal diagnosis. It is safe to conclude, therefore, that the women in our sample knew more than an average pregnant woman. Results of our study certainly overestimate the knowledge of pregnant women.

Most women in our sample overestimated the prevalence of delivering a child with Down syndrome. This was true for all age categories. As a result, the women of our sample may have been more worried about the risk of a Down syndrome child than justified by the facts. One explanation for this result could be the availability heuristic (Keller, Siegrist, and Gutscher 2006; Tversky and Kahneman 1982). Women were well informed about pregnancy, and they knew, therefore, quite a bit about the risk of delivering an unhealthy child. This awareness may have increased participants' risk estimates and may have resulted in an overestimation of having a Down syndrome child.

Almost all participants were aware that screening tests could provide false positive or false negative test results. Participants did not have a clear picture of the size of these errors, however. Diagnostic tests like amniocentesis or CVS have a much higher sensitivity and much lower false positive rate than the screening tests. However, a majority of the participants erroneously believed that these diagnostic tests are deterministic tests and that they are not associated with decision errors.

Amniocentesis or CVS involve the risk of a miscarriage. Knowledge about this risk was prevalent among participants. A clear majority of the participants estimated the size of risk as reported in brochures informing about pregnancy and prenatal tests.

Some limitations of the present study must be addressed. Our sample included pregnant women as well as women who had already given birth, the latter of which may have exhibited a hindsight bias. In other words, these subjects may have underestimated both the uncertainty they were originally faced with during the pregnancy and the risks of delivering, e.g., a Down syndrome child because they had the knowledge that a healthy child was delivered.

Altogether, results of the present study are disillusioning. We interviewed women, who were well educated and interested in prenatal diagnosis. All of them had experienced a screening or a diagnostic test. A majority of these women, however, did not have the knowledge needed to make an informed decision. Based on the sample examined, we cannot know the knowledge of an average pregnant woman. It seems safe to say that the average pregnant woman does not possess the knowledge necessary to make a sound decision. Future research should examine the effects of different risk communication strategies on knowledge and on perceived risks. We acknowledge that it is important to develop better diagnostic tests with fewer risks. It is equally important, however, to improve and standardize risk communication, so that parents can make decisions that are in line with their values.

Acknowledgement

The article is based on work supported by the EC Contract No LSHB-CT-2004-5053243 (NoE SAFE).

References

Abramsky, L., and O. Fletcher. 2002. Interpreting information: What is said, what is heard - a questionnaire study of health professionals and members of the public. *Prenatal Diagnosis* 22: 1188–94.

Al-Jader, L.N., N. Parry-Langdon, and R.J.W. Smith. 2000. Survey of attitudes of pregnant women towards Down syndrome screening. *Prenatal Diagnosis* 20: 23–9.

Beyth-Marom, R. 1982. How probable is probable? A numerical translation of verbal probability expressions. *Journal of Forecasting* 1: 257–69.

Binkert, F., M. Mutter, and A. Schinzel. 2002. Impact of prenatal diagnosis on the prevalence of live births with Down syndrome in the eastern half of Switzerland 1980–1996. *Swiss Medical Weekly* 132: 478–84.

Bogardus, S.T., E. Holmboe, and J.F. Jekel. 1999. Perils, pitfalls, and possibilities in talking about medical risk. *Journal of the American Medical Association* 281: 1037–41.

Cosmides, L., and J. Tooby. 1996. Are humans good intuitive statisticians after all? Rethinking some conclusions from the literature on judgment under uncertainty. *Cognition* 58: 1–73.

Edwards, A., G. Elwyn, and A. Mulley. 2002. Explaining risks: Turning numerical data into meaningful pictures. *British Medical Journal* 324: 827–30.

Georgsson Öhman, S., S. Saltvedt, U. Waldenström, C. Grunewald, and S. Olin-Lauritzen. 2006. Pregnant women's responses to information about an increased risk of carrying a baby with Down syndrome. *Birth* 33: 64–73.

Gigerenzer, G., and A. Edwards. 2003. Simple tools for understanding risks: From innumeracy to insight. *British Medical Journal* 327: 741–4.

Gigerenzer, G., and U. Hoffrage. 1995. How to improve Bayesian reasoning without instruction: Frequency formats. *Psychological Review* 102: 684–704.

Hoffrage, U., S. Lindsey, R. Hertwig, and G. Gigerenzer. 2000. Communicating statistical information. *Science* 290: 2261–2.

Hunfeld, J.A.M., A. Leurs, M. De Jong, M. Lesnik Oberstein, A. Tibben, and J.W. Wladimiroff. 1999. Prenatal consultation after a fetal anomaly scan: Videotaped exploration of physician's attitude and patient's satisfaction. *Prenatal Diagnosis* 19: 1043–7.

Jaques, A.M., L.J. Sheffield, and J.L. Halliday. 2005. Informed choice in women attending private clinics to undergo first-trimester screening for Down syndrome. *Prenatal Diagnosis* 25: 656–64.

Keller, C., M. Siegrist, and H. Gutscher. 2006. The role of the affect and availability heuristics in risk communication. *Risk Analysis* 26: 631–9.

Klinik für Geburtshilfe. 2004. *Ratgeber für Schwangere*. Zurich: Universitätsspital Zürich.

Marteau, T.M., G. Saidi, S. Goodburn, J. Lawton, S. Michie, and M. Bobrow. 2000. Numbers or words? A randomized controlled trial of presenting screen negative results of pregnant women. *Prenatal Diagnosis* 20: 714–8.

Miny, P. 2001. Medizinische Genetik 2001: Pränatale Chromosomendiagnostik. *Schweizerisches Medizinisches Forum* 51/52: 1281–3.

Müller, H., P. Imhasly, and M. Leuthold. 2005. Pränatales Screening und pränatale Diagnostik. *Schweizerisches Medizinisches Forum*, 696–7.

Neuhaus, G. n.d. *!Schwanger?* Zurich: Verein Ganzheitliche Beratung und kritische Information zu pränataler Diagnostik.

Santalahti, P., E. Hemminki, A.-M. Latikka, and M. Ryynänen. 1998. Women's decision-making in prenatal screening. *Social Science & Medicine* 46: 1067–76.

Sedlmeier, P., and G. Gigerenzer. 2001. Teaching Bayesian reasoning in less than two hours. *Journal of Experimental Psychology: General* 130: 380–400.

Siegrist, M. 1997. Communicating low risk magnitudes: Incidence rates expressed as frequency versus rates expressed as probability. *Risk Analysis* 17: 507–10.

Smith, D.K., R.W. Shaw, and T.M. Marteau. 1994. Informed consent to undergo serum screening for Down's syndrome: The gap between policy and practice. *British Medical Journal* 309: 776.

Smith-Bindman, R., W. Hosmer, V.A. Feldstein, J.J. Deeks, and J.D. Goldberg. 2001. Second-trimester ultrasound to detect fetuses with Down syndrome. *Journal of the American Medical Association* 285: 1044–55.

Thomson, R., A. Edwards, and J. Grey. 2006. Risk communication in the clinical consultation. *Clinical Medicine* 5: 465–9.

Tversky, A., and D. Kahneman. 1982. Availability: A heuristic for judging frequency and probability. In *Judgment under uncertainty: Heuristics and biases*, ed. D. Kahneman, P. Slovic, and A. Tversky, 163–89. Cambridge: Cambridge University Press.

Lead is like mercury: risk comparisons, analogies and mental models[1]

Ann Bostrom

Daniel J. Evans School of Public Affairs, University of Washington, Seattle, USA

This paper reviews evidence for the use of analogies and comparisons to understand risks and conflation of different risks, from mental model studies of lead paint hazard, global climate change and smallpox disease and vaccine. For each study participants use analogies with other risks explicitly, and often draw inferences based on their experiences or knowledge of those other risks. In the case of lead paint, study participants judged options for testing and mitigating lead paint by analogy with risks as diverse as radon and mercury, with corresponding differences in their assessments of proposed strategies. Mental models of smallpox disease and vaccine are often explicitly based on analogies with chickenpox, which is much more familiar to study participants. Many studies of climate change have demonstrated conflation of stratospheric ozone depletion with global warming from the greenhouse effect. Some study participants adopt other frames, such as weather, which affects their inferences in predictable ways. These findings are discussed in light of theories on the role of analogy and metaphor in thinking and inference. Taking into account the prevalence, structure and pitfalls of comparisons and analogies should help risk communication designers better realize their potential.

Introduction

Two broad classes of goals for written risk communications are education and advocacy. Those communicators aiming to support informed consent and decision-making lean toward education, by providing people with new information, or correcting misinformation. On the other hand, public health communicators aim often to persuade their audiences to adopt healthy behaviors, or abandon risky behaviors, using the communication tools and practices that they believe are most persuasive. Understanding analogy and metaphor better may help achieve both.

Analogies (e.g., 1 is to 3 as 3 is to 9) are comparisons based on alignment and mapping of the underlying structure of two represented situations (Gentner et al. ch. 6, in Gentner, Holyoak, and Kokinov 2001). Many metaphors are analogies. While some metaphors are based instead on common attributes, comparison and alignment processes are nonetheless fundamental in understanding metaphors. Structural alignment and mapping are also evident in similarities (Gentner and Markman 1997). Further, people consider similes (e.g., lead is like mercury) more metaphorical than metaphors generally (Bowdle and Gentner, cf. Gentner, Holyoak, and Kokinov 2001, 231). Thus, in how they are understood and used, both metaphor and similarity are like analogy.

Gentner and colleagues have shown in numerous studies that analogical mapping has three components (Gentner, Holyoak, and Kokinov 2001, 200): 1) one-to-one correspondence between the mapped elements in the two situations (the base and target), 2) parallel connectivity between the elements, that is, the arguments of corresponding predicates also correspond, and 3) higher systematicity is preferred in the selection of alignment – e.g., a system of causal relations would be preferentially aligned over a system of independent matches. Structurally aligning two situations can help people induce domain-general problem schemas.

Analogy also plays a central role in learning. There are salient examples of analogies leading to scientific inferences and discoveries, such as the analogy Keppler drew between light and the motive power that pushed the planets around the sun (*Astronomia Nova*, 383 – as described in Gentner and Markman 1997). Most of our conceptual structure is organized by spatialization metaphors (Lakoff and Johnson 1980, 1999), such as 'happy is up'. Also 'up' is 'healthy', 'rational', and 'control'. These metaphors are coherent systems, derived from our physical and cultural experiences of the world (Lakoff and Johnson 1999).

Comparisons and analogies are widely used in and advocated for risk communication (e.g., Covello, Sandman, and Slovic 1988; Edwards 2003; Monahan and Steadman 1996), though they can backfire (Freudenburg and Rursch 1994). Further, experts have had difficulty predicting when they will backfire, and when they will help achieve risk communication goals (Roth et al. 1990; Johnson 2002, 2003, 2004a, b; Slovic, Kraus, and Covello 1990).

To better understand the role of analogy, metaphor and comparisons in people's thinking about risk, evidence is provided here from: a mental models study of lead paint hazards to children; a mental models study of smallpox vaccine and disease; and a summary of previous studies on mental models and metaphors of climate change. The lead paint study included specific risk comparison questions asked of couples that suggest some ways in which risk comparisons can produce conflict; these are also reported here.

Methodology

Mental models researchers come from a variety of backgrounds – including but not limited to anthropology, psychology, and education – and apply methods ranging from ethnographic interviews to psychological experiments and home and classroom field observations. The risk research reported here on mental models of hazardous processes applies multiple methods adapted from decision research, cognitive and cultural anthropology, cognitive science, and cognitive aspects of survey methodology. It draws heavily on approaches taken in earlier studies in related domains, for example Kempton (1987), and Jungermann, Schutz, and Thuring (1988).

For the risk mental models studies reported here, the studies begin with two projects: (1) an expert analysis of the risk mitigation decision, and (2) an open-ended interview designed to reveal the decision-makers' mental model(s) of the underlying hazard(s), followed by increasingly structured questions designed to probe their risk-related beliefs and perceptions. The latter is called a mental models interview. Subsequently, interview results are compared to the expert decision model. The results of this comparison are used to design a survey instrument, in order to allow collection of a larger sample and permit more robust conclusions. Further description of mental models research on risk perceptions for risk communication design is provided elsewhere (e.g., Bostrom, Fischhoff, and Morgan 1992; Morgan et al. 2002).

This approach assumes that a verbal protocol, that is – saying one's thoughts aloud (Ericsson and Simon 1993) provides valid insights into the mental model(s) held by the speaker. There is also the implicit assumption that risk analysis is likely to improve risk mitigation decisions, based on the framework developed by Hohenemser, Kasperson, and Kates (1985; see also Kates 2001). While this assumption could be interpreted as what Kempton (1987) describes as assigning higher prestige to 'institutional' models over folk models, two attributes of the decision models used for comparison suggest otherwise. The first is that the decisions and context modeled are those the intended communication recipients face. The model is a composite model based on expertise from several domains, rather than from a specific discipline or institution. Second is that the method allows identification of weaknesses in the institutional models based on assessment of folk models (here called mental models), as well.

The excerpts and analyses from the lead paint hazards and smallpox vaccine and disease studies are chosen to exemplify similes, metaphors, and analogies volunteered by those interviewed. While the samples presented here are not exhaustive, they represent the majority of cases found by a single reader reading the transcripts from the two studies.

Lead paint hazards

(with Vic Adamowicz, Sandra Hoffmann and Alan Krupnick)

Study methodology

This section draws on a set of in-person interviews with 35 married couples (70 individuals) conducted for a study designed to examine how parents think about and value their children's environmental health. The study drew from the population of two-parent households in Atlanta, Georgia, with children under the age of 7, living in housing built before 1979. We limited the population to owner-occupied housing, to keep control options relevant, and to two-parent households, as modeling dyadic decisions was of particular interest to the study authors. US Census of Housing data were used to identify neighborhoods in the Atlanta, Georgia Metropolitan Statistical Area with housing stock built before 1979. Households were sampled from phone number lists by the Survey Research Center at the University of Georgia and screened for appropriate characteristics in initial phone contacts. The first 15 interviews were conducted by research assistants at a central location at Georgia Institute of Technology. Because of difficulty in recruiting couples to travel to the interviews, the remaining interviews were conducted in couples' homes, by the Survey Research Center interview staff.

The study employed three strategies to assess parental decision-making: characterization of direct statements by parents of how they make decisions; analysis of responses to closed-ended questions about decision-making and factors hypothesized to affect decision-making in the literature, and finally; examination of hypothetical decision-making about lead paint mitigation.

A semi-structured interview protocol was used to investigate parental decision-making behaviors and their mental models of lead paint risks. Prior to the interviews, each spouse was asked to write down three recent major children's health decisions. Interviewers selected the highest-ranking jointly mentioned decision as a focus for the first part of the individual interviews. The individual interviews began with open-ended questions exploring the decision-making process involved in the couples' most recent major children's health decision. The results reported here are from the second section of the interview, which dealt with children's environmental health problems and focused on parental awareness and level of concern about lead paint hazards compared to other environmental hazards. After this, each spouse was presented with a hypothetical lead paint decision scenario and asked to talk through what they thought their family would do. Follow-up prompts were used to assure that information on the information desired, factors considered and role of cost in the family decision was collected. After a break, spouses were interviewed as a couple. The results reported here from the couple's interview followed the same protocol as the individual interviews. Following this, the spouses were again separated and asked to fill out a written questionnaire, from which the demographic information reported here is drawn.

The oral interviews were transcribed verbatim and coded. Nineteen of these (38 individual interviews and 19 couple interviews) were complete and available for analysis and inclusion in this paper. Two coders first coded four interviews, then met to discuss and revise the coding scheme and coding instructions. A single coder then coded all of the interviews (blind to the hypotheses). A third coder's independent coding of four of the sets of interviews demonstrated a high level of reliability (less than 1% disagreement, out of 124 responses per interview set of two individual interviews and one couple interview, for the responses analyzed here).

The observations here are derived from these parents' mental models of lead paint hazards, as elicited during the open-ended portions of the interviews. The value of mental models interviews with parents and others lies in their semi-structured elicitation of parents' mental models, allowing them to shape their responses as they wish, thereby providing insight on both the specific content and general structure of their mental models.

Results

These parents' mental models of lead paint hazards were surprisingly varied, along multiple dimensions. One of the interviewees had taught lead paint workshops for an environmental agency, and another was described by her spouse as an environmental educator. Others reported knowing nothing or virtually nothing about lead paint. One couple had a child who had recently tested high for blood lead. Many appeared to rule out lead paint as a risk for their children, despite living in housing that allowed that possibility. In one couple, it was evident from the individual interviews that the father was aware of a lead risk in their house, which he had not revealed to his wife.

Mental models of hazards include measures of exposure – source and pathways, effects, and mitigation (Bostrom, Fischhoff, and Morgan 1992; Morgan et al. 2002). Risk communication theories such as the extended parallel process model (Witte 1994), suggest that effective decision-making about a risk requires understanding the basics of both exposure and effects, as well as mitigation concepts, in order to be able to assess the threat posed by the hazard as well as to assess response and self efficacy (for further discussion, see Bostrom 2003).

Within the category of exposure are more specific concepts of where lead paint can be found (source) and how it reaches children (pathway). In virtually all interviews, both individual and couple, at least one exposure concept, general or specific, was volunteered by the interviewees. Age of the house was mentioned as a risk factor by 58% of mothers and 53% of fathers. The source or location of lead paint in the home was mentioned as a risk factor by almost all mothers, and by 79% of fathers. More specifically, the 'type of surface' was mentioned as a risk factor for lead paint by about a third of respondents, also by more mothers (37%) than fathers (26%), while paint condition was mentioned by about half, and by more fathers (58%), than mothers (47%). Parents also volunteered several other sources of lead in the home, including in pipes or water (47% of mothers, 32% of fathers mentioned), in soil (11% of mothers, no fathers), and single mentions of mini-blinds, cribs, certain Mexican ceramic dishware, and pencils (the latter erroneously).

A large list of effects was included by the experts we consulted to construct and review our decision model (see Bostrom et al. 2005), although only a few attracted much attention in the interviews, including IQ loss, which was mentioned by half of parents, mental retardation (26% mothers, 21% fathers, and 42% of couples), and deficits in school performance, mentioned by 21% of mothers, 11% of fathers, and about a fourth of couples (26%). Many effects of children's exposure to lead paint were mentioned by few or no parents, including gastrointestinal pain (none mentioned), visual system deficits (11% of couples, fewer individuals), impaired hearing (auditory function and processing), anemia (mentioned by one mother, no fathers), and a range of other ill effects mentioned by no one (vomiting, colic, irritability, lethargy and others). Death, however, which would be extremely rare, was mentioned by 5% of women and 21% of men. Parents described harm in very general ways that did not map readily onto the expert model, including mentions of 'developmental effects', 'stunted growth', and other bad effects (by a total of 47% of mothers, 58% of fathers).

Like the radon tests

Among the analogies or parallels referenced by parents in their descriptions of lead paint hazards were radon, asbestos, and mercury. These analogies dominated in different contexts. The radon analogy appeared primarily in the context of testing, or test kits.

By analogy with their thoughts about radon, some parents appear to have concluded that home test kits would be available for lead paint, and would be a good way to test for lead paint. In response to the interviewer asking 'How could a person find out if there is lead paint in or around his or her home?' one father said:

> If I am not mistaken, there is probably some kind of test that they can do. Whether or not you have lead paint, that would probably increase the amount of lead in the air, like the radon tests. I don't know if that exists or not, but if it doesn't, it should. Also do things like take paint samples somewhere to get it tested. [20Male]

This illustrates a multiple models strategy for thinking about testing, and the parent's explicit uncertainty about the radon test analogy.

Although home test kits are available for lead paint, EPA advises against their use, in contrast to its policies and practices for radon testing.

Lead is like asbestos

By analogy with asbestos, some parents concluded that sanding (mechanical removal) of lead paint would be dangerous (which it is), and that lead paint could be inhaled, and its harmful effects might be mitigated with better ventilation or air filtration, which in reality probably would accomplish very little.

In response to a question about what should be done, if anything to reduce children's exposure to lead paint, one father responded:

> Keeping the woodwork painted and reducing the amount of peeling. If you see flaking or peeling chips, it should be removed and the wood repainted. Usually if you keep a good enamel coat on top of the lead based paint, then there is no exposure. It is a lot like asbestos. If you have asbestos there, the worst thing that you can do is make it breathable, and break it up. [14Male]

This father, however, mentioned development delays, learning difficulties, and aggression in small children as health risks from lead paint, earlier in the interview.

Another father described lead paint similarly, when asked about lead in the environment and any risk from lead:

> [...] I may be mistaken, but I think that lead is like asbestos, as long as you are not messing with it, sanding it, whatever, or eating it, then it is not going to hurt you. [20Male]

The interviewer then asked how a child might be exposed to lead paint in his or her home, to which he responded:

> Lead paint has been out of the market for a while, I don't know how long it has been. I would assume that older homes would be more at risk. With our home being older, I am sure it has had that at one time or another. We also used to have asbestos siding as well, that we took off the house. But, you know, I would assume that that would be a factor. Just being around it, does that... I don't know if that is bad or not, or if you just have to be messing with it. I think you have to be messing with it, but I am not sure of the extent. [20Male]

Again this analogy is about exposure processes, and is explicitly uncertain ('I am not sure of the extent').

When asked where lead might be found in a house, one couple [25Couple] responded as follows:

> Father: I am aware of paint. That is the only way that I have ever heard of it. I may have heard something in plumbing, but those are the only two ways that I have ever heard.
> Mother: Is asbestos similar to that?
> Father: Not the same.
> Mother: And what about wood, I did not realize wood...
> Father: It is in paint. I know paint and lead they are the two things that I heard. I know that my grandmother just moved out of her house because of the kids and the paint.

Here the wife introduces the asbestos analogy, demonstrating great uncertainty as she does so, and drops it when challenged by her husband.

Is lead like mercury?

By analogy with mercury, some parents thought lead would accumulate and stay in the body which it does, though in bones and other reservoirs, rather than blood, over the longer term).

> I don't know what happens. Is it like mercury, where it is there forever? [14Couple]

In one example, a father described having general concerns about poisons in the environment when asked about environmental hazards in the home, as well as wall coverings, the paint, how old the paint is, whether or not the HVAC was in working order. He followed with:

> You know the different gases, radon gases and things of that nature are very prevalent in Georgia. Also the hot water heater and things being in good working order. [29Male]

When asked a follow-up question regarding any concerns about lead in the environment and any risk from lead, the father responded:

> That is why I mentioned paint. That is a major concern for, any heavy metals really, whether it is mercury, you name it. [29Male]

This father described the risks from lead paint as thwarting a child's development and affecting motor skills.

> Heavy metals in general, your body is not able to fight and they generally stay in your soft tissues, but mainly in ones that are non-fibrous – brain tissue and things of that nature. [29Male]

Like the others, this analogy to mercury and general inference about lead from heavy metals is also focused on health effects.

One mother, when asked if she had any sense of whether there are environmental hazards in the home that might put children at risk, in particular lead in the environment or any risk from lead, volunteered the following about her daughter:

> Because of her delay with the mercury, which is a neurotoxin, which just breaks down the brain. The lead is also a neurotoxin. So if she is possible to have a content of mercury, she could also have been exposed to lead, easily. [21Female]

This mother was analogizing with her daughter's mental and verbal delays after receiving the MMR (measles, mumps and rubella) vaccine. In her own words:

> My concern was that, from doing my own research, a lot of the MMR, especially in the 1990s, contained what is called mercury. Then the federal government got involved because of complaints from parents and a few class action lawsuits. They said they had to remove the mercury from the vaccine, yet I found out a few years ago that they used another derivative of mercury called ferorwaso [sic – thimerosal?], which is a derivative of mercury. Actually, they did not totally take the mercury out. Now my daughter had the vaccine the early part of last year. I am still uncertain if the mercury is still in there. I think 2004 they started removing that, but anyway, she was just tested for mercury, mercury levels, actually, about three days ago. I have not gotten the results back, but I will be following up with the physician. [21Female]

The analogies with mercury focus on health effects, and appear to heighten concern about lead paint risks.

Paint fumes

Some parents also drew inferences from their general knowledge of paint, as illustrated by this quote from one mother in response to, 'Can you tell me about any risks from lead paint?' Note that here, too, explicit statements of uncertainty accompany the analogy:

> To children, yes. Well, I have heard about lead poisoning in children, the fumes from the paint if it contains lead. I am not sure exactly the effects to children, what that would lead to, but I have heard it mentioned before. [19Female]

This mother went on to describe exposure in a manner consistent with this, in response a question about how a child might be exposed to lead paint in his or her home:

> Perhaps a house, if you move into an older home, and you don't strip the paint, and that paint still contains the lead. Breathing in that paint, being in the room that was painted with that kind of paint could pose a problem. I am not sure what kind of effects that would have. I can imagine it would lead to breathing problems, sinus problems. I don't think it could lead to asthma, I am not sure. [19Female]

In these cases, the ready analogy with exposure to potential toxins in paint fumes (unnamed - possibly volatile organic compounds) appears to trigger inferences about respiratory health effects.

In one couple interview [22C], when asked about health risks to children from lead, the mother mentioned that

> We had a relative, her children, one of her children had lead poisoning, or something, and they could not figure it out because she was in a relatively new home [...] I thought about that because of the age of our house. In the past I have heard stories about lead paint causing damage, so I did have a concern. When I mentioned it to my husband, he said 'I think you have to ingest it for it to be a problem.' He did not really see it as a problem, so I just, when he said that, I felt relieved. I don't think she (daughter) will just be eating the paint.

The father [22C] responded that he had forgotten the conversation his wife described, but a few minutes later said about lead paint that

> If it was wet, the fumes might cause some kind of brain damage.

Again a discussion and analogy based on exposure processes appeared to trigger inferences about health effects.

Interview-induced risk comparisons

A unique feature of the lead paint study is that we interviewed both parents individually and as a couple. Both interview protocols included two risk comparison questions. The first asked 'How do you feel lead paint hazards compare to other environmental risks a child might face?' The second asked for a comparison of the risk that children face from lead paint to that they face from air pollution.

In general, the responses from individuals were fairly straightforward, but provide some insight regarding how these parents thought about risk comparisons. For example, one mother responded to the first that:

> I think the more things I worry probably are the growth hormones and what they are eating. Stuff that I know are happening to them instead of the things that I can only fear are happening to them.

And in response to the second:

> Air pollution would be worse. I would think it is more prevalent. It is more, I like to think that my house is all safe and any problems can be covered up. [4Female]

The response illustrates two dimensions to comparisons: the focus on exposure processes as a determinant of risk, according to which there is certain exposure to air pollution given that everyone breathes and uncertain exposure to lead paint; and explicit expressions regarding control over safety in one's house.

When this couple [4Couple] was interviewed together, they interrupted one another in response to the first comparison question:

> Father: I think it is the chances are less likely, the risk is lower than air pollution especially in a city like Atlanta with such congestion and air pollution.
> Mother: There are growth hormones and other things that are bad for them too (talking over her husband).
> Mother: I don't remember what I said when we were talking about that one.

Not only do these parents interrupt one another, their responses do not map onto one another at all.

Many couples conflicted when asked about the risk comparisons (air pollution comparison, [12C]):

> Mother: hmm… I think it's a less risk, less… because they're going to breathe.
> *Interviewer: So less risk with the lead paint, you said?*
> Father: I think it would be more.
> Mother: You think so?
> Father: Because to be exposed to it, they have to eat the lead, once it gets into your system…

The above response illustrates again that for many respondents, exposure was key to how much the parent worried. In response to 'How would you compare the risk children face from lead paint and from air pollution?' [15Female]:

> Not every child I don't think faces risk from lead paint, depending on where they spend their time. But all children breathe the air. So, um, I think lead paint can have more, if a child is exposed to lead paint though, then that is more of a problem for that child than I think air pollution. So, it depends on how much exposure the child would have to lead paint, how it would compare to air pollution.

But when one parent was focused on exposure and the other on the severity of the consequences for exposed children, they ended up talking over one another, often failing to reach a consensus ranking, as illustrated by the second comparison from [25Couple]:

> *Interviewer: And what about air pollution? How would you compare the risk that children face from lead paint to air pollution?*
> Father: The lead paint is much more dangerous. You could put me outside, and put someone else in a room with lead paint, and you wouldn't know.
> Mother: I said just the opposite (in the interview). I said because you don't have people who live in those older homes. I said air pollution because we are breathing it and everything, as opposed to older homes that people might not live in.
> Father: Come on…
> Mother: You might not have kids in an older home, but we are breathing air every day.
> Father: But she says the comparison. You are outside and I am in this room.
> Mother: Oh, well, in that case, it would be definitely. I thought she meant…
> *Interviewer: The risks that children face from both, compare the risks…*
> Father: Oh, if there is poison and death in affected areas… I think if you are going to poison people…

Not all couples conflicted, though there is still some hesitancy in the middle of this duet (comparison with lead paint hazards to other environmental risks, [20Couple]):

> Father: I would say that it is not at the top of my list, not something that I worry about every day.
> Mother: Yeah, I think that pollution and chemicals…

Father: Air quality.
Mother: Air quality, and that crap in our food.
Father: And water.
Mother: And water, probably is far more at risk than... But direct exposure... see it is just... I mean...
Father: But you have no control, you have little control over air quality.
Mother: We have control over 'if I take the sander to get that spot'
Father: I worry less about things that I have control over.
Mother: I agree.
Father: I have control over lead in my house.
Mother: Yes, so.
Father: As far as environmental hazards
Mother: Less than other things, but we certainly think about it.

When both parents were focused on exposure, and in this case on control, they conveyed agreement on the relatively lower risk to their children from lead paint as compared to other environmental risks. Note that the mother still made a point of emphasizing at the end that lead paint concerned them, whereas the father initiated their response by deprioritizing lead paint hazards.

A final dialogue concerning how to compare lead paint hazards to other environmental risks to children illustrates the central role of analogy when there is great uncertainty or lack of knowledge, and again the kinds of conflicts seen above appear in this dialogue as well [21Couple]:

Interviewer: How do you feel lead paint hazards compare to other environmental risks a child might face?
Father (whisper): Go ahead, you can answer.
Mother: Well, I don't know, it could be just as bad.
Father: You said how do you feel...?
Interviewer ... lead paint hazards compare to other environmental risks.
Mother: I said just as bad [in individual interview]
Father: Just as bad? But what, just as bad as what?
Mother: Other environmental things.
Father: Like?
Interviewer: Compare it to air pollution, for example?
Mother: Yeah, you can get asthma.
Father: Can you get blood poisoning out of air pollution?
Mother: I want to say yeah...
Father: You can? You can get blood poisoning out of air pollution?
Mother: I want to say yeah, because you can inhale, you know.
Father: No, I don't know. That is why I am asking.
Mother: You can inhale, like lead, and umm
Father: Lead is not air pollution. These are two different things.
Mother: Okay, well, chlorine is harmful, strong and in the air.
Father: Well, I am trying to say, if we can...
Mother: Chlorine can kill brain cells.
Father: Listen to what I am saying... I am just trying to say of all those things that you said before that can happen when a person
Mother: I just said just as bad, can you please just leave me, you don't know...
Father: I am just saying, out of all those things that you listed earlier that can be caused by lead poisoning, they are greater than these other things? I think lead is important, not just, asthma is important.
Mother: I would say just as bad.
Father: Okay, I tried.

The conflicts generated between parents by the risk comparison task were not just conceptual, but as this illustrates, also emotional.

Smallpox vaccine and disease

The events of 9-11 and consequent advent of a renewed, strong interest in bioterrorism, as well as a new smallpox vaccine policy from the Bush administration motivated this study of smallpox vaccine and disease mental models, trust, and policy attitudes (Bostrom and Atkinson 2007).

Study methodology

The smallpox study interview protocol opened with a few very general, open ended questions, then became more structured, to explore what mental model(s) the interviewee might be using to make decisions about smallpox vaccine and disease risks. For purposes of comparison and a reliability check, this study also included several of the questions from Blendon et al. (2003; see Bostrom and Atkinson 2007).

We interviewed 24 Georgia Tech students enlisted in the Psychology experiment pool in the spring of 2003. The ages of our participants ranged from 17 to 26, averaging 21. Of our 24 respondents, 11 were male and 13 were female; all were current students; none of the 24 respondents had any children; 20 had never had any training in public health and four had; five respondents had family members who were in the medical profession; 22 respondents were American citizens and those who were not had lived in the US for an average of six years; and the most common ethnicity was Caucasian, followed by Asian, other (Burmese/Myanmar, Caucasian and middle-eastern, and mixed), and one respondent each who was African-American, Asian-American, and Hispanic.

Results: smallpox is like chickenpox

A single analogy dominated the smallpox interviews: chickenpox. At the time of these interviews, when experts were comparing monkeypox to smallpox, to use the familiar to draw inferences about a more novel disease, young adults were relying on still somewhat prevalent and controversial disease, chickenpox, to explain smallpox, which was eradicated in 1977 (see Bostrom 2003 for additional discussion). Results are presented in two separate sections, although some interviewees use the disease to make inferences about the vaccine, or vice versa.

Smallpox disease

The following excerpts demonstrate that metaphors and analogies highlight similarities and differences, as predicted by structural alignment and comparison processes. An explicit appeal is made to the parallels in causal processes for chickenpox (the base, and more familiar disease) and smallpox (the target):

> *Interviewer: What happens when a person is infected with smallpox?*
> [20]: I know it's similar to the chicken pox as far as, um, severe flu symptoms, high fever, cough, and probably congestion, sneezing, everything. But also, um, as far as the sores go, I know they're characterized by smaller sores and probably more numerous.
> [...]
> *Interviewer: And, do you know how the smallpox disease causes these effects?*
> [20]: Not specifically, but I know it's like, related to chickenpox, a lot. I know the causes and effects through – it's a bad disease, as far as the severe effects, you know. Fever would mainly cause a person to get brain damage or harm their organs, and um, that's about it.

Note that the following respondent is explicit about uncertainty (below), and gets to the point of saying 'I don't want to say anything because I don't know'. The chickenpox analogy conflicts directly with stylized facts about death rates from smallpox that this respondent [26] had heard on TV, which produced hesitation and discontinuity in the response.

> *Interviewer: How might a person get smallpox?*
> [26]: I would guess just from another person with smallpox.
> *Interviewer: So, like, person-to-person contact, or airborne? ...*
> [26]: Like, yea, I guess it's kind of like chickenpox where it's like airborne. Somebody can cough, you can touch ... it's really contagious.
> [...]
> *Interviewer: And how does smallpox cause these harmful effects that you've mentioned?*
> [26]: Well, I think the way chickenpox works is that, like, the actual, like, chickenpox cell, like virus, can fill the rest, like actual, like working body cells, and then replicate itself and like cause rash and ruptures. [...] I'm guessing smallpox works the same way where it just takes over your healthy body cells, and reproduces [...]
> [26]: I've seen pictures on TV of it. I think they're, I guess, smaller than chickenpox and you get more of them. And then, I don't know [...] I think I know of it, like, killing a lot of people.
> *Interviewer: Okay, so you can die from it.*
> [26]: I think so.
> *Interviewer: Okay. Anymore? Can you think of any long-term effects of smallpox?*
> [26]: Maybe... I don't want to say anything because I don't know.
> *Interviewer: Okay, and how does smallpox cause these harmful effects that you've mentioned?*
> [26]: Well, I think the way chickenpox works is that, like the actual, like chickenpox cell, like virus, can fill the rest, like, actual, like, working body cells and then replicate itself and cause rash and ruptures. So there are lots of little [unintelligible] all over your body that can latch on to the other working body cells. So I'm guessing smallpox works the same way where it just takes over your healthy body cells, and reproduces the [unintelligible].

This final excerpt illustrates again the same kind of comparative thinking as the above, as well as the direct and explicit appeals to uncertainty and lack of knowledge.

> *Interviewer: What happens when a person is infected with smallpox, what does the disease do to his or her body?*
> [29]: I really do not know.
> *Interviewer: Okay, what are short term effects of smallpox?*
> [29]: Short term effects? I would think your actions would be inhibited, you wouldn't be able to do much. It's related to the chickenpox, so... I don't know, then again chickenpox is a very mild version of it. I'd, when I had the chickenpox I was very young, I wasn't able to do anything. I don't know if smallpox has that itchiness, but...
> *Interviewer: Okay, what about any long term effects of smallpox?*
> [29]: I don't believe there are any, or I haven't heard of any long term effects, except for death, but I, full recovery. I haven't heard of any long term effects.

Once again, the analogical leap provides a basis for several concrete suggestions about effects, but the respondent is unsure of the extent of the parallels between the two situations.

Smallpox vaccine

The results suggest that the analogy with chickenpox led the young adults interviewed to believe that they had been vaccinated for smallpox. However, it is not possible to assert this with any certainty. The rapid increase in the number of vaccines administered routinely to children over the last few decades complicates the task of knowing which vaccines a youngster has received or not. Here also uncertainty appears to accompany the biggest analogical leaps.

Interviewer: At what age do you think it is appropriate to receive the vaccination?
[24]: I would say, I'm not sure, I would… I really don't know. I would put it probably in the context of as young as, maybe, some people develop chickenpox, when they get the vaccine for that. I guess an early age, depending on – then again, it really all depends on how much is needed. So I guess without me really knowing how much is needed it's hard to say at what age.
Interviewer: Ok.
[24]: But if it's needed, then I would definitely say as early as possible.

With regard to the vaccine and possible harm from it, [26] appears to be analogizing from public dialogues around chickenpox and other vaccines that are based on altered or killed disease viruses, including the issues of differences in vulnerability to the effects of such vaccines, and forecasting which influenza vaccine should be produced any given year, based on the influenza strains circulating before flu season. This generates an appropriate concern for how sensitive subpopulations might vary in their responses to such a vaccine, even though smallpox vaccines have historically been produced from *vaccinia* rather than the smallpox virus itself. The second [26] and third [31] interview excerpts below illustrate how interviewees analogize with chickenpox vaccine, expressing a general model of how vaccines work as a basis for inference about how smallpox vaccine works and its potential effects.

[26]: I guess if the disease is fatal it could kill you, but I don't know if the vaccine could really do that.
Interviewer: So could someone die from the smallpox vaccination? That's the next question.
[26]: Theoretically, yes.
Interviewer: Okay, Do you know how the smallpox vaccine causes these harmful effects?
[26]: Okay, I read that this one time, but I don't remember. I think like, with the chickenpox vaccine, what can actually happen is that when it alters to a certain state it actually makes it more likely to infect certain body types, but just as general. It's no good for the population. So maybe with smallpox it's much more sensitive […] like you can only use just a little bit, when it's actually going to end up hurting other people.

[31]: The vaccine? I know that, as I said earlier, that it could actually be dangerous.
Interviewer: Okay.
[31]: Like any vaccine, there's always a chance of actually contracting the virus. People might have mild symptoms, this is just purely guess, they might have mild symptoms of actual smallpox.
Interviewer: Okay
[31]: That it will, if it's given accurately and it works effectively, it's supposed to just, your body will build immunity […]
Interviewer: Okay, can you say anything about the benefits of, like how long the benefits of the vaccine might last?
[31]: Purely guess, I think, I thought it would last indefinitely, because once your body builds immunity to it, it's always going to be there.
Interviewer: Okay
[31]: It's like if you had chickenpox for the first time, you get immunity to it, and most people don't get it again.

Even [36] who describes smallpox disease and vaccine in considerable detail, hesitates when faced with a conflict between an analogy with vaccines in general and specific stylized facts about smallpox vaccine:

> *Interviewer: Tell me about smallpox.*
> [36]: What I know about smallpox is that it is a virus; that they have been developing vaccinations for it; that it could be considered a biological weapon; and vaccinations are highly recommended to the public and especially medical workers. And I know a lot of them have already gotten vaccinations, and also military personnel. That is what I know.
> [...]
> *Interviewer: Ok. Tell me about the smallpox vaccine.*
> [36]: The vaccine is a very weakened strain of the regular smallpox. I think it might be like cowpox or something like that, that is [unintelligible] much weaker, and you're supposed to be able to develop your own defenses against that, and that will help you if you actually come in contact with the real, much stronger version of smallpox. I know that there can be adverse consequences to just the vaccine.

Policies can also be taken as metaphors for scientific fact, as the following illustrates. While in fact smallpox vaccine has not been a childhood vaccine for several decades, this respondent thought it still was, which appears to be the basis for the respondent's explicit inference about potential harm:

> *Interviewer: Who is likely to be harmed by the vaccine?*
> [37]: Harmed? I think there's none.
> *Interviewer: And why do you say none?*
> [37]: Because everybody, I think, growing up, I mean in any way, in like, you know, for example in the United States or Europe, or any country that's not a third-world country knows that you have to be vaccinated for it.
> *Interviewer: Ok*
> [37]: And so there is no harm.

Smallpox vaccine is also riskier than most vaccines on the market. Some vaccines were harmed and a few died from the vaccine after the Bush administration implemented its smallpox vaccine policy (see Bostrom 2002).

Climate change

A handful of explanations dominate those people have volunteered over the last decade and a half when asked to describe or explain global warming or climate change. These include stratospheric ozone depletion, tropospheric pollution, and weather (Bostrom and Fischhoff 2001; Bostrom and Lashof 2007; Böhm and Pfister 2001; Kempton 1991, 1997; Löfstedt 1991). Additional, less prevalent explanations have included, for example, explaining climate change in terms of photosynthesis (Kempton 1997), or spacecraft puncturing the ozone layer on takeoff.

Although the evidence is substantial that a majority of the US public already supports CO_2 emission-reduction policies, many people still confuse weather with climate, and attribute or link global warming to stratospheric ozone depletion, which leads to relatively inappropriate or ineffective actions, including avoiding aerosols sprays, blaming chemical industries, and ignoring energy efficiency (Bostrom and Lashof 2007; Kempton, Boster, and Hartley 1995). In one tri-state US study, only 26% of those surveyed believed that energy use causes global warming; about the same proportion were willing to pay more for utilities to avert climate change.

Common climate change analogies include not only ozone holes, pollution and weather, but also those analogies proposed and perpetuated by scientists and researchers themselves, such as the greenhouse that retains heat. People do not refer spontaneously to the greenhouse effect, leading some researchers to conclude that the greenhouse analogy has failed (Aubrun and Grady 2001). In one experimental study, a 'thickening blanket of CO_2, warming the planet' outperformed other metaphors for greenhouse gases. People who saw a 'heat trapping' or 'CO_2 heat trapping' explanation, as compared to no explanation, a more detailed scientific explanation, or an explanation that focused on impacts, were more likely to mention relevant concepts when asked to describe global warming (Aubrun and Grady 2001; see also Bostrom and Lashof 2007).

'*The key determinant of behavioral intentions to address global warming is a correct understanding of the causes of global warming*' (Bord et al. 2000, 205). Having a simple, correct mental model or analogy of the causal process can promote effective action.

Conclusions and discussion

Based on his ethnographic studies of how people think about and use thermostats, Kempton describes two 'folk models' of thermostats, as simple analogies. For example, some people have a 'valve' model of thermostats (Kempton 1986). The higher they turn the thermostat, the wider open the valve, the faster the room will heat. The analogy influences behavior in predictable ways (Kempton 1987). Kempton is careful to describe the merits of such folk models, which by-and-large serve people remarkably well.

The results reported in this paper suggest several ways in which simple analogies or 'folk models' can help people reason about a hazard, even one that is fairly well known. The parallels parents drew between mechanical abatement methods for lead paint and asbestos highlight risks that have been central to lead paint mitigation research controversies in the US (Hirsch and Bozeman 2005). The analogy between lead and mercury provides a basis for drawing inferences about the neurotoxicity of a heavy metal that may be less familiar from one that may be more widely discussed. This analogy was also one of the rationales for the choice of lead paint as the topic of the family decision-making research project itself. Note that the analogy with mercury was not elicited in the context of mitigation, however, and did not appear to help parents reason about what they could, should, or might want to do about lead paint risks. Similarly, the greenhouse analogy appears ineffective by comparison with an analogy that better supports causal thinking that incorporates CO_2 emissions.

The results also illustrate how such analogies can be dysfunctional. The analogy with radon is misleading in several ways, in part because the accuracy of home tests varies depending on the specific technology of the test, in part because of the physical differences between human exposure processes for radiation and lead paint dusts. The analogy between chickenpox and smallpox is powerful, as it is based on systematic parallels between causal processes, from exposure processes to some aspects of effects. This analogy appears strong enough to produce a generalized problem representation that leads many interviewees to conclude that smallpox is a childhood disease, and that people are still vaccinated for this disease. Findings would likely have differed had we interviewed those targeted for vaccination under the Bush smallpox vaccination policy. Nevertheless, the results suggest that hazard analogies can be misleading specifically when the analogies are based on parallels for exposure processes, but are then used for inferences about health effects or risk.

A similar conclusion can be drawn from a study of folk models of *Pfeisteria piscicida* outbreaks by Kempton and Falk (2000). Estimated recreational fishing losses of $4.3 million in Maryland followed *Pfeisteria* outbreaks. These appeared to be attributable at least in part to the folk or 'cultural models' of *Pfeisteria* held, including: a disease in fish, a parasite in fish, a water pollutant, and a toxin or poison (Kempton and Falk 2000), all of which suggest routes for exposure that are inconsistent with scientific models of *Pfeisteria* hazards. The experienced economic losses far exceeded the losses that would have occurred had people only avoided areas where there were outbreaks (Kempton and Falk 2000), as an expert model of the hazardous process would prescribe.

Exploratory studies like those reported in this paper do not support firm conclusions about the prevalence of phenomena in general, or even in specific contexts other than the study contexts. Nevertheless, it is notable that the use of analogical thinking and metaphor was pervasive in the smallpox study, where a single comparative, chickenpox, dominated. By comparison, the lead paint interviews produced fairly sparse use of analogies and metaphor, and no single comparative dominated. Kempton points out that people use multiple models, for the most part, when they explain phenomena (e.g., Kempton and Lave 1983). This is evident from the excerpts presented here too; for lead paint a single father [20] used different analogies to explain different parts of the hazardous process. Both studies elicit spontaneous comparative thinking, which suggests its importance in thinking about risks. However, dyads demonstrated difficulty negotiating an imposed shared risk comparison task.

It is tempting to conclude with an appeal to the kinds of advice given by experts on risk comparison over two decades ago (Covello, Sandman, and Slovic 1988), regarding the relative merits of different kinds of comparisons and analogies. A growing body of empirical literature shows that the advice was an oversimplification and does not hold (Freudenberg and Rursch 1994; Johnson 2002, 2003, 2004a, b; Roth et al. 1990; Slovic, Kraus, and Covello 1990). The results presented here suggest that the advice fails in part because it ignores that structural alignment increases salience of both similarities and dissimilarities. Further, as Gentner and Markman (1997, 47) point out, predicting the outcome of a comparison requires knowing 'the person's current psychological construal of the things being compared, including goals and contextual information as well as long-term knowledge'.

The discussion here has focused primarily on cognition and inference. However, experimental evidence from marketing suggests that metaphor can also be more persuasive than literal statements (Hitchon 1997). The Hitchon study also hypothesized that general affect transfers from the base to the target in metaphors, but found no affective transfer of that nature. Studies of this sort for hazard analogies and risk comparisons would be helpful for risk communicators. The heightened conflict between couples apparent in the lead paint risk comparison task suggests that it was not the affective attitude held toward the base that generated emotional responses to the comparison, but dyadic differences in structural alignment processes.

Good analogies, metaphors and comparisons can simplify communication and improve learning and inference. This paper adds to the small body of research illustrating several characteristics of their use in risk contexts, with an eye toward informing and improving how they are deployed in risk communication.

Acknowledgements

I gratefully acknowledge the collaborators listed in each study section, and National Science Foundation support for the climate change studies, US Environmental Protection Agency support for the lead paint research through Resources for the Future, and Georgia Tech for supporting Emily Atkinson's work on the smallpox study. Any errors in this paper are solely the author's responsibility.

Notes

1. Prepared for New Perspectives on Risk Communication: Uncertainty in a Complex Society: A Multi-Disciplinary International Research Conference, 31 August–2 September, 2006, Göteborg, Sweden.

References

Aubrun, A., and J. Grady. 2001. *The missing conceptual link: Talkback testing of simplifying models for global warming*. Providence, RI: Cultural Logic.

Blendon, R.J., M. Catherine, P.H. DesRoches, J.M. Benson, M.J. Herrmann, K. Taylor-Clark, and K.J. Weldon. 2003. The public and the smallpox threat. *The New England Journal of Medicine* 348, no. 5: 426–32.

Bostrom, A. 2002. *Risky diseases, risky vaccines, risky policies. Research horizons, fall 2002*. Atlanta, GA: Georgia Institute of Technology. http://gtresearchnews.gatech.edu/reshor/rh-f02/column.html.

———. 2003. Future risk communication. *Futures* 35, no. 6: 553–73.

Bostrom, A., B. Fischhoff, and M.G. Morgan. 1992 Characterizing mental models of hazardous processes: A methodology and an application to radon. *Journal of Social Issues* 48, no. 4: 85–100.

Bostrom, A., S. Hoffmann, A. Krupnick, W. Adamowicz, R. Goldman, M. McWilliams, and J. Varner. 2005. Parental decision-making and children's health. Working paper presented at EPA NCER Star grantees workshop, November, in Washington, DC. Available from the authors.

Bostrom, A., and E. Atkinson. 2007. Trust and risk in smallpox vaccination. In *Trust in cooperative risk management: Uncertainty and scepticism in the public mind*, ed. M. Siegrist, T.C. Earle, and H. Gutscher. New York: Earthscan.

Bostrom, A., and D. Lashof. 2007. Weather it's climate change? In *Creating a climate for change: Communicating climate change and facilitating social change*, ed. S.C. Moser and L. Dilling. Cambridge: Cambridge University Press.

Covello, V.T., P.M. Sandman, and P. Slovic. 1988. *Risk communication, risk statistics, and risk comparisons: A manual for plant managers*. Washington, DC: Chemical Manufacturers Association.

Edwards, A. 2003. Communicating risks through analogies. Editorial letter. *British Medical Journal* 327, no. 27: 749.

Ericsson, K.A., and H.A. Simon. 1993. *Protocol analysis: Verbal reports as data*. Cambridge, MA: MIT Press.

Freudenburg, W.R., and J.A. Rursch. 1994. The risks of "putting the numbers in context": A cautionary tale. *Risk Analysis* 14, no. 6: 949–58.

Gentner, D., K.J. Holyoak, and B.N. Kokinov. 2001. *The analogical mind: Perspectives from cognitive science*. Cambridge, MA: MIT Press.

Gentner, D., and A. Stevens, eds. 1983. *Mental models*. Hillsdale, NJ: LEA Associates.

Gentner, D., and A.B. Markman. 1997. Structure mapping in analogy and similarity. *American Psychologist* 52, no. 1: 45–56.

Bozeman, B., and P. Hirsch. 2005. Science ethics as a bureaucratic problem: IRBs, rules, and failures of control. *Policy Sciences* 38, no. 4: 269–91.

Hitchon, J.C. 1997. The locus of metaphorical persuasion: An empirical test. *Journalism & Mass Communication Quarterly* 74, no. 1: 55–68.

Hohenemser, C., R.E. Kasperson, and R.W. Kates. 1995. Causal structure. In *Perilous progress: Managing the hazards of technology*, ed. R.W. Kates, C. Hohenemser, and J.X. Kasperson, 25–42. Boulder, CO: Westview Press.

Johnson, B.B. 2002. Stability and inoculation of risk comparisons' effects under conflict: Replicating and extending the "asbestos jury" study by Slovic *et al. Risk Analysis* 22, no. 4: 777–88.

———. 2003. Are some risk comparisons more effective under conflict? A replication and extension of Roth *et al. Risk Analysis* 23, no. 4: 767–80.

———. 2004a. Risk comparisons, conflict, and risk acceptability claims. *Risk Analysis* 24, no. 1: 131–45.

———. 2004b. Varying risk comparison elements: effects on public reactions. *Risk Analysis* 24, no. 2: 103–14.

Jungermann, H., H. Schutz, and M. Thuring. 1988. Mental models in risk assessment: Informing people about drugs. *Risk Analysis* 8, no. 1: 147–55.

Kates, R.W. 2001. Queries on the human use of the Earth. *Annual Review of Energy and the Environment* 26: 1–26.

Kempton, W. 1986. Two theories of home heat control. *Cognitive Science* 10: 75–90. [Reprinted and expanded 1987 in D. Holland & N. Quinn (Eds) *Cultural models in language and thought* (Cambridge: Cambridge University Press)].

———. 1987. Variation in folk models and consequent behavior. *American Behavioral Scientist* 31, no. 2: 203–18.

———. 1987. Two theories of home heat control. In *Cultural models in language and thought*, ed. D. Holland and N. Quinn. Cambridge: Cambridge University Press.

———. 1991. Lay perspectives on global climate change. *Global Environmental Change* June: 183–208.

Kempton, W., J.S. Boster, and J.A. Hartley. 1995. *Environmental values in American culture*. Cambridge, MA: MIT Press.

Kempton, W., and J. Falk. 2000. Cultural models of pfiesteria: Toward cultivating more appropriate risk perceptions. *Coastal Management* 28: 273–85.

Kempton, W., and J. Lave. 1983. Review of 'Mental Models', by Dedre Gentner and Albert L. Stevens (eds). *American Anthropologist* 85: 1002–4.

Lakoff, G., and M. Johnson. 1980. The metaphorical structure of the human conceptual system. *Cognitive Science* 4: 195–208.

———. 1999. *Philosophy in the flesh: The embodied mind and its challenge to Western thought*. New York: Basic Books.

Löfstedt, R. 1991. Climate change perceptions and energy-use decisions in northern Sweden. *Global Environmental Change* 1: 321–4.

Morgan, M.G., B. Fischhoff, A. Bostrom, and C.J. Atman. 2002. *Risk communication: A mental models approach*. Cambridge and New York: Cambridge University Press.

Monahan, J., and H.J. Steadman. 1996. Violent storms and violent people: How meteorology can inform risk communication in mental health law. *American Psychologist* 51, no. 9: 931–8.

Roth, E., M.G. Morgan, B. Fischhoff, L. Lave, and A. Bostrom. 1990. What do we know about making risk comparisons? *Risk Analysis* 10, no. 3: 375–87.
Slovic, P., N. Kraus, and V.T. Covello. 1990. What should we know about making risk comparisons? *Risk Analysis* 10: 389–91.
Witte, K. 1994. Fear control and danger control: A test of the extended parallel process model (EPPM). *Communication Monographs* 61, no. 2: 113–34.

The public meeting as a theatre of dissent: risk and hazard in land use and environmental planning

Åsa Boholm

CEFOS, Göteborg University, Sweden

Land-use and environmental planning typically involves diverse actors such as representatives of government authorities and administrations, special interest organisations, scientific experts, lawyers, stakeholders and concerned citizens. During the course of the planning process, communication is not limited to written documents. The Swedish Environmental Code also calls for public meetings and consultation efforts with stakeholders. Considering that public meetings have a prominent role in environmental planning, there is surprisingly little research on the form, content and development over time of face-to-face communication and social interaction taking place. This article addresses communication and interaction at public meetings concerning a facility siting project of considerable longevity – the building of a railway tunnel through the Hallandsås in southern Sweden. Using data from participant observation of local public consultation meetings regarding the Hallandsås railway tunnel through a period of several years, this paper analyses and discusses communication about environmental risk. The theoretical foundation derives from pragmatic communication theory, Niklas Luhmann's distinction between risk and hazard, and social anthropological theory on agency and conflict generation in communication.

> Communication presupposes beings that exist independently, with their own environments and their own information-processing apparatuses. Every being sifts and processes what he perceives for himself. Under such circumstances, how is communication, that is, co-ordinated selection, possible at all? (Luhmann 1995, 157)

Public consultation: risk as a topic of argumentation

Public meetings are one of several formats used by government agencies, municipalities and companies to inform the public and stakeholders about local projects or planning ventures. These meetings facilitate the dissemination of planning alternatives and their reasons, goals and objectives, including risks and benefits associated with decision alternatives. There can be manifold objectives for encouraging participation in land-use planning, depending on the category of participants and what roles they have in the planning process (Petts 1999). A developer, a governmental authority, a local resident, a stakeholder group or an environmental group have different interests and motives. Also, the scope for participation may vary, from being solely a recipient of information about a project to becoming a participant in decision-making about planning issues in ways that actually influence the outcome of a project. Moreover, to be consulted is not the same thing as to have power (Soneryd 2003).

According to the Swedish Environmental Code,[1] consultation meetings are mandatory in the planning process of environmental projects. An environmental impact assessment (EIA) and evaluation must include public involvement, called 'consultation' (*samråd*), at several stages of the project. Other authorities and municipalities that are affected, the county administrative board, private individuals and organizations likely to be affected must be consulted in the process. The Environmental Code does not specify what form consultation should have; it only states that consultation should be held and that it should be documented at certain junctures in the planning process. The Code distinguishes between early and late consultation; the rationale for early consultation is that it should give an input into the EIA required in order to obtain necessary permits.

In Sweden, public trust in government is relatively high (Löfstedt 2005). The public administration favours a consensual style of risk regulation where expert advice has a key role in decision-making. The public has a limited role in the planning process, and the consultation process prescribed by the Environmental Code mainly serves as a forum for negotiation between different authorities (Päiviö and Wallentius 2001; Rönnborg 2006). Only to a limited extent does consultation serve as a political arena for deliberation between government and citizens. The common format is to stage consultation as a public meeting where information is delivered by authorities and project managers. The public and identified stakeholders are invited to comment, discuss, ask questions, criticise, and voice consent or dissent. Also, according to the Planning and Building Act, there are requirements for consultation, but at this stage plans are usually advertised so that people may take part in the scheduled presentations and deliver commentaries in writing before a set time.

Consultation meetings in EIA contexts regarding infrastructure development and facility siting projects as a rule include a multitude of risk issues. Risks pertaining, for example, to a new road, railway line, wind farm, cell phone mast, or airport extension may be identified in such diverse areas as construction technology, groundwater flows, bird and marine life, landscape aesthetics, and human health. An EIA investigation potentially can actualise a multitude of risk issues: What are the risks? How big are they? How could the risks be assessed? How should the risks be evaluated? What priorities should be made? Which values are at stake? How should values at stake be prioritised? Is there to be more or less concern about birds at risk as compared to flowers at risk? Is physical damage to humans worse than psychological trauma or stress? These kinds of questions tend to amass when major projects are planned in a local setting. It is not only the sheer number of issues and perspectives that adds to the complexity. Knowledge of all these issues is necessarily knowledge from a certain standpoint (van den Hove 2006) and the standpoints diverge, sometimes quite dramatically. Such 'coexistence of irreducible standpoints' (van den Hove 2006, 11) has two possible solutions: either one standpoint establishes a hegemonic position vis-à-vis all others (which goes against the principle of democratic governance in society) or the parties with the diverse standpoints have to communicate and exchange views, arguments and facts in order to arrive at some form of agreed solution to a common problem.

The multitude of specific consequences and risks of policy decisions on land use and environmental planning and natural resource management offers rich nourishment for heated local debate. When locations need to be found for infrastructure facilities or industrial plants, the voicing of conflicting interests and values, demands and wishes regarding the uses and meanings of a particular landscape follow suit (Boholm and Löfstedt 2004). Planning sets in motion processes of counter-planning which may readily turn into rhetorical battles between opposing forces, each using arguments to destroy the other side in a power contest over what the decision alternatives are, and what decisions should or should not be made. Ultimately the legitimacy of power over the local environment is at stake (Binde and Boholm 2004; Corvellec 2001, 2002; also Sidaway 2005).

Time and energy are spent on arguing about what specific planning alternatives are the best, for what reasons, and what the stakes are for the various options. In heated and long-drawn disagreements sometimes continuing for many years, values are weighed and prioritized, arguments come and go, and shifts are made between broad and specific arguments. Cause-specific arguments are embedded in a broader discourse on trust and mistrust, responsibility and accountability of authorities and regulatory bodies (Binde and Boholm 2004; Boholm 2000, 2005a; Corvellec 2001). Such debates display a rhetorical organisational logic very different from any ideal model of decision-making characterised by '...perfect dialogues between rational actors convincing each other on the mere virtue of their arguments' (Corvellec 2001, 37–8). As Healey and Shaw (1994) put it:

> Through processes of argumentation, the different values, policy theories, and ideas about cause and effect can be brought into play, with instrumental reasoning being used alongside moral reasoning and aesthetic sensibility. [...] In this way, the values held by individuals or groups can be drawn into a judgment, so long as a clear line of reasoning is established. (435)

Characteristic of meetings as a form of risk communication is that people engage in face-to-face interaction during a set period of time. They spend time together in the same room and communicate verbally and non-verbally, the course of interaction unfolding as interplay among the participants.

Aim and methodology

Although public meetings and other forms of face-to-face discussions and interactions such as public hearings or workshops are a prominent arena for risk communication and environmental management often used by governmental agencies in Sweden and elsewhere (Chess and Purcell 1999), there is a surprising lack of research on the subject, particularly on deliberate efforts to involve the public in democratic processes (Ryfe 2005). This was noted by Daniel Fiorino (1990), while McComas (2003a, b) made the observation 13 years later that not much had happened in the field.

The literature is mainly prescriptive and normative, discussing how ideals such as dialogue and participation – taken to be conducive to trust-building and confirming the legitimacy that government agencies yearn for – could be achieved in practice (Renn 2006). Many public meetings fail due to open conflict, distrust and sometimes lack of interest (very few citizens show up). Noted problems include the substance of the meetings and what is accomplished during them (McComas 2003a, b). Public meetings as a forum for consultation with the public have been criticised both for over- and under-representation. Influential and strongly-voiced stakeholders or citizen interests can use the opportunity to exercise undue influence. However, in the literature there are very few empirical studies of actual meetings and the communication taking place at them. In one study (McComas 2003a, b) of participants' views of meetings held regarding two proposed local waste sites in upstate New York, a main result was that most participants felt the meetings to be quite meaningless (2003a) and that those attending, especially the more active participants (2003b), tended to distrust the organisers.

This article offers to fill part of this gap by means of a case study of communicative interaction at local public meetings concerning the Hallandsås railway tunnel project in southern Sweden. The study was conducted during a period of almost seven years and during this period the tunnel project was followed continuously from early 1998 to mid-2004. Field observations were made by the author during a 4.5-year period (from January 2000 until June 2004) of a substantial number of official public meetings arranged by the National Rail Administration's project organisation. There were basically three types of local meetings, although similar in organisational structure: consultation meetings proper in accordance with the Environmental Code, information meetings called ('consultation forum'), and court hearings held by the Environmental Court and the Supreme Environmental Court.[2] Notes were taken at the meetings to document what key participants were present, what topics were introduced and discussed, and also to record statements, arguments, verbal expressions, emotional tone and attitude. Social arrangements and other features pertaining to the context of communication were also noted. Most meetings were taped[3] so that the notes could be compared with audio recordings. In addition, the meetings were officially documented by a secretary and these minutes were later published on the Hallandsås tunnel project's home page on the internet.[4] Also other sources have been used, including informal talks with, and spontaneous questions to, participants at the meetings (officials, experts, stakeholders and citizens). It also happened that I questioned actors about specific occurrences at a meeting later, on some other occasion. Another source of information has also been the local media[5] and its reporting from the meetings. It should be added that research on the Hallandsås tunnel project has been conducted over an extended period of time by a small team of researchers at the Centre for Public Sector Research at Göteborg University, Sweden. This meant that we often had three to four researchers present at the same meeting and we therefore had good opportunities to check and compare our observations, notes and impressions.[6]

Before we proceed to the presentation of the case a brief outline of the theoretical framework is needed. Communication about risk at consultation meetings is approached from a pragmatic perspective addressing the context of communication, including assumptions, knowledge and experience, intentions and actions of participants (Boholm 2007). Communication is understood as a situated social activity (Allwood 2000). At consultation meetings people gather for different reasons. A public official goes to a meeting to present a proposal for a new landfill site since this is part of his work while a local resident goes to the same meeting because she lives nearby the proposed site and will become its neighbour. So at the very start the participants have very different roles in the ensuing communication that will unfold at the meeting. The public official represents the decision-making and regulatory authority, and the resident/citizen is a stakeholder, someone who will be affected by the decision (see Luhmann 1993, ch. 6). Therefore we need to address not only the factual content and the manner of presenting that content to an audience. Pragmatics asks us to consider communication as a complex social process of meaning creation and interpretation situated in real life. This includes the participants (who they are and what roles they have), the structures and procedures that condition communication, the motivation of participants and their rationality, and how 'relevance' in communication is established – that is to say, what relationships are meaningful, and how interconnections between messages are made (Allwood 2000; Sperber and Wilson 1986/2003). A pragmatic research approach further demands detailed ethnographic case studies of natural situations of communication.

This study addresses how co-operation is established and maintained in communicating about risk, how conflicts are managed, and how participants with different roles and professional status interact – as experts and laypersons, decision-makers and stakeholders – who cannot be expected to share assumptions and knowledge and who may have diverging interests and communicative intentions and goals.

Back on track – a case from reality

The railway tunnel project[7] has been going on since 1992 and it has been, and still is, strongly contested for a number of reasons. The project has encountered several technical difficulties due to the hydrogeological conditions of the Hallandsås, a ridge consisting of severely fissured and porous material and huge deposits of ground water. The Hallandsås is partly a nature reserve and partly an agricultural area with many farms specialising in agriculture, dairy and animal husbandry.

The Hallandsås tunnel belongs to a category of mega-projects (Flyvjbjerg, Bruzelius, and Rothengatter 2003), namely multi-billion investments in complicated infrastructure (tunnels, bridges, airports, railways or roads) that systematically becomes tremendously more expensive than originally budgeted. In most cases, such works are publicly funded and promise to offer huge benefits for society, in terms of growth, quality of life, regional development, national or international integration, etc. Such projects are characteristically technically and organisationally complex and depend on the mobilisation of political-technocratic-industrial networks of actors over considerable periods of time. They also draw heavily on symbolism and give rise to masses of sense-making argumentation to legitimate the project and to prove its merits and worth (Boholm 2005a; Darian-Smith 1999; Latour 1996).

Building of the tunnel started in October 1992 and soon encountered considerable technical problems. In 1995–96 ground water leaked uncontrollably from the excavated tunnels far beyond the permission granted to the National Rail Administration in a verdict from the Water Rights Court. Sealing did not work properly, and in 1996 the dramatic sinking of the ground water table began to be noticed above ground. Wells in the area above the tunnels started to dry out. This caused disturbances for local residents and farmers depending on a reliable supply of water for irrigation, cleaning purposes and drink for their animals. In order to deal with the leaking tunnels, a chemical sealant containing acrylamide was used which spread into the ground water, wells and streams (see Baier 2003; Boholm and Löfstedt 1999; Löfstedt and Boholm 1999; Sjölander-Lindqvist 2004).

The project was stopped in October 1997 after the toxic leak and the dramatic losses of ground water. Several government commissions and investigations were launched to figure out what had gone wrong. Also several court proceedings were initiated to investigate various legal aspects of the project and to determine legal responsibility for the violation of the water permit, the toxic leak, and the tunnel workers affected by acrylamide poisoning (Heiefort 2004). In 2001 the Swedish government decided that, under the condition that necessary permits were given by the relevant authorities, the tunnel should be built despite the past problems and the fact that it had become tremendously more expensive. The new water permit deemed necessary by the Rail Administration to build the remaining tunnels was granted by the Environmental Supreme Court in 2003, and Båstad municipality granted the building permit. Building under the contractors Skanska (which had been in the project before) and Vinci (a new company) started in 2004–2005 using new technology including deep freezing of parts of the ridge and state-of-the-art tunnel boring technology by means of a new, specially-designed tunnel drilling machine.

The effects of leaking ground water from the tunnel, starting in 1995 and culminating with the toxic leak in the autumn of 1997, have been a source of concern for local residents. Many are worried about how the surrounding environment and the local landscape will be effected in the long term, and how agriculture and farming will be affected in the future (Sjölander-Lindqvist 2004). In order to assess the risks and decide what risks are 'acceptable', environmental investigations and environmental court proceedings have been undertaken. The tunnel has produced an abundance of oral and written material in the form of information brochures; political, administrative and legal documents; environmental impact assessments; expert reports; newspaper articles; information meeting oratory; and informal commentaries. While some institutions and actors are eager to build the tunnel, others are hesitant and still others are decidedly against it, especially and understandably so those who have been most severely affected, namely local residents and farmers. Since the tunnel is being built in a democratic society, dialogue and consultation among various stakeholders and interests are essential.

A great number of organisational actors have been involved in the process – government agencies, companies, experts, members of NGOs, local citizens, and the news media who report about the tunnel project in terms of news stories. Communication has been profuse and enlivened by means of an array of media (public meetings, printed matter, exhibitions, media stories, narratives and even film and photography). Goals, aims and reasons for decisions have been formulated, deliberated, explained and criticised. Procedures and plans have been developed, presented and debated over many years. Decision alternatives have been suggested, motivated, and reconsidered. Since 1998, the National Rail Administration's Hallandsås project has arranged a number of local consultation meetings at which information about the project has been presented, the project has been explained, and experts have been called in to give their views on various aspects of the project; the general public has had the opportunity to discuss the project, ask questions, and express viewpoints.

Public consultation – forums for communication

During the building phase from October 1992 until the toxic leak in October 1997, the Rail Administration had not arranged any specific meetings or information activities for the local public. House owners and farmers who had problems with their wells going dry, a problem that began in 1996, were dealt with individually; however, there were no organised efforts to address the issues of adversely affected local people or environments or to incorporate these issues into the definition of the project (see Sjölander-Lindqvist 2004). Directly after the toxic leak in October 1997 the municipality of Båstad declared parts of Hallandsås a risk area and initiated a crisis group to handle residents' worries. Turbulent days followed. The Environmental Assessment Group (Miljögranskningsgruppen, MGG) composed of experts in hydrogeology, chemistry and ecology, which had been commissioned by the Rail Administration to assess the environmental impacts of the tunnel project soon assumed the role of communicators of facts and risk assessments while the Rail Administration's project management stepped into the background. Trust in the Rail Administration was low (Boholm, Löfstedt, and Stranberg 1998; Palm 1998) while MGG and Båstad municipality faired better in trustworthiness. In their role as experts, communicating their findings and assessments to the public and to various concerned groups of residents, between October 1997 and January 1998, MGG had a number of formal and informal contacts with individuals, organisations, interest groups and local residents.[8] Most of these contacts were in the form of information meetings arranged in cooperation with stakeholder organisations like the local branch of the Swedish Environmental Protection Society (SNF), residents organised in three road associations on the northern side of the ridge, and residents on the southern side represented by their lawyer.

Issues discussed at the meetings ranged from the effects tunnel building might have on water resources, the risks of acrylamide contamination and further spread of the chemical, to complaints by residents about how the situation was organised by the authorities, and matters regarding water distribution. These meetings and information efforts were in the beginning not publicly announced and were targeted to affected groups to deal directly with the issues and problems relevant to their situation. The value of having a more institutionalised forum for providing information about various aspects of the tunnel project to those affected was acknowledged by MGG in January 1998, when it noted that the tunnel project was gradually moving from a stage of acute mitigation to a stage of planning for continuation.[9] The group also noted that a continuation of the tunnel project would need a consultation process (*samrådsprocess*) in accordance with the demands of the Environmental Code (§§ 4 and 5); MGG considered the contacts and meetings held during the three last months of 1997 as a good start.[10]

Regarding the continued consultation process (*utökat samråd*) ahead of the eventual restart of the tunnel building project, MGG noted 'a great need of public transparency and opportunity to influence'.[11] The group considered various models for how such a process of public consultation could be organised. One model was the open public meeting (*stormöte*) including a panel of experts. The stated advantage of this format was that no-one was excluded from participation. But drawbacks were also noted, namely that the experts might come to play a too prominent role, and that aggressive voices and extreme opinions might dominate the discussion at the cost of more moderate opinions. Another option discussed was the 'limited consultation' (*begränsat samråd*) which was to be publicly announced and which would require special invitation. The rationale for such restricted meetings was that representatives of the Rail Administration, the municipality, the building contractor and specially invited experts could then communicate with affected local residents and other stakeholder groups. However, this suggested format for a more exclusive lobby network of various stakeholders and interests was never realised.

The first consultation meeting, called a Consultation Forum (*Samrådsforum*) was held on 9 March 1998 in the municipality office in Båstad. Sixteen citizens were present (several of them are still active in attending meetings and following the tunnel project) together with representatives from Båstad municipality, two local journalists and a representative from MGG. The minutes from the meeting record the following: MGG should be the organiser of the meetings; MGG should see that minutes are taken; the Rail Administration and Skanska (the building contractor) can be represented but should not be the organisers; the meeting should be based on dialogue and not on one-way dissemination of information; there should be printed announcements for the meetings; and mass media should not be allowed to attend. In addition, the meeting defined a number of tasks for MGG to undertake: investigate the amount of remaining acrylamide in a local stream and in private wells; assess the risks of damage to buildings due to the lowered ground water table; assess the risks of altered flows of groundwater when the tables are restored; and answer questions about whether local residents had any means to ask for the water permit to be quashed.[12]

The next meeting was held in a school in Båstad three weeks later (30 March 1998).[13] This meeting was attended by the same individuals with the addition of a few newcomers, including representatives from the municipality and the Rail Administration. The formal procedures and aims of the meetings were restated and the issues discussed concerned acrylamide in agricultural produce and the legal status of the water permit. The Rail Administration's newly recruited project manager declared that he was new on the job and that he had limited experience of what had happened before October 1997. Noting that the relationship between the Rail Administration and the local public and the municipality was not very good, he advised that 'we' should now look forward. The project manager declared that tunnel construction in the future would continue only on the condition that it would not harm nature, the environment or residents in the area. He explained that the Rail Administration had learnt from 'the events' (referring to the acrylamide incident). Several people at the meeting wanted to know what guarantees there were that new mistakes would not be made in the future. To this the project manager answered that the project would be certified and have external audits. A number of questions were then raised regarding aspects such as: Who will be responsible for judging future risks? How big is the actual area of potential effect? How should the water permit of 33 litres of water discharge per second (33 l/sec) be interpreted? How will the water quality in wells be affected? What about alternative routes for the new railway? How will streams, springs and ground water be affected? Have wild animals and farm animals been affected by drinking contaminated water, and if so, how? These themes of communication at the first two meetings recurred in many of the ensuing consultation meetings for years to come.

In 2000 the Rail Administration undertook an Environmental Impact Assessment of the tunnel project aimed to serve as a basis for the Swedish government in its decision whether the project should continue, and if so, under what terms, conditions and budgetary limits. The local community was now included in the project.[14] In total, over 30 consultation meetings were held from October 1997 until June 2003. Six of these meetings were consultation meetings held as part of the process of the Rail Administration's EIA process in 2000, presented as a report to the government.

Another major arena for communication about the tunnel project has been the hearings in Båstad held by the Environmental Court (which had replaced the old Water Rights Court) and the Environmental Supreme Court on three occasions during 2002–2003. The matter for the proceedings in the Environmental Court was that in order to resume tunnel building, the Rail Administration needed a new and considerably more generous water permit (the old one permitting 33 l/sec had proved to be technically impossible to live up to). The Rail Administration had therefore applied to the Environmental Court for a new permit of 100 l/sec. These hearings were held in public and included numerous stakeholders, public as well as private. The majority of private stakeholders consisted of local residents and farmers, strongly opposed to the proposed new permit. At these hearings the applicant and the various stakeholders were allowed to present arguments for or against the application and with special focus on the environmental consequences of the 100 l/sec proposal. The hearings were very well attended including, apart from the members of the court, personnel from the Rail Administration, concerned local residents, interested citizens, various experts called in to give statements, lawyers, representatives of national and local NGOs, and media reporters. During the hearings, the environmental aspects of the tunnel were debated both in minute detail and in general terms. In addition to debates and presentations, the hearings also included *in situ* visits to a number of selected sensitive local environments or biotopes so that the court could get a more 'practical' view of what was actually discussed (Boholm 2005b).

The staging of a public meeting

The consultation meetings concerning the Hallandsås rail tunnel gather together a number of organisational actors (representatives of government agencies and authorities, and various kinds of experts – law, ecology, biology, risk assessment, or engineering) with various stakeholder groups, interested parties, local residents, and journalists. A more or less intensive process of communication takes place during a meeting. People speak, argue, explain, use slide presentations including maps and animations; decisions and decision alternatives are demonstrated and explained by authorities and experts, then discussed, questioned and criticised by members of the audience. The meetings tend to have a standardised structure divided into three consecutive stages:

(i) The opening: staging the perspective of the authorities (especially the Rail Administration) and experts. An obligatory part of this stage is to present arguments to prove that the tunnel project contributes to the 'greater good of society'.
(ii) A coffee break in the middle of the meeting.
(iii) The questioning: staging various arguments against the project; voicing dissent, making counter-arguments, making claims that arguments deriving

from other perspectives must be considered. This stage may include speeches of local residents victimised by the project.

If one looks at the meetings over time – from early 1998 to mid-late 2004, a period of nearly seven years – there is surprisingly little variation. Of course, new topics emerge, for example numerous technical matters referring to the intricacies of tunnel building technology; individual experts as well as specific evidence come and go, but overall the themes of communication remain stable and so does the conflict line in the debate. The thematic issues concern the legality of the project, the status and interpretation of the old (and new) water permit, the reasons for the project, the history of the project, the credentials and credibility of the Rail Administration, risk assessments and judgements, what is included in the project environment (or what should be excluded from it), future scenarios, risk management operations and routines, and of course, many technical matters.

The opening of the meeting

A meeting always opens with an official representative of the project (such as the project manager representing the Rail Administration) who explains the project from the point of view of the concerned authority. The national perspective and broader policy goals are made explicit. The argumentation builds on a number of assumptions about rail traffic, the environment, good governance, and the role of the Rail Administration in society: the railway is a safe and environmentally friendly means of transport; the Rail Administration promotes railway and therefore it follows (seemingly in a logical fashion) that the Administration gives priority to environmental values; it makes wise decisions based on a meticulous scrutiny of decision alternatives, and an office-holder and employee of the Administration works for a just cause ('contributing to a better society'). Another cornerstone in the argumentation is that the national objectives have been democratically decided by the government. Decision-making about the Hallandsås tunnel is embedded in a broader context of more substantial decisions within the context of national environmental policy and the decision to upgrade the West Coast Line. An obligatory part of the opening of the meeting is the assertion that the decision to build the tunnel is a just and rational one (the tunnel is needed) and that the Rail Administration and project management is trustworthy (having the necessary competence and a suitable frame of mind). Here is the opening statement of the project manager at the April 2002 hearing:

> The West Coast Line is a highly prioritized railway route in Sweden. The reason for this is that capacity on rail traffic must be increased, linking Norway with the European continent. Traffic must be relocated from road to rail. Never before have so many people travelled by rail; there is a five percent increase per year of rail travel in Sweden. There is a lack of capacity on the rail. In the Mälarregion, for example, there has been a tremendous expansion, and we have seen a decrease in road traffic. We must counteract the increasing road traffic on the E6 motorway. With the Hallandsås tunnel we will have an increased capacity – 24 trains per hour. The single track meanders over the ridge and is a typical railway of the nineteenth century that does not live up to the requirements of today. The Hallandsås tunnel will offer a long term opportunity to direct traffic from road to railway. This will improve the environment and road safety. (Project manager, introductory address to the court, Environmental Court hearing, Båstad, 22 April 2002, notes taken by author)

And one year later, the same project manager states:

> The project started ten years ago and it has been standing still for over six years. The lorry traffic has doubled on the E6 motorway and will double one more time. It will be difficult to accomplish the government's environmental goals. Hallandsås is the biggest bottleneck. We have a functioning railway net and we should be anxious to get the railway system in order. We must benefit from investments already made. The building project will be carried out as a well planned activity, with concern for the environment. We have discussed shortcomings. But this does not mean that nothing works. It is healthy to be really self-critical from time to time. This also holds for the contractors. The investigations underlying the decisions have high quality... The environmental impact is acceptable. The entire investigation is reliable and has been commended by *Naturvårdsverket* (the Swedish Environmental Protection Agency) and the *Statens Geologiska Institut* (Swedish Geotechnical Institute). (Project manager, concluding address to the court, Environmental Supreme Court hearing, Båstad, 5 September 2003, notes taken by author)

The next step is that experts (representing sciences such as geology, hydrogeology, ecology, biology, chemistry, or the technologies) make presentations of subjects in relation to the project, in terms of technical matters, risk identification and risk management. These presentations tend to be structured as factual, technical expositions of states of affairs and causal relationships (technical, biological, scientific or managerial) presented by experts or officials (from the Rail Administration and other governmental authorities, as well as from the county board and the municipality of Båstad). They are sometimes interrupted by attacks of counter-argumentation from members of the audience who question a particular fact, or a suggested causal relationship, or who disagree on what the facts are, which facts matter and how they should be interpreted.

The coffee break

Coffee is served with sandwiches or cakes during the break at the expense of the organiser, and the assembled meeting splits up into an informal gathering of smaller groups talking, discussing and exchanging greetings. Food has social meaning and eating together is a marker of belonging and inclusion in most societies. In a Swedish cultural context the coffee break (*kaffe*) is an institution at workplaces, at home, with neighbours, visitors, and friends. Drinking coffee together opens up an informal way of interacting with those you know, as well as with strangers that promotes small talk, personal comments and gossip. The informality of the event does not, however, mean that it lacks capacity for social structuring. The coffee break serves to 'bond', making visible the borders of a social group or constellation, differentiating those who belong from those who do not belong, and which actors constitute the core of the group. Serving coffee with sandwiches or cakes also expresses hospitality on the part of the host when one party 'invites' guests to coffee (for example at a public meeting). Such hospitality may be understood as a token of social respect and recognition that people have devoted time and effort to come to the meeting.

The questioning

After coffee, the meeting re-assembles and the concluding part consisting of discussion and questions follows. This part characteristically displays a number of polarised themes: factual evidence versus emotions; risks versus benefits; experts versus locals; scientific knowledge versus local experience; economic versus environmental values; and trust versus suspicion. While the authorities and experts are the most active during the first part, the members of the audience are the most active during this last part of the meeting. People in the audience raise questions, make statements, report contrary evidence, suggest alternative solutions, and engage in emotional outbursts of personal frustration, anger or despair. Some contributions receive spontaneous applause. Critical speakers tend to focus on their unique competence and knowledge as local residents, for example:

> The former speaker [a university professor called in as an expert] does not live here. I do. We should learn from history as history repeats itself. The residents are very worried. Trust in the Rail Administration is very low as they exceeded the old water permit… (Resident, Environmental Court hearing, Båstad, 25 April 2002, notes taken by author)

Especially notable is a recurrent form of speech that expresses the point of view of the 'innocent victim'. At almost every meeting, one or more local residents deliver speeches with a strong moral and emotional tone focusing on their sufferings, the destruction of unique values (losses of water, landscape and local nature), and their lack of trust in the Rail Administration and the tunnel project. These monologues may be more or less elaborated, occasionally prepared in advance and read aloud as statements. The core message is always the same: that local residents have been mistreated by the authorities. The tone is quite emotional and personal and brings to the fore the suffering and despair felt by those who experience themselves to be victims of the decisions made by the authorities. The themes focus on perceived lack of control, lack of trust, suffering, and a feeling of vulnerability. For example:

> The environment, fish and water, has been on the agenda. This is good. But where is the individual person? The residents? They are shut away in the background. I want to tell about our experiences. Maybe it is a bit tedious. But it is needed. How we have been living now for several years. It has been great pressure since 1996. A foreign power has occupied our yard. Move, or we will drive over you! During the summer of 1997 when the ground water was at its lowest we did a questionnaire covering six square kilometres including many objects, dams, brooks and wells. We compiled the answers and made a map with dates. It was a terrible sight. There was no water! We approached the county board for help. Nobody bothered. In October 1997 it stopped. The Rail Administration had exceeded the water permit for several years by 80–100 litres per second. That was done deliberately. Otherwise the project would have been stopped. The great lowering of the groundwater – that is the big crime! As a consequence of the toxic leak we ended up living in a risk area. We were not allowed to deliver milk and animals for slaughter for a period of six weeks. We were lucky; we did not have to put down any animals. Then there was the water tank. Twice a day we had water delivered to our farm. It was an endless traffic of experts, people to take samples, and workmen. Meetings after meetings. Still poor water, it was called 'serviceable but with comments', from a new drill. Not one week without contact with the Rail Administration during these years. … We cannot take additional problems. We want our fresh water back and then we want to be left in peace. (Farmer, concluding address to the court, Environmental Court hearing, Båstad, 25 April 2002, notes taken by author)

Such speeches, putting emotions into words, are strongly applauded by members of the audience expressing solidarity with the speaker. Confronted with such counter-rhetoric in the form of criticism, questioning of project logic and reason, and emotional self-victimisation, the Rail Administration project management and the summoned experts usually adopt a deferring position. They tend to repeat project 'reason' by presenting again the factual arguments and statements, explicating project legitimacy in broad terms, and the political mandate to act, but without answering directly to any emotional or personal misgivings or grievances. Nor do they go into any discussion of planning solutions and options already excluded from the decision process. These meetings can thus be described as basically two sets of monologues, separated by the coffee break. We are indeed very far from the participatory dialogues prescribed in the risk communication literature as being the ideal format for 'rational discourse'.

Victims versus decision-makers

From attending a considerable number of consultation meetings on the subject of the Hallandsås tunnel over a period of five years, a recurrent observation is that the meetings have a strong built-in element of conflict. This is also an often-reported feature of contested facility siting projects where over time various involved actors tend to become increasingly polarised in a continuing debate about the pros and cons of a project (Binde 2000; Binde and Boholm 2004; Corvellec 2001). A facility siting case can give rise to a myriad of conflicts, for numerous reasons and regarding many issues, including colliding interests and goals, diverging priorities and experiences, disputes about evidence, what the facts are and what facts are needed and should count. But what I want to discuss here is not conflict in terms of opposing interests, values, or standpoints deriving from knowledge, whether of an expert or lay person. Behind the conflict there is – or at least can be in some cases – a much more profound divide between (lacking a better term) modes of existence. These modes have to do with how individuals and groups perceive themselves in terms of agency (or lack of it) in relation to the outside world, including other people (Stoffle and Evans 1990).

Action is linked to intentionality, the capacity to act and to make things happen as planned in the world. The concept of agency distinguishes 'happenings' (caused by physical laws) from 'actions' (caused by prior intentions) (Gell 1998):

> An agent is one who 'causes events to happen' in their vicinity...An agent is the source, the origin, of causal events, independently of the state of the physical universe. (16)

To have agency means that a being has a capacity to do things and get things done; there is a self-movement that is monitored from the inside ('the internal self'). It should be noted that agency is not an absolute feature of a being. It is a quality that is attributed by an observer (observing himself or others) according to conventionalised expectations:

> The idea of agency is a culturally prescribed framework for thinking about causation, when what happens is (in some vague sense) supposed to be intended in advance by some person-agent or thing-agent. (Gell 1998, 17)

Since agency is directed (intentional), it has a goal or object outside the acting being. This other object or being, which is acted upon, is the 'patient' of the actor (Gell 1998, 21–3). Agency and patiency constitute two relationally constituted states; the one who acts and the one acted upon (Karlsson 2002; Gell 1998). Agency and decision-making are related, since to make a decision presupposes agency. Without agency there can be no decisions. A decision is a specific form of human action and reasoning in which alternatives are assessed and computed in a systematic and deliberate process. Without agency and the capacity to choose deliberately there would be no risk. Risk can therefore be understood as a by-product of decision-making (Luhmann 1993). Decisions, however, not only relate to the actual decision maker since decisions often have consequences that affect other people. Those affected find themselves in a different position from the decision-maker; we could say that they are patients of the agency of the decision-maker's agency (his/her capacity to act).

Luhmann's distinction between risk and hazard (1993) relates to agency and lack of agency. To attribute something as a hazard implies that it has an external cause; a hazard is imposed from outside. A risk, on the other hand, is attributed to have an internal cause since it presupposes agency, i.e., choice between decision alternatives. 'One is exposed to dangers, but one takes a risk', in Luhmann's vocabulary. The distinction of risk and hazard, and the associated distinction between agency and patiency, depends on attribution. An observer (observing self, others, and the relationship between self and other) identifies agency, its source and direction. Whether or not something is attributed as a risk or a danger depends on the stance of the observer and '…if we wish to know which is which, we must observe the observer and if necessary develop theories on the conditioning of his observation' (Luhmann 1993, 27).[15]

In the modern welfare state many decisions on public affairs which also affect us as individuals are delegated to other people: such as politicians, public officials, and regulators. The relation between decision-maker and those affected by a decision (whom we may call 'stakeholders') is asymmetrical (see Table 1). Stakeholders find themselves in a different position from the decision-maker; they are affected by the decision which might endanger their interests, values, identity or assets. The stakeholder is affected and exposed, in terms of danger, to the consequences of decisions made or not made (Luhmann 1993, ch. 6). For the stakeholder the consequences of a decision are hazards while for the decision-maker/regulator the negative consequences are risks. This means that it is not the issue as such – the substance of a decision, such as the building of a railway tunnel – that defines whether or not something counts as a hazard or a risk, but on who the observer is: 'One man's risk is another man's danger' (Luhmann 1993, 153).

Risk and hazard are distinctions within the same schema (Luhmann 1993) – risk is contingency in the agency mode; hazard is contingency in the patiency mode. This means that something can oscillate from risk to hazard and back again depending on how a relationship is framed in terms of agency–patiency, depending on the salience of decisions (contingency) or the saliency of determination: to be patient under someone else's agency (Gell 1998):

Table 1. Distinction between regulator/decision-maker and stakeholder positions.

Regulator/Decision-maker	Stakeholder
Make decisions	Affected by decisions
Agent	Patient
Calculated contingency (intentionality)	Causality
Risk mode	Hazard mode

> Marking risks then allows danger to be forgotten, whereas marking dangers allows the profits to be forgotten that could be earned if risky decisions are made. In older societies it was thus danger that tended to be marked, whereas modern society has until recently preferred to mark risk, being concerned with optimizing the exploitation of opportunity. (Luhmann 1993, 24–5)

Let us return to the consultation meetings where the Rail Administration's project managers, other officials, experts, the local public and the affected stakeholders meet and communicate. An obligatory part of those meetings, we have noted, is a demonstration on the part of the Rail Administration and its contractors of their capacity to act, by means of making rational and well-founded decisions, politically legitimated as democratic decisions made by the Swedish government. Such demonstrations further include a display of facts to be assessed in the decision-making process for the implementation of the building of the tunnel, which requires a myriad of technical, legal and organisational decisions. During a meeting, numerous options and assessments which are included in this decision process are indicated. The focus on risk aspects makes tunnel building transparent as a decision process. Risks are identified, named, calculated, isolated, and deliberated, while strategies for coping with them are dealt with in terms of risk management decision procedures and strategies. The consultation meetings can be seen as instances of 'performative governance' (Futrell 1999):

> Performative governance is a situation in which impressions of committed governance are displayed by officials yet effective inclusion of citizenry in the decision-making processes – the principle objective of the occasion – is negligible. (1999, 502)

The Rail Administration and its contractors make an effort to convince the audience about their capacity to make qualified decisions and to act competently based on the decisions made. They claim competence and agency as rational decision-makers, as is necessary for a sustained 'license to operate' as tunnel builders (Corvellec 2007). The other obligatory part of the consultation meetings involves the stakeholders in the audience, who use the 'concluding questions and discussion' part of the meeting to voice their lack of trust in the Rail Administration and the tunnel project, and their long-term sufferings from the project. The stakeholders in the audience are in a very different position from the Rail Administration, the officials, the contractors and the expert(s) who make their presentations during the first part of the meeting. They are affected by the decision made by the Rail Administration and the contractors. The stakeholders emerge as the suffering party; they are the ones acted upon. They have no licence to operate regarding the building of the tunnel and have little authority regarding how their own land and property is being affected (or used) by tunnel building and tunnel builders. They identify themselves as victims of the operations. For them the tunnel is a hazard, and not a risk.

Conflict generation in communication

Schismogenesis is a term introduced by social anthropologist Gregory Bateson (1935) in his analysis of ritual interaction and interpersonal communication. It refers to processes of social differentiation and opposition (including phenomena such as contestation, hierarchies and domination) where individuals' reactions to the actions of other individuals influence and modify their actions (Bateson 1935, 1958, 1973;

Nuckolls 1995). For Bateson, interaction is interplay of coordinated action or behaviour. Individual A reacts to the actions or behaviour of B, and A's reactions modify and influence B's actions or behaviour in the future. Bateson distinguished analytically between two possibilities: either A and B respond to each other by the same behaviour (for example, anger leads to anger to more anger) or they respond by functionally complementary behaviours (for example, by means of contrast pairs such as dominance – submission).[16]

The consultation meetings regarding the Hallandsås tunnel develop according to a process of complementary schismogenesis involving risk and hazard – agency and patiency modes of existence, which differentiates regulators from stakeholders. The more the Rail Administration emphasises its agency through its decision-making capacity and its claims to sustained license to operate, the more the affected stakeholders respond by demonstrating their patiency, their lack of agency and their consequent victimisation. The discourse of risk which the Rail Administration uses to display its agency mode is responded to by stakeholders using a discourse of hazard. When stakeholders argue against the tunnel project in terms of their lack of agency, being victims from the tunnel as a hazard, the Rail Administration responds by intensifying their risk discourse in an effort to convince the audience about their capacity to act and to make wise decisions. Such demonstrations of agency and decision-making capability however only make stakeholders more vulnerable, and exposed to the consequences of the decision-making of others. They are deprived of agency and decision capacity vis-à-vis their own land and their own local and everyday environment, which add to frustration and increasing hostility towards the entire project (see Sjölander-Lindqvist 2004).

In fact, the schismogenetic structuring of the consultation meetings resulted in a breakdown on 27 October 2003 when the most vocally critical stakeholder group of residents at a meeting read a statement summarising their feelings of victimisation at the hands of the Rail Administration's actions and the tunnel project. At this meeting the Rail Administration manager of the tunnel project reported that the Environmental Supreme Court had decided to approve the application of the Rail Administration for a new water permit. Then followed an account from the project management and the contractors about how an especially difficult section of the planned tunnel (with extremely difficult geological conditions) could be dealt with according to different technical solutions (one decision alternative being deep freezing of the section in question). After this matter had been accounted for at the meeting, the group of stakeholders read a statement ending in a declaration that, from then on, they were not going to attend any more meetings. The group, consisting of about 30 people, then got up and marched out of the room in the middle of the meeting. The statement was also delivered as a written document signed by the action group *Rädda Lya* (Save Lya).[17] The first part of the statement consisted of a summary of the concerned stakeholders' experiences of the tunnel project:

> Let us summarize our experiences of the tunnel project together with our conclusions. There is probably no other building project in our time that has treated local residents so badly over such a long period. We have lived with the Rail Administration's illegal sinking of ground water, their betrayed promises and legal tricks since 1995, when Skanska took over the project. We have endured toxic contamination, dried-up wells, wetlands, streams and springs, diminished yields from agriculture, processes of compensation, difficulties in planning our economy and our future, recurrent intrusion and negotiations with the Rail Administration. Our lives are a continuous worry over

> what the Rail Administration will do next. These meetings are called Consultation Forums. But consultation requires a little bit of humility. But that has not been the attitude of the Rail Administration. They have presented biased information on decisions already made. The Rail Administration's Consultation Meetings are just play for the galleries. We have no influence on the decision process. (Spokesperson from *Rädda Lya*, MGG PM 296, 2003, translation by author)

After the stakeholder group had declared that they no longer intended to participate as patients to the Rail Administration's decision, the meetings lost legitimacy as consultation meetings. The incident made the Rail Administration project management rename the meetings *Forum Hallandsås*, rather than *Samrådsforum Hallandsås* (Consultation Forum). This shift of naming, however, did not essentially change the content of the subsequent meetings, apart from the fact that they lost much of their earlier schismogenetic dynamic since the opposition between agency and patiency modes ceased to be articulated at the meetings. In a conflictive interaction system, each new move – action or communication – is framed within the conflict; in an escalating conflict the move will trigger a conflictive response. Meanings are integrated so that events, actions, and communications in an escalating development are incorporated and interpreted in the light of the conflict – which means that it is nearly impossible to solve conflicts from within such a system (Luhmann 1995, ch. 9).

By way of conclusion

Within risk communication as a practical field, it is often taken for granted that other people's ideas about risks can be monitored by means of skillfully designed dissemination of information, by means of pedagogical presentation of facts (e.g., Lundgren and McMakin 2004). Such efforts, however, are likely to fail since the context of communication, the concepts and assumptions, the knowledge, experiences, expectations and social and cultural conventions that guide speaker and hearer in their understanding of the meanings of messages are not sufficiently taken into account. A pragmatic perspective on risk communication shifts focus from the technical aspects of information transmission – what is to be communicated, to whom, and in what way – to communication as action and interaction (Allwood 2000). Communication is not something apart from interaction systems or social systems (Luhmann 1995).

Communication involves assumptions that people have about themselves, other people and the world. Assumptions steer the interpretation of messages; many assumptions are shared but not all. Messages can therefore be understood differently. There is no neutral, assumption-free position from which information is interpreted. This means that a risk communicator has little control of the ways in which messages are understood unless the assumptions that guide a hearer's interpretation of meanings are known (Chess, Burger, and McDermott 2005). Since risk is a relational concept, risk issues are inherently open to interpretation from diverse standpoints. Risks are observed and judged; there are no absolute risks as such. A risk is judged in relation to another state where the risk could be bigger or smaller. The concept of risk is a *comparative* term. Another feature of the modern notion of risk is that it is a *temporal* construct. Risks are assessed with reference to possible states in the future, which are unknown and contingent. Risk is also an *epistemological* category constituted by what we know and do not know regarding potentially harmful influences between entities (Reith 2004).

Participation is often advocated as a road towards mitigating conflicts in risk communication by means of creating shared expectations, consensus and respect for other viewpoints. It is assumed that participation works through communication: by intensifying and broadening communicating, inviting more participants to have their say and by encouraging dialogue (Renn 2006). However, it is hardly communication as such that has a capacity to make people more cooperative and disposed to agree on a common characterisation of the world. Participation works, not by enforcing people to communicate, but by altering or modifying existing agency–patiency relationships (Luhmann 1993, 152 ff). Participation procedures in public policy (for example, in the implementation of decisions having considerable consequences for local groups who are not the decision-makers) can 'empower' stakeholders with a capacity to act and make decisions on matters of concern to them. Such empowerment enables stakeholders to move from a hazard mode to a risk mode of relating to an issue. But in order for this to occur, decision power has to be transferred from regulators to stakeholders (Stoffle and Evans 1990).

Communication alone cannot be expected to abate schismogenetic differentiation of existence modes. In order for communication about risk to result in mutual understanding (or an approximation of it), stakeholders and regulators must come to share fundamental assumptions about the context at hand – something that will require a transition of decision power from regulator to stakeholder. The transition weakens the contrast between a risk and a hazard mode and between the regulator and a stakeholder positions as adversaries. For these positions to converge (by deliberative and participatory processes), decision power must be diffused. A somewhat provocative conclusion, therefore, is that risk communication leading to understanding, for this reason can never be fully reached in a democratic society which requires the rule of the majority's view (and its interests) over the minority. The reaching of understanding between regulator and stakeholder through risk communication is therefore a democratic paradox. Sometimes the majority may want to make certain decisions even though the decisions may have negative side-effects for a minority, and without allowing the affected minority agency in relation to the matter decided upon (wildlife conservation, and natural resource management abound with such examples). The (democratic) principle of concentration of decision-making power in the majority means that there will always be minorities deprived of agency over matters of concern to them. No matter how much effort is spent on communication, understanding cannot be fully achieved as long as the existence modes of the parties involved fundamentally differ – one party making the decisions and the other being exposed to the consequences of those decisions.

Acknowledgements

The research on which this article is based was funded by the Swedish National Rail Administration, and the Swedish Research and Communications Board. For research time to finish the article for publication the author is indebted to the Swedish Research Council, the Swedish Rescue Services Agency, the Emergency Management Agency, the Swedish Governmental Agency for Innovation Systems, the Swedish Road Administration, and the Swedish Maritime Administration. The author wishes to thank Max Boholm, Gustaf Kastberg, Vicki Johansson, and Ragnar Löfstedt for commentaries on an earlier version at a session during the conference "New perspectives on risk communication" held in Göteborg in August–September 2006. Also to thank the participants at this session for their suggestions, questions and commentaries. The author is also grateful for constructive criticism from the anonymous reviewers at the *Journal of Risk Research.*

Notes

1. Miljöbalk (SFS 1998, 808) (ch. 6, sect. 4–6).
2. I have been a participant observer of 15 out of a total of 16 'consultation forum' and 'consultation' meetings. I was present during all three Environmental Court Hearings regarding the tunnel project: in April 2001 (the Environmental Court), in November 2001 (the Environmental Court) and in September 2003 (the Supreme Environmental Court). Each of those hearings extended for several full days.
3. The Environmental Court Hearings were not taped since audio recording or photographing is not allowed in the court room during proceedings. The Court, however, tapes the hearings and since the tapes are official public documents they have been obtained.
4. http://www.banverket.se/sv/Amnen/Aktuella-projekt/Projekt/1869/Hallandsas.aspx (accessed 10 July 2007).
5. Nord Västra Skånes Tidningar (NST) has reported regularly from the meetings.
6. Another study coming out of the Hallandsås tunnel research project is Sjölander-Lindqvist (2004). Also Grimes (2005) partly concerns the tunnel case.
7. Two parallel tunnels, 8.6 km long. To date since building started in 1992, one-third of the tunnel has been built. The Hallandsås is a horst, a geological formation created 70–100 billion years ago by movements of the earth's crust. The interior of the ridge consists partly of solid rock, partly of gravel and clay, intersected by huge water-filled crevices. It contains enormous amounts of ground water, and at certain places at 170–180 m above sea level, water is emerging naturally through the ground in ponds, brooks and springs. The area above ground is very fertile and ideal for farming. The landscape is varied, partly pastoral with grazing animals, fields with vegetables or grain; there are pastures, farmsteads intersected by small roads, areas of beech or alder forest, and moor. The north side of the ridge is a nature reserve, and on the Hallandsås there are several dispersed Natura 2000 habitats protected by the European Union.
8. MGG PM 015, 1998-01-08.
9. MGG PM 024, 1998-02-25, p. 3.
10. MGG PM 015, 1998-01-08, p. 7.
11. MGG PM 024, 1998-02-25, p. 3.
12. MGG PM 107, 1998-03-09.
13. MGG PM 108, 1998-03-30.
14. Banverket, Miljökonsekvensbeskrivning, 2000, p. 14.
15. People's judgements of the causes of other people's behaviour (in terms of inner dispositions like intentions, or external factors and what responsibility they have for their actions) have been researched in social psychology since the late 1950s (Heider 1958; for an overview, see Fölsterling 2001).
16. Bateson's concept has been used to explain conflicts in land-use planning and wildlife management (see Binde 2000; Binde and Boholm 2004; Brox 2000).
17. *Aktionsgruppen rädda Lya*, Lya is a little village on the ridge above the tunnel.

References

Allwood, J. 2000. An activity-based approach to pragmatics. In *Abduction, belief and context: Studies in computational pragmatics*, ed. H. Bunt and W. Black. Amsterdam: John Benjamins Publishing Company.

Baier, M. 2003. *Norm och rättsregel: En undersökning av tunnelbygget genom Hallandsås*. Lund: Sociologiska institutionen, Lunds universitet.

Banverket. 2000. *Projektutredning Hallandsås: Miljökonsekvensbeskrivning*. Borlänge, Sweden: Banverket.

———. 2004. *Årsredovisning*. Borlänge, Sweden: Banverket.

Bateson, G. 1935. Culture contact and schismogenesis. *Man* 35: 178–83.

———. 1958. *Naven: The culture of the latmul people of New Guinea as revealed through a study of the "naven" ceremonial*. Stanford, CA: Stanford University Press.

———. 1973. *Steps to an ecology of mind: Collected essays in anthropology, psychiatry, evolution and epistemology*. Herts: Paladin.

Binde, P. 2000. Risk discourse: The rhetoric of railway hazards in siting controversy. In *Foresight and precaution. Procedings for ESERL 2000, SARS and SRA-Europe annual conference, Edinburgh, Scotland, UK, 15–17 May 2000*, ed. M.P. Cottam, D.W. Harvey, R.P. Pape, and J. Tait. Rotterdam: Balkema.

Binde, P., and A. Boholm. 2004. Schismogenesis in a Swedish case of railway planning. In *Facility siting: Risk, power and identity in land use planning*, ed. Å. Boholm and R. Löfstedt. London: Earthscan.

Boholm, Å., ed. 2000. *National objectives – Local objections: Railroad modernization in Sweden*. Göteborg: CEFOS, Göteborg University.

———. 2005a. Greater good in transit: The unwieldy career of a Swedish rail tunnel project. *Focaal* 46: 21–35.

———. 2005b. Riskbedömningars ontologi och epistemologi: Hallandsåsen och dess vatten. In *Risk och det levande mänskliga*, ed. I. Brinck, S. Halldén, A.-S. Maurin, and J. Persson. Nora: Nya Doxa.

———. 2007. Risk communication: A pragmatic approach. In *Communication – Action – Meaning. A Festschrift to Jens Allwood*, ed. E. Ahlsén, P.J. Henrichsen, R. Hirsch, J. Nivre, Å. Abelin, S. Strömqvist, and S. Nicholson. Göteborg University, Department of Linguistics.

Boholm, Å., and R. Löftsedt. 1999. Issues of risk, trust and knowledge: The Hallandsås tunnel case. *Ambio* 28, no. 6: 556–61.

Boholm, Å., and R. Löfstedt, eds. 2004. *Facility siting: Risk, power and identity in land use planning*. London: Earthscan.

Boholm, Å., R. Löfstedt, and U. Stranberg. 1998. *Tunnelbygget genom Hallandsås: Lokalsamhällets dilemman*. Göteborg: CEFOS, Göteborg University.

Brox, O. 2000. Shismogenesis in the wilderness: The reintroduction of predators in Norwegian forests. *Ethnos* 65, no. 3: 387–404.

Chess, C., and K. Purcell. 1999. Public participation and the environment: Do we know what works? *Policy Analysis* 33, no. 16: 2685–92.

Chess, C., J. Burger, and M.H. McDermott. 2005. Speaking like a state: Environmental justice and fish consumption advisories. *Society and Natural Resources* 18: 267–78.

Corvellec, H. 2001. Talks on tracks – debating urban infrastructure projects. *Studies in Cultures, Organisations and Societies* 7: 25–53.

———. 2002. *På tal om tredje spåret vid Riddarholmen*. Göteborg: Bokförlaget BAS.

———. 2007. Arguing for a licence to operate: The case of the Swedish wind power industry. *Corporate Communications – An International Journal*, 12, no. 2: 129–44.

Darian-Smith, E. 1999. *Bridging divides: The channel and English legal identity in the new Europe*. Berkeley, LA: University of California Press.

Fiorino, D.J. 1990. Citizen participation and environmental risk: A survey of institutional mechanisms. *Science, Technology & Human Values* 15, no. 2: 226–43.

Flyvbjerg, B., N. Bruzelius, and W. Rothengatter. 2003. *Megaprojects and risk*. Cambridge: Cambridge University Press.

Fölsterling, F. 2001. *Attribution. An introduction to theories, research and applications.* London: Taylor and Francis.

Futrell, R. 1999. Performative governance. Impression management, teamwork, and conflict containment in city commission proceedings. *Journal of Contemporary Ethnography* 27, no. 4: 494–529.

Gell, A. 1998. *Art and agency. An anthropological theory*. Oxford: Oxford University Press.

Grimes, M. 2005. *Democracy's infrastructure. The role of procedural fairness in fostering consent*. Göteborgs Studies in Politics 97. Kungälv: Department of Political Science, Göteborg University.

Healey, P., and T. Shaw. 1994. Changing meanings of 'environment' in the British planning system. *Transaction of Institute of British Geographers* 19, no. 4: 425–38.

Heider, F. 1958. *The psychology of interpersonal relations*. New York: John Wiley & Sons.

Heiefort, R. 2004. When complexity becomes a problem: 'Law' and 'fairness' on separate tracks in Sweden. In *Facility siting: Risk, power and identity in land use planning*, ed. Å. Boholm and R. Löfstedt. London: Earthscan.

Karlsson, M.M. 2002. Agency and patiency? Back to nature. *Philosophical Explorations* V, no. 1: 59–81.

Latour, B. 1996. *Aramis or the love of technology*. Cambridge, MA: Harvard University Press.

Löfstedt, R. 2005. *Risk management in post-trust societies*. New York: Palgrave.

Löfstedt, R., and A. Boholm. 1999. Off track in Sweden. *Environment* 41, no. 4: 16–44.

Luhmann, N. 1993. *Risk: A sociological theory*. New York: Aldine de Gruyter.

———. 1995. *Social systems*. Stanford, CA: Stanford University Press.

Lundgren, R., and A. McMakin. 2004. *Risk communication*. Columbus: Battle Press.

McComas, K.A. 2003a. Trivial pursuits: Participant view of public meetings. *Journal of Public Relations Research* 15, no. 2: 91–115.

———. 2003b. Public meetings and risk amplification: A longitudinal study. *Risk Analysis* 23, no. 6: 1257–70.

Nuckolls, C.W. 1995. The misplaced legacy of Gregory Bateson: Toward a cultural dialectic of knowledge and desire. *Cultural Anthropology* 10, no. 3: 367–94.

Palm, L. 1998. *Hallandsåstunneln som tvistefråga, kris och förtroendeproblem*. Stockholm: Styrelsen för psykologiskt försvar.

Päiviö, J., and H.-G. Wallentius. 2001. The Hallandsås railway tunnel project. In *EIA, large development projects and decision making in the Nordic countries*, ed. T. Hilding. Stockholm: Rydevik, Nordregio Report.

Petts, J. 1999. Public participation and EIA. In *Handbook of environmental impact assessment*, ed. J. Petts, 145–77. Oxford: Blackwells.

Reith, G. 2004. Uncertain times: The notion of 'risk' and the development of modernity. *Time & Society* 13, nos. 2/3: 382–402.

Renn, O. 2006. Participation processes for designing environmental policies. *Land Use Policy* 23: 34–43.

Ryfe, D.M. 2005. Does deliberative democracy work? *Annual Review of Political Science* 8: 49–71.

Rönnborg, P. 2006. *Finding the right place: The story about an offshore wind power project*. Göteborg: School of Business, Economics and Law, Göteborg University.

Sidaway, R. 2005. *Resolving environmental disputes. From conflict to consensus*. London: Earthscan.

Sjölander-Lindqvist, A. 2004. *Local environment at stake: The Hallandsås railway tunnel in a social and cultural context*. Lund: Department of Human Ecology, Lund University.

Soneryd, L. 2003. Public involvement in the planning process: EIA and lessons from the Örebro airport extension, Sweden. *Environmental Science & Policy* 7: 59–68.

Sperber, D., and D. Wilson. 1986/2003. *Relevance: Communication and cognition*. 2nd ed. Malden, MA: Blackwell Publishing.

Stoogle, R.W., and M.J. Evans. 1990. Holistic conservation and cultural triage: American Indian perspectives on cultural resources. *Human Organization* 49, no. 2: 91–9.

van den Hove, S. 2006. Between consensus and compromise: Acknowledging the negotiation dimension in participatory approaches. *Land Use Policy* 23: 10–17.

What environmental and technological risk communication research and health risk research can learn from each other

Ragnar E. Löfstedt[a] and Perri 6[b]

[a]*King's Centre for Risk Management, King's College London, London, UK;* [b]*Graduate School of Social Research, Nottingham Trent University, Nottingham, UK*

The study of risk perception and communication has emerged in several disciplines and fields, and there has not been strong convergence between them, especially in Europe and specifically in Britain. The lack of connection between, for example, the fields of health risk and environmental and technological risk, and still more with respect to fears of crime or old age poverty or privacy violation, has meant that theories have not been shown to be adequate, and practical prescriptions for conflict containment and communication have not been made robust. This article examines the roots of the fragmentation, examining the particular case of the relationship between health and environmental/technological risk studies, considering why the flow of ideas and methods has been so limited, and much of it in a rather surprising direction. The paper concludes by offering some principles that could inform a strategy for more integrated research on risk perception and communication.

Introduction

Academic research has always been fragmented by discipline and by empirical field. As the research community expands, and as more specialist journals form, researchers themselves help promote this fragmentation. As they are inundated with specialist information, they have little time to conduct research outside their respective fields, knowing full well that in order to climb the academic ladder of success, they need to specialise in the 'cutting edge' issues in their own particular area. This has benefits, of course. But it also brings problems. These include lack of learning, theories developed in one area that are never tested in others where they would have to work if they were valid, lack of well-designed comparative research between fields, and mutually incompatible perspectives between the disciplines.

The fields of research into risk perception and risk communication present a particularly severe case of the problems of fragmentation. Environmental-technological researchers focus on their particular fields of risks, the health scholars on theirs, and criminologists on theirs, privacy and information risk scholars on theirs, financial services and social policy researchers on theirs, and so on; but there is little learning across fields and even less overarching theory development. No genuinely unified body addresses the concerns of all the risk scholars or professionals. For example, over the past years, the conferences organised by the

Society for Risk Analysis-Europe have mainly attracted social scientists doing technological and environmental risk perception research, while the engineers have been attending the European Safety and Reliability Association meetings. The Conferences of the Society for Behavioural Medicine provide a platform for many papers on patterns of individual risk perception around diet, alcohol, illicit drugs and exercise, but health psychologists concerned with collective public health communication[1] rarely collaborate with environmental and technological risk perception researchers. Much more research in public health is devoted to health-related behaviour than to emotions such as fears or insouciance about disease. In psychiatry, there has been much focus recently on the management of risks presented by very small proportions of those with mental health problems who present risks to the wider public, and there has been a little research on the risk perceptions of both professionals and lay people in this field. However, this work has largely ignored other fields of risk study, except for occasional literature comparisons between, for example, siting studies from the technological risk field and siting conflicts over medium secure facilities for people with mental health problems. As a result, the field of risk perception and communication research paradoxically remains in a state of frenetic and diverse activity but weak in theoretical and methodological maturity.

The situation may be somewhat better in the United States, where the Society for Risk Analysis has a larger membership from public health schools, and where public health schools run programmes in environmental health. Nevertheless, as we argue below, methodological and theoretical integration between the fields even in the United States remains limited. However, much of the argument that follows focuses particularly on the situation in Europe and in the UK.

There is a clear need for the risk perception and communication research communities to integrate their work across the various fields of risk. For if theories of how social fears develop or die away, and theories of how they might be deliberately smoothed or stirred, are valid in one area, they ought to generalisable, and the effort to generalise them between fields of risk should constitute a central part of theory-testing. Moreover, many of the theories have been developed in relation to particular kinds of risk. Therefore, comparative research into their relative strengths and weaknesses now needs to be done, which must involve examining their usefulness across wider ranges of types of risk. For practitioners in risk communication this fragmentation leads to poorly grounded understanding of the conditions under which techniques will succeed, prove counter-productive, or simply have no appreciable effect.

There are radical and deep disagreements about just what risk perception is, and these reflect the fragmentation between disciplines. Cognitive psychologists have tended to define risk perceptions as beliefs, which may have emotional charges attached to them (dread, excitement, etc). By contrast, anthropologists and sociologists have tended to define them as social fears, or as emotions rooted in social practices which may include particular beliefs which are used in (and functionally explained by their role in) blaming others, praising others, binding solidarities with others, and so on (6 2001). While risk communication cannot – at least yet, partially as a result of its fragmentation – be defined as an independent discipline, it is perhaps best described as 'the flow of information and risk evaluations back and forth between academic experts, regulatory practitioners, interest groups and the general public' (Leiss 1996, 86).

In this article, we examine how far two elements in the risk perception and communication community remain disconnected, and review what has been done to reconnect them, before proposing ways forward. We focus on two particular subsets of the risk perception and communication field, namely, environmental/technological risks and public health risks. We could, of course, have picked out others, such as research on perception of and communication about risk of crime, or personal financial problems, or information and rights risks such as privacy violations. However, it is particularly appropriate and useful to examine and compare the disconnections and the linkages between environment/technological and health risk areas for at least four reasons:

a. Unlike the continuing lack of connection between studies on perceptions of risks of crime, of personal financial problems and of information risks such as privacy violations, there have been some recent attempts by researchers to bridge the fields of health and environmental risk perception studies. For example, approaches using mental models that were originally developed better to understand public concerns toward electromagnetic fields as well as environmental radon have recently been used to understand American adolescents' perceptions toward sexually transmitted diseases (Downs et al. 2004; Fischhoff, Dows, and Bruine de Bruin 1998; Fischhoff 2005; Morgan et al. 2001); however, this remains more a diagnostic tool than a sophisticated causal theory well linked with wider social science theories of organisation and bias.
b. There is a causal relationship between the two types of risks: many technological and environmental risks do present health hazards. Therefore, to the extent that the causal links risks are recognised by the public, perceptions may be systematically related.
c. Both are concerned with risks associated with mortalities and morbidities rather (in the first instance) than loss of property, or poverty, or loss of rights. Therefore, to the extent that the intrinsic character of the harm or damage involved influences perception or the possibilities for deliberate communication, there ought to be common factors.
d. The two fields both seem to exhibit failures in programmes of risk communication. At its best, risk communication is not a top-down form of communication from expert to the lay public, but rather a constructive dialogue between all those involved in a particular debate about risk. The outcomes of the various risk communication programmes implemented in Europe and the United States have been mixed (Fischhoff 1995; Fischhoff, Bostrom, and Quadrel 2002; Freudenburg 2004; Kasperson and Palmlund 1987; Lofstedt 2003, 2005; McComas 2006). The research suggests that the proximate reasons for which these programmes have not been more successful, particularly with regard to the siting of noxious facilities such as waste incinerators, have to do with factors such as social amplification of risk, decline in public trust toward policy makers and industry and the inability of risk communication practitioners to adequately engage in public dialogue (Cvetkovich and Löfstedt 1999; Fischhoff 1995; House of Lords 2000; Leiss 1996; Löfstedt 1996a, b, 1999, 2005; Renn and Levine 1991; Renn, Webler, and Wiedemann 1995; Renn, Webler, and Kastenholz 1996; RCEP 1998). These factors seem to be common to the two fields.

This article first sets out the reason why integration is important. It then reviews the main traditions in two fields. Then it considers the extent of commerce between them, first from environmental to health risk and then in the reverse direction. We examine the following puzzle: given the historically much greater authority of the health field, why is it that recently, most of the (meagre) intellectual traffic between them has been from the environmental and technological risk fields to the health risk community? Then, we present and consider four hypotheses about what might lie behind this puzzling trend. Finally, we set out four principles or maxims. We call not only for greater cross-fertilisation between these two fields but also for an integrated research strategy by which social fears can be understood. This should represent neither the renascent hegemony of old public health approaches nor the imperialism of environmental risk models.

The rationale for integration

There are three reasons, we shall argue in this section, to welcome the few modest initiatives in cross-fertilisation between the two fields cited below, and to call for more efforts of this kind.

Social fear is basically a single phenomenon, that can best be understood through comparisons between fears in different empirical contexts

Social fears about risky behaviours of others, and insouciance about one's own risky behaviour are explained in, for example, anthropology, psychology and political science by basically the same kind of risk factors, whether in the field of health fears or in that of fears about environmental or technological risk, or indeed anywhere else (6 2001). Simple information processing models and psychometric accounts of fear and insouciance explain these phenomena by reference to competence in performing implicit calculations of probabilities on the balance of available information, as distorted by particular biases and blinkers, whatever the content of the fear or recklessness (Slovic 2000a). Anthropological theories are similarly symmetrical, treating the forms of social organisation and the institutional processes of blame and conflict for which certain types of fear and fearlessness are functional, as operating in every sphere of life, spanning the most intimate matters of personal health and the most public and distant matters (Caplan 2000; Douglas 1992). Social psychological models using social representation theory, in the same way, deny any suggestion that the dynamics by which representations are produced and internalised are fundamentally different in different arenas of human alarm (Breakwell 2001; Breakwell and Canter 1993; Joffe 1999; Moscovici 2000). If the one claim these otherwise rival approaches agree upon is even broadly right, then it makes little sense to conduct separate studies on each type of fear or initiative to calm or whip up those fears.

Unfortunately, hitherto, theories studies of risk perception and social fear have been developed in a single field, albeit for understandable reasons for cost and, sometimes, difficulty of defining comparable operational measures. Psychometric studies have covered a wide range of risks, but most of the risk ranking exercises, for example, encompassed only risks of mortality arising from technologies. Some theorists such as Douglas (e.g., 1992), have written about several kinds of risk between the same covers, but have not undertaken comparative empirical research on fears of quite different kinds of risk. Yet the only way in which any of these theories can therefore be tested effectively, is to examine their performance, the variance explained, their goodness of fit, parsimony, collinearity of variables, fundamentality, etc, by conducting empirical work that affords comparison between fears in several fields of fear or perception of risk.

The fields of risk communication practice all face the same underlying challenges

Risk communicators in government and business, and for that matter in social movements, face essentially the same problems, whether the risks in question are those of chemical pollution, bovine spongiform encephalopathy, sexually transmitted diseases, old age poverty through lack of saving, violation of privacy, or acquisitive or violent crime. For example, the Phillips enquiry into BSE ranged over policy issues affecting both public health and technological risk, albeit without the assistance of much academic research crossing those boundaries (Phillips, Bridgeman, and Ferguson-Smith 2000). Governmental, professional and business risk communicators face problems of insecure authority, of distinguishing lack of understanding from lack of trust, of fragile trust among at least some sections of their target publics in their credibility or at least that of their message, of complexity and uncertainty of messages, and they must also face the risk of litigation should their communication turn out to be misguided. The trust problems that, for example, the British public health community is facing over the combined measles, mumps and rubella vaccine are no different, *structurally*, from those faced by the US civil nuclear waste processing industry over their storage plans. In the field of medical risk research, there have been studies for many years on trust by individual patients in their own personal physicians and their advice, but this raises rather different issues from the problem of understanding trends in collective trust between citizens and the profession, especially with respect to the nexus of institutions that bring together the professional bodies with government ministries for public health advice. Moreover, research on trust in physicians has been much less well connected with the study of social movements and their impact on citizens' attitudes than has the work on trust in environmental and technological fields (Alazewski and Horlick-Jones 2003). The public understanding of science tradition has insisted that public ignorance and incompetence with probabilities is the key problem in environmental and technological risk communication (Noelle-Neumann 2000; Warner 1998) just as many public health scholars argue that failures to accept or follow health risk advice are rooted in patient ignorance. In both fields, this assumption is being challenged by those who argue that the public are not generally stupid, but distrusting for explicable reasons (in environment, see e.g., Wynne 1992, 1996; in public health, see Ball, Evans, and Bostrom 1998; Bogardus, Holmboe, and Jekel 1999; Edwards and Prior 1997; Edwards et al. 1998; Fischhoff 2000a, b; Fong, Rempel, and Hall 1999; Patrick 1999; Rothman and Salovey 1997; Thornton, van den Borne, and de Bruijn 1996). Complexity in the communication of multiple, competing and differently uncertain risks is a central problem both for public health researchers (Banks et al. 1995; Bottorff et al. 1998; Cody and Lee 1990; Fischhoff 1999; Kreuter 1999; Maibach 1999), and in the environmental and technological fields (Fischhoff 1995; Renn, Webler, and Kastenholz 1996; Slovic 2001). Problems of trying to explain benefits at the same time as explaining risks occur in public health (Black, Nease, and Tosteson 1995; Davidson, Degner, and Morgan 1995; Fischhoff 1999) as much as in privacy risk issues in relation to the benefits of joined-up public services (6 2002). The problem of authority is different only in degree for public health authorities vis-à-vis their citizens and for public officials vis-à-vis residents in neighbourhoods near to industrial waste storage centres, or, for that matter, psychiatric care facilities unpopular with fearful and prejudiced residents (Dear 1992; Gaber 1996; Repper 1996; Takahashi 1997; Takahashi and Dear 1997; Cowan 2002). The well known fear of damages in civil suits that leads to defensive medicine leads also to increased readiness to recall pharmaceutical products on something not unlike a precautionary principle. This is not different in kind from the defensiveness that infects contemporary official and commercial communications about any risks that mobile phone use may one day be found to present to children. The blunt disbelief of many young people in the public health messages of the late 1980s and early 1990s about the dangers of ecstasy and amphetamines is not different in the formal structure of reasons or even necessarily in its associated social organisation from the blunt disbelief of many people of all ages in the assurances of the nuclear power industry about the safety of its waste storage facilities. It is by comparison between ways of approaching these problems that these fields hope to make progress.

Risk communicators in different fields are now converging on the same techniques

A plethora of experiments with stakeholder conferences citizens' juries and the like have been tried in both environmental technological and public health fields (Renn 2001; Renn, Webler, and Wiedemann 1995; Renn, Webler, and Kastenholz 1996; Renn et al. 2002). Those concerned with health risk related behaviours have tried to offer more 'neutral' presentations of 'just the facts', just as environmental and technological risk communicators have experimented with attempts at 'neutrality' (Fischhoff 1995). By comparing 'what works' between fields risk communication as a field can best be developed beyond an unstructured list of sometimes-helpful, sometimes-misguided maxims and anecdotes. One promising methodological tool for transcending both research areas is the work taken from wider communication studies to develop the analysis of mental models. It is already being used in both areas and it is likely that the popularity will continue to grow. For example, the Centre for Disease Control has funded Fischhoff's group at Carnegie Mellon to do risk communication research surrounding the MMR vaccine using mental models (Casman et al. 2000; Fischhoff, Bostrom, and Quadrel 2002).

The nature and roots of disconnection between the two fields

The academic study and the professional practice of environmental technology and public health risk communication, and the understanding of risk perception that underpins both, have been surprisingly disconnected from each other.

Mainstream health risk researchers either use predominantly linear models of cognition (Conner and Norman 1995) and call for people to worry more about things they seem lackadaisical about (such as smoking or drinking too much or lack of exercise), or else they disdain the whole business of risk communication in health promotion as insufferably paternalistic, victim-blaming and authoritarian (e.g., Lupton 1995). By contrast, environmental and technological risk researchers now generally eschew linear models, flock to mental mapping techniques instead (Morgan et al. 2001) and call for people to worry less about things that seem to exercise them greatly (such as nuclear power), or else they explain fear anthropologically by reference to the social function that fears play in sustaining certain types of social organisation (e.g., Douglas 1992).

The environmental and technological risk and health risk fields developed separately for several reasons. As we shall document in more detail in the next two sections, those reasons can be grouped as follows.

Origins in different periods and in different types of profession and authority

The roots of explicitly organised bodies of knowledge about how professionals should communicate with the public about health risks go back centuries. Public health is an ancient discipline, and has had its specialists at least since mediaeval times, and the practice can be found, albeit with rather different professional organisation, in antiquity. The basic priorities and practices of risk amplification, of public persuasion, focusing upon both inducement to comply with lifestyle regimes and on encouragement or even the requirement to present oneself for professional medical attention, can be found in well developed forms at the time of the great plagues, and in well organised form by the time of the great plague in London in the seventeenth century. The great debates between quarantine and civil engineering of the nineteenth century were conducted by bodies of professionals with formal bodies of knowledge and with already defined government resources to secure (Porter 1999). By contrast, environmental and technological risk perception was hardly professionalised until the twentieth century, and in many developed countries, not until well into the second half of the century (Covello 1983; Renn 1998). Academic disciplines almost invariably develop after professions emerge, in order to support them, rather than in advance.

The quite different historical roots of the professions have meant that their bodies of knowledge were institutionalised in different ways, to different paradigms. Public health risk communication has long been institutionalised within the context of the project of building social and even political authority for the public health professions, and the historically subordinate status of public health promotion to clinical public health reflects this. By contrast, professional risk communicators in the environmental risk field could not rely on authority built with care and difficulty over at least the last two centuries, but had to borrow such authority as politicians and business leaders and civil servants could lend them.

Different academic traditions

Although both fields have drawn upon psychology, they have tended to draw upon that discipline quite differently. In the field of public health, more individual-level psychological approaches have tended – at least since the second world war – to be used, in order to explain behaviours that are represented as failures of 'compliance' with régimes prescribed by public health authorities. The importance of randomised controlled trials, experiments and large scale quantitative studies in medical research has led some public health scholars to be suspicious of the kinds of survey or qualitative research by which social scientists measure social fears. Moreover, linear models, because they can be represented by readily soluble equations in algebraic form and which can therefore be explored using standard regression techniques, have made psychological researchers fearful that the use of models that do not lend themselves so readily to such treatment would be considered less respectable.

By contrast, the environmental and technological risk fields have tended to look to social psychological work using survey methods based on attitudes detected by survey instruments or qualitative research schedules among groups or populations (Gilovich, Griffin, and Kahneman 2002). In the field of health risk, and no doubt partly because of the dominance of medical conceptions in the developed world, the anthropological critique of the limitations of psychological approaches could only find a foothold in the field of 'exotics' or studies of cross-cultural variations in the recognition of what counts as health. By contrast, anthropological critiques of psychological studies of environmental and technological risk perception and communication were made in the core of what was a young, fragile and insecure discipline (consider, for example, the continuing brouhaha over Douglas and Wildavsky's 1982 book). Sociological work, including within the Foucaultian tradition, has also offered functional explanations of social fears, as means by which solidarities are maintained, and exclusion of rejected groups sustained (a body of this work is reviewed in Lupton 1999a, chs. 7–8).

Interests seeking opposed directions of persuasion

By and large, interest groups in the public health field are concerned to make us worry more about our lax health-related behaviours (Brush and floss! Vaccinate your children! Boil all water until further notice in these districts!). By contrast, the interest groups in the environmental and technological risk field want us to worry less about their behaviours (This nuclear power plant is quite safe; please stop demonstrating outside!). Only in the very small field of encouraging green consumption are mainstream commercial interests willing to amplify risk perception (Buy our environmentally friendly washing-up liquid!); only in respect of assuring sceptical minorities about the safety of such things as vaccines are public health professionals principally concerned with attenuation. Although in both fields, there are common problems of public trust, government and business risk communicators have adopted very different approaches, reflecting not only the generally distinct requirements of communicating negative and positive messages but also very different assumptions about reasons for citizens' reluctance to behave in the desired ways. The health risk amplifiers have focused mainly on individual risky behaviour and are much more concerned with explaining and countering what they see as individual deviance – conceived as a destructive form of cognition (Alaszewski and Horlick-Jones 2003; Bellaby 2003). By contrast, the environmental and technological risk attenuators facing social movements hostile to what they seek as corporate and governmental recklessness tend to argue that these movements represent some kind of social breakdown, although they are often forced to concede later that the results of their campaigns have nevertheless sometimes proven creative (Cvetkovich and Löfstedt 1999; Leiss 1995).

The result of these three factors has been a very particular kind of fragmentation of the study of risk perception and risk communication. Yet, as we shall note in the next two short sections of this article, there has not been much conflict between the two traditions. Rather, aside a small number of exceptions, they have largely passed each other by. Rarely do environmental and technological risk researchers test in their field the linear models used by many public health scholars. Likewise, until quite recently, it was just as rare for mainstream public health risk scholars to use mental mapping models and similar techniques developed in the environmental and technological risk community.

Given that the public health tradition is now centuries old, the continuing social authority (albeit recently somewhat tarnished in some countries) of the public health professions (Moran and Wood 1993; Gabe, Kelleher, and Williams 1994), and the deeper public policy commitment to health than to environmental action, it might have been expected that, if and when there was any cross-fertilisation between these fields, this would have been an intellectual imperialism by public health. Certainly, the newer field could have been expected to develop new approaches, but the weakness of influence from public health remains surprising. Yet, as we shall see, what appears to have happened is almost the reverse. Before examining this puzzle, we present very brief reviews of some of the key strands of thought in the two fields as they have developed in recent years.

Environmental and technological risk communication research

The environmental and technological risk communication field developed originally from psychometric studies of risk perception. In psychology, the study of risk perception has two origins. First, work by White in the 1940s and that of Fischhoff, Lichtenstein and Slovic in the 1970s (for a major retrospective, see Slovic 2000a) was influential in establishing the dominance of a quantitative social psychological approach, using statistical methods to analyse survey data. This work tended to stress collective experience and attitudinal explanations – in effect, giving greatest weight to proximate variables. White's work (especially 1945) on public perception of flood risk, for example, led researchers to theorise that reactions to hazards were based on experiential factors.

On these accounts, human groups with different social, economic and cultural background characteristics, living in diverse geographical areas, would perceive risks in ways reflecting their knowledge and environment (Burton and Kates 1964a, b; Burton, Kates, and White 1978, 1992; White 1945, 1961). People's responses were found to be dependent both on their perception of the hazard and their knowledge of the ability of society and individual, to deal with that hazard (Burton, Kates, and White 1978, 1992). The second impetus for the development of the field came from psychologists studying technological hazards.

Their focus was on the measurement of individual perceptions of risk and judgements associated with different risk-related options. The research findings illustrated the importance of contextual factors such as perceived controllability, catastrophic potential and dread as factors of risk perception (Fischhoff et al. 1978; Renn 1990; Slovic 1987). At this stage most psychological research on risk perception was focused on public perceptions of civil nuclear power, contaminated land (a big issue following the Love Canal scandal in the US) and chemical plants. Little attention was given to health-related risks, with the notable exception of smoking (Fischhoff et al. 1981) until the late 1990s (Slovic 2001; Viscusi 2002). Later, the study of environmental and technological risk was influenced by quite different traditions from anthropology, which took their inspiration from social anthropologists such as Evans-Pritchard (1937) who examined social fears as functional elements in social systems, examining the mobilisation of fear as a strategy of social authority, resistance, discipline and blame. The best known strand of anthropological theory of risk, developing this approach, stems from the work of Douglas (e.g., 1982, 1992; also Adams 1995; Wildavsky and Dake 1990; Grendstad and Selle 1997; Ellis and Thompson 1997; Grendstad 2001).

The outcome of understanding environmental and technological risk perception in these ways was the development of risk communication 'methods' for different potential hazards – that is, strategies for public persuasion in order deliberately to manipulate levels and distributions of fear and insouciance in order to further public policy goals or secure better 'public understanding of science'. It was during the 1980s that significant numbers of researchers began to experiment with the application of some of the findings of risk perception research to risk communication (National Research Council 1989; Stern 1991).

Most of the work on risk communication has focused on siting of perceived noxious facilities such as waste incinerators, the clean-up of contaminated land sites, nuclear waste dumps and chemical plants and food-related risks (e.g., BSE and GMOs) (Eldridge et al. 1998; Eldridge and Reilly 2003; Frewer et al. 1996, 1997; Powell and Leiss 1997; Renn, Webler, and Wiedemann 1995; Renn, Webler, and Kastenholz 1996; Wynne and Dressel 2001). Most of the risk communication studies in this field have used case study methods and surveys of target community awareness and recognition of messages.

Public health risk communication

There is a large and multi-disciplinary body of British research on the perception and communication of health risk related behaviours, including alcohol, tobacco, illicit drugs, and unprotected sex. Some of this has been organised through formal linear psychological models of cognition, such as the health belief model (Rosenstock 1966, 1974) and the more recently developed variants in the pure psychological tradition (Conner and Norman 1995; Rutter and Quine 2002). Most of these approaches have some kind of basis in bounded rational choice theory. For example, the protection motivation theory (Boer and Seydel 1995) treats perceived severity and perceived costs of 'maladaptive behaviour' as straightforward, and is only interested in deviations from scientifically correct probabilities. Causal determination of perception and behaviour is represented by linear cybernetic models only, and only simple feedback mechanisms are recognised. The effects of social processes and structures are given limited weight, appearing mainly as sources of misinformation about probabilities. Planned behaviour theory allows a wider range of social factors and some personality traits to enter the model, but again only as a rather unstructured set of sources of information and misinformation about probabilities or severity, which then drive the model in a fairly conventional linear cognitive fashion.

Other studies have used attribution theory in psychology which stresses the internalisation of ascriptions especially about responsibility (e.g., Booth Davies 1992), which has some links with sociological theories of identity. However, much British work on risk perception and communication in such fields as substance misuse and sociology of the fear of crime has been done without much reference to work on risk in other fields, using fairly straightforward empirical survey methods (e.g., the studies reported in Bean 1993). Some studies in this tradition draw more on sociological theory, including the theory of moral panics or collective processes of positive feedback among structurally semi-marginalised groups (Cohen 1987 [1972]; Ben-Yehuda 1986). Insofar as this work has connections with other fields, it tends to be with the sociology of deviance and with older traditions in criminology (Cohen 1987 [1972]), where risk perception studies have more recently come to be dominated by cognitive behavioural traditions (with some use of psychoanalytic models, e.g., Hollway and Jefferson 2000).

One important resource that work in this area brings, which is missing in the environmental and technological risk perception and communication studies, is a rich integration with social network theory, especially in the studies on peer effects among adolescents and young people as structures for risk communication and the shaping of risk perception (e.g., Cotterell 1996) and with identity theory (e.g., Hendry et al. 1993, ch. 8). Work on risk communication in relation to illicit drugs has drawn heavily on these approaches (see e.g., Dickerson and Stimson 1995 and Power et al. 1995). Other studies of risk perception in relation to health-related behaviours have drawn upon bodies of criminological theory, such as theories of norm internalisation and reputation management (e.g., Emler and Reicher 1995). In public health generally, much recent work on risk perception and risk communication has been influenced by the arguments that priority for personal heath, awareness of risk perception and openness to risk communication – and hence capacity for sustaining health – is shaped powerfully by the social network situation.

This has been conceptualised in terms of 'social capital', and has guided a number of studies sponsored by the former Health Education Authority and its successor the Health Development Agency, including those by Cooper et al. (1999), Campbell with Wood and Kelly (1999) and Swann and Morgan (2002). This is in line with wider arguments in public health epidemiology about the importance of institutional context, social situation and relations in shaping the distribution both of health risk and (relatedly) awareness of risk and motivation to act (see e.g., Blane, Brunner, and Wilkinson 1996 and Wilkinson 1996). However, this view is opposed by the critique of the public health promotion movement from the Foucaultian perspective in the sociology of public health, which has been influential in recent British and Australian work (Bunton, Nettleton, and Burrows 1995; Lupton 1995; Petersen and Lupton 1996; Peterson and Waddell 1998; Lupton 1999b). This tradition argues that public health promotion through the manipulation of fear and insouciance about, for example, health consequences of dietary choices, represents steadily more invasive and insidious processes by which institutions of power exercise control over people's private lives.

Risk communication research within the public health world has often drawn upon 'compliance' models of acceptance of individual medical advice in the physician-patient relationship, and used randomised or semi-randomised methods, including asking subjects to keep diaries of health risk related behaviours (Miller and Philo 1995).

The influence of environmental risk studies in the health field

Despite the historic separation, recently there has been some cross-fertilisation of approaches between the two fields. In this and the following section, we briefly review these attempts, beginning with the perhaps surprisingly greater influence of approaches drawn from the environmental and technological field upon the health field, before – in the next section – examining the much more meagre flow of ideas in the opposite direction. Here, we focus on only a few examples from the UK and the US.

It was not until the mid-1990s that British environmental and technological risk researchers became interested in health care risk communication in a serious way. One of the few major edited volumes published on the subject in this period was Bennett and Calman's (1999) collection entitled *Risk Communication and Public Health*. This was based on papers given at a 1997 conference discussing how public health professionals can best communicate risk to a public that is seen as distrustful. For this volume, some of the UK environmental and technological risk researchers contributed papers on issues in the health field but the contributors to this volume did not report on recent or current research in the health care communication area. Rather they suggested how work in the environment and technological field could be applied to the public health areas (e.g., Pidgeon, Kasperson, and Slovic 1999). Likewise a short report published by the Department of Health in 1997 discussing how one might communicate risks was also based on research conducted in the environment and technological area (DoH 1997). Once again, little attempt was actually made to integrate the two areas. The Bennett and Calman volume was essentially a 'cook book' – a volume of recipes on how practitioners in the health care areas should communicate risks that they are faced with, based on models and techniques developed in the environmental and technological risk community. Following the publications of these papers, two UK research councils (the Economic and Social Science Research Council and the Medical Research Council) commissioned two researchers, one a health expert (Andy Alaszewski) and the other a social scientist (Tom Horlick-Jones) to examine the further potential for applying findings from the social scientific risk literature (primarily from the environmental and technical fields) to practical public health issues (Alaszewski and Horlick-Jones 2003a). These efforts have led to a greater interest among the public health profession for environmental and technological risk research (e.g., Berry 2002, 2004; Calman 1998; Calman, Bennett, and Coles 1999), helped in no small part by the establishment of a new risk journal, *Health, Risk and Society*, that caters to this very audience.

In the US, there had been some attempts made over a longer period to integrate the environmental and technological risk issues with work on health care risks. One particular health area that has received significant attention among those with a background in environmental and technological research is that of smoking. Their research, to a large degree driven by law suits against US-based tobacco companies, has focused on whether adolescents are aware of the risk associated with smoking once they start to smoke. There are two schools of thought on this. Economist W. Kip Viscusi, who began his research looking at the costs associated with environmental and safety issues (Viscusi 1983, 1998, 2002), argues that smokers 'knew the risks and made informed, rational decisions to smoke' (Viscusi 1992, 149). He bases this argument on the fact that cigarette packages have warning labels on them stating that smoking is dangerous to one's health (Viscusi 1990, 1991a, b, 1992a, b). Against this, risk psychologists Paul Slovic and Neil Weinstein argue that adolescents who start smoking do not understand the consequences of their actions

(Slovic 1998, 2000a, b). Slovic writes 'When individuals decide to begin smoking, they rely on their feelings and not on analytic thinking' (Slovic 2001, x). Weinstein argues that the adolescents who start to smoke suffer from optimism bias, not being fully aware how addictive cigarettes are or how difficult it actually is to quit (Weinstein 1998, 1999).

There are legal and business interests at stake. For example, Slovic's (2001) collection on smoking grew out of an initiative to prove earlier work of Viscusi to be basically 'incorrect' showing that smokers do not behave rationally. This research was driven by US based law suits, in which individuals (usually dying smokers) but also the US government or US-states sue large US-based tobacco companies for in effect misleading them (e.g., the smokers state that they were not aware of the risks). These suits have led to the requirement of more research to prove the plaintiffs to be either right or wrong. Indeed to date in these suits, Viscusi has been pitted as the star expert witness representing a tobacco company (often Philip Morris: see Viscusi 2002) against the plaintiff's star witness which in many cases has been Paul Slovic, Neil Weinstein or Harvard law scholar John Hanson.

These findings of course also reflect their disciplinary biases for rational choice and for affectively-led or socially-led deviations from rationality. Both schools share the view – denied, for example, by many anthropologists – that rationality is a single phenomenon to which one either conforms or from which one deviates.

Another area that US-based environmental/technological risk researchers are working on is how best to communicate risks associated with waterborne diseases. Following a series of cryptosporidium outbreaks, environmental/technological risk researchers at Carnegie Mellon have helped water companies over the past four years or so to develop better communication tools (information leaflets and brochures) for the general public regarding such outbreaks (Casman et al. 2000; Fischhoff, Bostrom, and Quadrel 2002).

Also of importance is the work of Fischhoff and Bostrom in the health risk communication field on 'need to know' criteria in the health practitioner-patient interface, and on how to best communicate cancer risks (e.g., Fischhoff 1999, 2000). Bostrom has focused her attention on how to best communicate the benefits of vaccines in an era of distrust (e.g., Bostrom 1997, 1999).

In general, then, the influence of environmental and technological research approaches in public health has been to bring to bear rational choice economic or else cognitive psychological methods upon health risk behaviours and message acceptance, and some in the public health communities in the UK and the US have been at least moderately receptive to such approaches.

The influence of health risk studies upon environmental risk research

However, on the opposite side of the coin, there have been few attempts by the public health care risk experts to integrate their work into the environmental and technological area. Even the risk guidance book for the Department of Health was based on research carried out by environmental and technological researchers (DoH 1997). That said, some health practitioners are attempting to bring in research results from the environmental/technological area. For example, in the personal health area, discourses around issues such as dental amalgam, multiple chemical sensitivity, debate about Gulf War syndrome and the like incorporate elements of both fields

(Petrie et al. 2001; Petrie and Wessely 2002; Wessely and King's College Gulf War Research Unit 2001).

However, these initiatives are recent and few. On the whole many more environmental/technological risk researchers attend health risk meetings than vice-versa. The *Journal of Risk Research*, the official journal of the Society for Risk Analysis Europe and Japan, has, in its ten years of existence carried only five articles by health risk researchers. Mainstream public health methods using linear models such as the health belief model, the theory of planned behaviour, the theory of reasoned action, and the like have attracted no interest in the environmental and technological field.

Some Foucaultians who began their work in public health have shown some interest in a few technological fields. However, only a few public health scholars from the Foucaultian tradition have attempted much in the environmental field. The best known Foucaultian risk scholars have however concentrated on those risks that social policy scholars study or else on criminological risk and the fear of crime (e.g., Culpitt 1999; Burchell, Gordon, and Miller 1991).

The reasons for the lack of influence probably have little to do with any lack of recognition of the importance of communication studies in the health area. In fact, the opposite is rather the case. For example, in the year 2000 the US-based National Cancer Institute (NCI) funded 165 grants for research using communication as a component, process or study focus. NCI decided to fund so many studies because, in their view, '[t]here has been a significant deficit in the scope of NCI extramurally funded research program that focuses on explicating processes underlying cancer communications. Such research is desperately needed to inform communications practices throughout the cancer continuum' (NCI 2001). Other government bodies have funded health risk communication related research. For example, over the past few years the US Department of Defence has invested heavily in risk communication research, not least in a major call for proposals in 2000–2001 around Gulf War syndrome. In the years 2000–2001, it invested several million dollars in risk communication to find how best to handle public and soldier outcry about the Gulf War syndrome crisis as well as to handle the refusals of a large number of US troops to take the anthrax vaccine, even though refusal could lead to dismissal and court martial. The funding for this type of research has increased exponentially following the second Gulf War and terrorist attacks on the United States on 11 September 2001 (e.g., Lerner 2003; Schuster et al. 2001).

Explaining non-communication between risk communicators

Beyond these limited cases of influence in one direction or the other, there are very few cases of research projects in which perceptions of or communication about health and environmental risks are directly compared.

Why, then, has the flow of ideas been so much greater from the environmental and technological field to the health field than *vice versa*?

We briefly consider several hypotheses that might explain this puzzle. In general, there can only be a limited number of factors that could explain a phenomenon of this kind. Differentials in resources might be part of the story. In intellectual work, perhaps sheer weight of superior intellectual firepower might account for it. Alternatively, more micro-political factors to do with the relative confidence of the researchers and practitioners in the two fields might have something to do with it. We consider four hypotheses in turn. The first three are not very convincing.

Greater resources for environmental and technological risk research

Whatever might appear to be case, this hypothesis is not compelling. Although today, environmental and technological risk perception and communications studies are burgeoning, historically, public health was the much better funded field. Some of the major psychological and psychometric studies in the environmental and technological field in the 1960s, 1970s and 1980s in the US were funded on a rather hand-to-mouth basis. In Europe, only with the advent of panics around biotechnology did governmental and foundation funding pour into the field and major cross-national surveys begin (e.g., the Eurobarometer surveys; this was also the case with the rushed response to the emergence of bovine spongiform encephalopathy in the 1990s. E.g., Kjaernes 2006; UK Strategy Unit 2002). Although health risk work has probably not grown hugely in volume, it is hard to argue that greater funding for environmental research can be an important factor.

Greater intellectual cogency of environmental and technological risk models

Again, a dispassionate consideration of the two fields would hardly bear this hypothesis out. Much as environmental and technological risk researchers might like to boast of the power of their models, and to point to the weaknesses of the linear models still used by many public health scholars, in fact, their achievements are still limited and greatly contested. Even its leading practitioners have come to acknowledge that the 'psychometric paradigm' is sometimes problematic (Slovic 2000a). The amplification paradigm (Kasperson et al. 1988) may be moderately helpful as a simplified description, but it has little predictive value (Wåhlberg 2001; but see Breakwell 2007; Kasperson et al. 2003; Pidgeon, Kasperson, and Slovic 2003 for illuminating discussions). Much of the social constructionist work in effect offers few explanations but simply notes differences between groups in perceptions, and points to unspecified and unexplained social factors. While one of the present authors believes that anthropological work is much more theoretically promising, this remains a much smaller and less influential body of empirical research, even within the environmental and technological field, and in any case this has had almost no impact in health risk studies with the exception of some studies of sexual behaviour by some AIDS and HIV researchers. For practitioners, the body of knowledge in risk communication remains at the level of maxims and anecdotes, and the field has as many catastrophic mistakes to report ranging from Monsanto's failure to introduce GMO crops to Europe to siting mobile telephone masts in the UK as it has successful cases of persuasion to show, such as re-licensing hydropower dams in the US (Lofstedt 2003, 2005; Burgess 2004; Freudenburg 2004). Undoubtedly, the field has made great progress, and does have achievements to show. However, if its ideas are now being picked up by health risk scholars, it cannot be because the environmental and technological risk field can offer a solid body of empirical research with accepted, explanatorily powerful and adequately tested, predictive, and theoretically integrated models.

Intellectual imperialism of environmental and technological risk researchers

Again, this hypothesis has limited power. We have already noted the intellectual aspirations of both rational choice and cognitive psychological approaches in the debate about decisions to smoke. No doubt there will be those in every discipline who believe that their own disciplinary paradigms should largely replace those of other disciplines, and it would be surprising if there are not some in the environmental and technological risk perception and communication fields. However, with the exception of a few individual economists such as Becker, it is hard to document many cases of this (most economic studies on risk have focused almost entirely on financial risk beginning with Frank Knight, 1965). In general, most of the borrowing by health researchers seems to arise from 'pull' factors within the health risk community looking for fresh ideas, rather than 'push' factors from bored and over-confident environmental and technological risk scholars looking for new terrain to occupy. The Bennett and Calman collection, for example, arose from a health research initiative.

Loss of confidence among public health researchers

If 'pull' factors within the public health community have more to do with the trend toward greater borrowing than by environmental and technological risk researchers, and if neither resources nor intellectual cogency explain much of what is going on here, then what needs to be explained is the nature of this demand by public health scholars for approaches developed beyond their borders. We cannot here offer a full explanation. However, one promising line of enquiry might focus on the huge political challenge to the legitimacy of health promotion that has been made since the late 1960s in the UK and throughout western Europe, although perhaps with less resonance in the United States. Challenges from the left were that it effectively promoted middle class lifestyles, denigrated working class lifestyle choices, was patronising, victim-blaming, and concerned more with saving taxpayers' money (in countries with socialised systems of medical care) than with health status. In academic circles, these critiques were given a peculiar twist by the neo-Foucaultian sociological critics of health promotion in the 1980s (Lupton 1985, 1999a, b). From the neo-liberal right, too, there came a politically influential critique of public health preventive work (Le Fanu 1994; Fitzpatrick 2001), and of the overdone fears of the 'worried well' (Furedi 1997). Faced with this political and intellectual challenge to the legitimacy of their risk communication practice, and by a public increasingly anxious about food safety and also decreasingly willing to accept public health assurances about the safety of vaccinations and other procedures, it would not be surprising if public health promotion researchers and practitioners had begun during the late 1980s and 1990s to see whether they could borrow fresh approaches from other fields.

Toward a strategy for research integration

How might the research communities overcome some of the problems identified above of disconnection, fragmentation, lack of learning and lack of theory testing? In this section, we briefly suggest four principles that would, if adopted, represent an important step toward tackling the problems identified above, and would constitute the bare bones of a research agenda. We do not set out particular empirical questions for collaborative work between healthy and environmental or technological risk researchers – there are too many that cry out for attention. Our aim here is to set out the principles for an integrative research agenda.

Clearly, lack of comparison between fields makes for weaknesses in the work done in either of these two, or in any of those that make up risk perception and communication studies. In our view, the crucial first step is the encouragement of the acceptance across all the fields of risk perception and communication research of the following methodological maxim:

> Maxim A: *Any theory of risk perception or of risk communication should not be considered adequate until it has been tested comparatively in a single research design, and not just in separate studies, across a sufficiently wide range of substantive types of risk.*

But what is 'sufficient'? Table 1 provides a breakdown of what are, in our view, the main substantive categories. Risk are distinguished by the nature of the harm – one of death, illness or injury, lack or loss of resources, of injury to social relations or status. Next, they can be distinguished by empirical field, often allocated typically to different disciplines. In each cell, examples are given of the kinds of risk. The table can be used to give content to the concept of sufficiency in Maxim A, yielding the following refinement:

> Maxim B: *A research design will be sufficiently wide to provide a test of the adequacy of a theory of risk perception or communication, to meet the standards proposed in Maxim A, if either*
>
> i. *It involves testing on risks covering all four types of harm, and at least (say) three fields that exhibit variety on key variables in the theory; or*
> ii. *The theory itself offers some theoretically empirically credible reason why its application should and can only be restricted to one or a few substantive categories of risk.*

Adequacy is not a strong standard, but a necessary one. It is reasonable to suppose, for example, that economic rational choice and the closely related public health linear models, cognitive psychological models using mental mapping methods, the Douglasian anthropological institutionalist theory developed from Durkheim and Evans-Pritchard, could be shown – with sufficient comparative research – to meet at least this adequacy requirement. More than this is needed. This suggests at least the following:

> Maxim C: *High priority should be attached to comparative research across fields of risk that attempts to measure the relative explanatory power of more than one putatively adequate theory at once, in the same research design.*

Theories of risk perception and communication are sought, not only for reasons of pure social science, but also for applied and practical purposes of designing interventions, designing communications, managing or at least containing conflicts over risks, making decisions over such matters as siting controversial facilities or authorising or mandating certain public health procedures such as vaccinations or investing in public information campaigns. As we have seen, the practical challenges are common to all the fields of risk. This suggests another general principle for an integrative research strategy:

> Maxim D: *The recommendations derived from a theory that yields practical prescriptions about what governmental, business or social movement decision makers ought to do in risk communication, or indeed what individual consumers or citizens ought to fear or be*

Table 1. Four categories of substantive harms, studied by different academic disciplines of risk research, with examples from nine fields of public policy.

	Bodily		Quasi-economic	Social / Relational
Type of damage υ Field of policy ξ	Mortality	Morbidity	Resource loss or disadvantage	Social situation or status change
Health [Health care, public health, health risk behaviour studies]	Accident, injury, fatal poisoning, some genetic conditions	Infectious disease, some genetic conditions	Uninsurability	Stigma (e.g., mental illness, AIDS)
Environment [Environmental studies, geography]	Toxic exposures	Toxic releases	Rising sea levels due to climate change	Stigmatisation of areas (e.g., siting of contentious facilities)
Technology [Engineering]	Unsafe design	Unsafe design	Technology failure; erosion of values of nearby assets	Stigmatisation of technologies
Occupational health and safety [Occupational engineering, human resources management]	Some working conditions	Occupational exposures and accidents, disablement	Loss of job or income or promotion opportunities from disablement, stress etc	Stigmatisation of certain workers e.g., those working in sewerage processing
Crime (victimisation) [Criminology]	Murder, manslaughter	Assault	Acquisitive crime	Crimes intended to humiliate e.g., rape, racist violence
Material thriving [Social policy]	Starvation	Malnutrition, chronic disease	Poverty, unemployment, homelessness, inadequate pension provision, falling housing asset values, famine, drought	Social exclusion, unjust discrimination
Information [Information law and policy]	n/a	n/a	Deprivation of intellectual property rights, 'passing off', information torts e.g., software viruses and Trojan horses	Loss of privacy, defamation, obscenity, blasphemy, censorship
Business [Economic, business studies, management studies]	n/a	n/a	Project risk, business failure, bankruptcy: economic loss	Stigma associated with bankruptcy, arbitrary use of state of social movement power (political risk), restrictive anti-competitive practices, abuse of monopoly power

insouciant about in relation to any given risk, should not be considered robust until the likely consequences of following those recommendations have been examined against at least as wide a range of substantive types of risk as would be required for assessing the adequacy of the pure theory under Maxim B.

If these principles were to be widely accepted, the result would, by definition, be much greater collaborative and cross-disciplinary research not only between public health researchers and environmental and technological risk scholars, but more generally across the risk perception and risk communication fields. The argument of this article is that only thereby can any of the fields hope to make progress in either pure or applied studies.

Conclusion

The fragmentation of risk perception and communication studies by field and by discipline is understandable, but increasingly it represents a barrier to the development of any of the fields and disciplines. Overcoming this can only be the work of many scholars over many years. The fields of public health and of environmental and technological risk provide a particularly clear view of the depth of the problem, and, because of the similarities and links between the substantive risks, their disconnection and the limitations of the recent efforts to bridge them provide a good measure of what remains to be done and how important it is to pursue integration. However, this can only be a first step toward more ambitious programmes of developing fully integrated understandings of the nature of social fear and insouciance.

Acknowledgements

The authors are grateful to Baruch Fischhoff and Simon Wessely for invaluable comments on an earlier draft. We are also grateful to two anonymous referees for their helpful comments and suggestions. Naturally, they bear no responsibility for our errors, nor should they be accused of agreeing with our arguments. In writing this article the first author was supported by a grant provided by Pfizer Global Research and Development.

Notes

1. In this paper, our concern is almost entirely with the literature on *collective* public health risk communication, and not with individual advice from physician to patient: the individual relationship with a professional, invoked at a point of individual dependence or even need, raises quite different questions of risk perception, communication and trust from those at issue in the collective health and environmental contexts, and therefore we do not press our arguments for integration and cross-field learning in this area of specifically *medical*, as opposed to health risk communication.

References

6, P. 2001. Can there be a unified science of fear? Review article on "Risk and 'the other'", by Hélène Joffe, Cambridge University Press, Cambridge 1999. *Journal of Risk Research* 4, no. 4: 431–45.

———. 2002. *Strategies for reassurance: Public concerns about privacy and data sharing in government*. London: Performance and Innovation Unit, Cabinet Office. Available at http://www.cabinet-office.gov.uk/innovation/2002/privacy/report/papers/perri6.pdf.

6, P., K. Lasky, and A. Fletcher. 1998. *The future of privacy, volume II: Public trust in the use of private information*. London: Demos.

Adams, J. 1995. *Risk*. London: UCL Press.

Alaszewski, A., and T. Horlick-Jones. 2003. How can doctors communicate information about risk more effectively? *British Medical Journal* 327: 728–31.

Ball, L.K., G. Evans, and A. Bostrom. 1998. Risky business: Challenges in vaccine risk communication. *Pediatrics* 101, no. 3: 453–58.

Banks, S.M., P. Salovey, S. Greener, A.J. Rothman, A. Moyer, J. Beauvais, and E. Epel. 1995. The effects of message framing of mammography utilisation. *Health psychology* 14, no. 2: 178–84.

Bean, P., ed. 1993. *Cocaine and crack: Supply and use*. Basingstoke: MacMillan.

Bellaby, P. 2003. Communication and miscommunication about risk: Understanding UK parents' attitudes to combined MMR vaccination. *British Medical Journal* 327: 725–8.

Ben-Yehuda, N. 1986. The sociology of moral panics: Toward a new synthesis. *Sociological Quarterly* 27, no. 4: 495–513.

Bennett, P., and K. Calman. 1999. *Risk communication and public health*. Oxford: Oxford University Press.

Berry, C. 2002. Conceptions and misconceptions of risk. *In Royal Society of Medicine Conference, How risky is health? Abstracts and Biographies.*

Berry, D. 2004. *Risk, communication and health psychology*. Maidenhead, UK: Open University Press.

Black, W.C., R.F., Jr. Nease, and A.N. Tosteson. 1995. Perceptions of breast cancer risk with screening effectiveness in women younger than 50 years of age. *Journal of the National Cancer Institute* 87, no. 10: 720–31.

Blane, D., E. Brunner, and R. Wilkinson, eds. 1996. *Health and social organisation: Towards a health policy for the 21st century*. London: Routledge.

Boer, H., and E.R. Seydel. 1995. Protection motivation theory. In *Predicting health behaviour: Research and practice with social cognition models*, ed. M. Conner and P. Norman, 95–120. Buckingham, UK: Open University Press.

Bogardus, S.T. Jr., E. Holmboe, and J.F. Jekel. 1999. Perils, pitfalls, and possibilities in talking about medical risk. *Journal of the American Medical Association* 281, no. 11: 1037–41.

Booth Davies, J. 1992. *The myth of addiction: An application of the psychological theory of attribution to illicit drug use*. Chur, Switzerland: Harwood Academic Publishers.

Bostrom, A. 1997. Vaccine risk communication: Lessons from risk perception, decision making and environmental risk communication research. *Risk: Health, safety and environment* 8, no. 2: 173–200.

———. 1999. Who calls the shots? Credible vaccine risk communication. In *Social trust and the management of risk*, ed. G. Cvetkovich and R.E. Löfstedt. London: Earthscan.

Bostrom, A., B. Fischhoff, and M.G. Morgan. 1992. Characterising mental models of hazardous processes: A methodology and an application to radon. *Journal of Social Issues* 48, no. 1: 85–100.

Bostrom, A., C.J. Atman, B. Fischhoff, and G.M. Morgan. 1994. Evaluating risk communications: Completing and correcting mental models for hazardous processes, Part 2. *Risk Analysis* 14: 789–99.

Bottorff, J.L., P.A. Ratner, J.L. Johnson, C.Y. Lovato, and S.A. Joab. 1998. Communicating cancer risk information: The challenges of uncertainty. *Patient education and counselling* 33, no. 1: 67–81.

Breakwell, G.M. 2001. Mental models and social representations of hazards: The significance of identity processes. *Journal of Risk Research* 4, no. 4: 341–51.

———. 2007. *The psychology of risk*. Cambridge: Cambridge University Press.

Breakwell, G.M., and D.V. Canter, eds. 1993. *Empirical approaches to social representations*. Oxford: Oxford University Press.

Bunton, R., S. Nettleton, and R. Burrows, eds. 1995. *The sociology of health promotion: Critical analyses of consumption, lifestyle and risk*. London: Routledge.

Burchell, G., C. Gordon, and P. Miller, eds. 1991. *The Foucault effect: Studies in governmentality*. Hemel Hempstead, UK: Harvester Wheatsheaf.

Burgess, A. 2004. *Cellular phones, public fears, and a culture of precaution*. Cambridge: Cambridge University Press.

Burton, I., and R.W. Kates. 1964a. The perception of natural hazards in resource management. *Natural Resources Journal* 3: 412–41.

———. 1964b. The flood plain and the seashore: A comparison analysis of a hazard zone occupance. *Geographical Review* 54: 366–85.

Burton, I., R.W. Kates, and G.F. White. 1978. *The environment as hazard*. New York: Oxford University Press.

———. 1992. *The environment as hazard*. 2nd edn. New York: Oxford University Press.

Calman, K.C. 1998. *The potential for health: How to improve the nation's health*. Oxford: Oxford University Press.

Calman, K.C., P.G. Bennett, and D.G. Coles. 1999. Risks to health: Some key issues in management, regulation and communication. *Health, Risk and Society* 1: 107–16.

Campbell, C., R. Wood, and M. Kelly. 1999. *Social capital and health*. London: Health Education Authority.

Caplan, P., ed. 2000. *Risk revisited*. London: Pluto Press.

Casman, E., B. Fischhoff, C. Palmgren, M. Small, and F. Wu. 2000. Integrated risk model of a drinking waterborne Cryptosporidiosis outbreak. *Risk Analysis* 20: 493–509.

Cody, R., and C. Lee. 1990. Behaviors, beliefs, and intentions in skin cancer prevention. *Journal of Behavioral Medicine* 13, no. 4: 373–89.

Cohen, S. 1987 [1972]. *Folk devils and moral panics: The creation of the mods and rockers*. 3rd edn. Oxford: Blackwell.

Conner, M., and P. Norman, eds. 1995. *Predicting health behaviour: Research and practice with social cognition models*. Buckingham: Open University Press.

Cooper, H., S. Arber, L. Fee, and J. Ginn. 1999. *The influence of social support and social capital on health: A review and analysis of British data*. London: Health Education Authority.

Cotterell, J. 1996. *Social networks and social influence in adolescence*. London: Routledge.

Covello, V. 1983. The perception of technological risks: A literature review. *Technological Forecasting Social Change* 23: 285–97.

Cowan, S. 2002. Public arguments for and against the establishment of community mental health facilities: Implications for mental health practice. *Journal of Mental Health* 11, no. 1: 5–15.

Culpitt, I. 1999. *Social policy and risk*. London: Sage.

Cvetkovich, G., and R.E. Löfstedt, eds. 1999. *Social trust and the management of risk*. London: Earthscan.

Davidson, B.J., L.F. Degner, and J.R. Morgan. 1995. Information and decision-making preferences of men with prostate cancer. *Oncology Nursing Forum* 22, no. 9: 1401–8.

Dear, M.J. 1992. Understanding and overcoming the NIMBY syndrome. *Journal of the American Planning Association* 58, no. 3: 288–300.

Department of Health. 1997. *Communicating about risks to public health: Pointers to good practice*. London: Stationery Office.

Dickerson, J.W.T., and G.V. Stimson, eds. 1995. *Health in the inner city: Drugs in the city*, supplement to the *Journal of the Royal Society of Health*. London: Royal Society of Health.

Douglas, M., ed. 1982. *Essays in the sociology of perception*. London: Routledge and Kegan Paul.

———. 1992. *Risk and blame: Essays in cultural theory*. London: Routledge.

Douglas, M., and A. Wildavsky. 1982. *Risk and culture: An essay in the selection of technological and environmental dangers*. Berkeley, CA: University of California Press.

Downs, J.S., P. Murray, W. Bruine de Bruin, J. White, C. Palmgren, and B. Fischhoff. 2004. An interactive video program to reduce adolescent females' STD risk: A randomized control trial. *Social Science and Medicine* 59: 1561–72.

Edwards, A., and L. Prior. 1997. Communication about risk: Dilemmas for general practitioners. *British Journal of General Practice* 47, no 424: 739–42.

Edwards, A., E. Mathews, R. Pill, and M. Bloor. 1998. Communication about risk: Diversity among primary care professionals. *Family Practice* 15, no. 4: 296–300.

Eldridge, J., J. Kitzinger, G. Philo, et al. 1998. The re-emergence of BSE: The impacts of public beliefs and behaviour. *Risk and Human Behaviour Newsletter* 3: 6–10.

Eldridge, J., and J. Reilly. 2003. Risk and relativity: BSE and the British media. In *The Social Amplification of Risk*, ed. N. Pidgeon, R.E. Kasperson, and P. Slovic. Cambridge: Cambridge University Press.

Ellis, R.J., and F. Thompson. 1997. Seeing green: Cultural biases and environmental preferences. In *Culture matters: Essays in honour of Aaron Wildavsky*, ed. R.J. Ellis and M. Thompson, 169–90. Boulder, CO: Westview Press.

Emler, N., and S. Reicher. 1995. *Adolescence and delinquency: The collective management of reputation*. Oxford: Blackwell.

Evans-Pritchard, E.E. 1937. *Witchcraft, oracles and magic among the Azande*. Oxford: Oxford University Press.

Fischhoff, B. 1995. Risk perception an communication unplugged: Twenty years of process. *Risk Analysis* 15: 137–45.

———. 1996. Public values in risk research. *Annals of the American Academy of Political and Social Science* 545: 75–84.

———. 1999. Why (cancer) risk communication can be hard. *Journal of the National Cancer Institute* 25: 7–13.

———. 2000a. Need to know: Analytical and psychological criteria. *Roger Williams University Law Review* 6: 237–61.

———. 2000b. Scientific management of science? *Policy Sciences* 33: 73–87.

———. 2005. Decision research sciences. *Health Psychology* 24: s9–s16.

Fischhoff, B., A. Bostrom, and M.J. Quadrel. 2002. Risk perception and communication. In *Oxford Textbook on Public Health*, ed. R. Detels, J. McEwen, R. Beaglehole, and H. Tanaka, 1105–23. Oxford: Oxford University Press.

Fischhoff, B., J. Dows, and W. Bruine de Bruin. 1998. Adolescent vulnerability: A framework for behavioral interventions. *Applied and Preventive Psychology* 7: 77–94.

Fischhoff, B., P. Slovic, S. Lichtenstein, S. Read, and B. Combs. 1978. How safe is safe enough? A psychometric study of attitudes towards technological risk and benefits. *Policy Studies* 9: 127–52.

Fischhoff, B., S. Lichtenstein, P. Slovic, S.L. Derby, and R.L. Keeney. 1981. *Acceptable risk*. New York: Cambridge University Press.

Fitzpatrick, M. 2001. *The tyranny of health: Doctors and the regulation of lifestyle*. London: Routledge.

Fong, G.T., L.A. Rempel, and P.A. Hall. 1999. Challenges to improving health risk communication in the 21st century: A discussion. *Journal of the National Cancer Institute Monographs* 25: 173–6.

Fox, C.R., and J.R. Irwin. 1998. The role of context in the communication of uncertain beliefs. *Basic and Applied Social Psychology* 20: 57–70.

Freudenburg, W.R. 2004. Can we learn from failure? Examing US experiences with nuclear repository sitting. *Journal of Risk Research* 7: 153–69.

Frewer, L.J., C. Howard, D. Hedderley, and R. Shepherd. 1996. What determines trust in information about food-related risks? Underlying psychological constructs. *Risk Analysis* 16, no. 4: 473–86.

Frewer, L.J., C. Howard, and R. Shepherd. 1997. Public concerns about general and specific applications of genetic engineering: Risk, benefit and ethics. *Science, Technology and Human Values* 22: 98–124.

Furedi, F. 1997. *The culture of fear: Risk-taking and the morality of low expectation*. London: Cassell.

Gabe, J., D. Kelleher, and G. Williams, eds. 1994. *Challenging medicine*. London: Routledge.

Gaber, S.L. 1996. From NIMBY to fair share: The development of New York City's municipal shelter siting policies, 1980–1990. *Urban Geography* 17, no. 4: 294–316.

Gilovich, T., D. Griffin, and D. Kahneman. 2002. *Heuristics and biases: The psychology of intuitive judgement*. Cambridge: Cambridge University Press.

Grendstad, G., and P. Selle. 1997. Cultural theory, postmaterialism and environmental attitudes. In *Culture matters: Essays in honour of Aaron Wildavsky*, ed. R.J. Ellis and M. Thompson, 151–68. Boulder, CO: Westview Press.

Grendstad, G. 2001. Nordic cultural baselines: Accounting for domestic policy convergence and foreign policy divergence. *Journal of Comparative Policy Analysis, Research and Practice* 3: 5–29.

Hendry, L.B., J. Shucksmith, J.G. Love, and A. Glendinning. 1993. *Young people's leisure and lifestyles*. London: Routledge.

Hollway, W., and T. Jefferson. 2000. *Doing qualitative research differently: Free association, narrative and the interview method*. London: Sage.

Hood, C.C., and D.K.C. Jones. 1996. *Accident and design: Contemporary debates in risk management*. London: University College London Press.

Joffe, H. 1999. *Risk and the 'other'*. Cambridge: Cambridge University Press.

Kasperson, J.X., R.E. Kasperson, N. Pidgeon, and P. Slovic. 2003. The social amplification of risk: Assessing fifteen years of research and theory. In *The social amplification of risk*, ed. N. Pidgeon, R.E. Kasperson, and P. Slovic. Cambridge: Cambridge University Press.

Kasperson, R.E., and I. Palmlund. 1997. Evaluating risk communication. In *Effective risk communication: The role and responsibility of government and non-government organisations*, ed. V.T. Covello, D.B. McCallum, and M.T. Pavlova. New York: Plenum.

Kasperson, R.E., O. Renn, P. Slovic, H.S. Brown, J. Emel, R. Goble, J.X. Kasperson, and S. Ratick. 1988. The social amplification of risk: A conceptual framework. *Risk Analysis* 8, no. 2: 177–87.

Kjaernes, U. 2006. Trust and distrust: Cognitive decisions or social relations? *Journal of Risk Research* 9: 911–32.

Knight, F.H. 1965. *Risk, uncertainty and profit*. New York: Harper and Row.

Kreuter, M.W. 1999. Dealing with competing and conflicting risks in cancer communication. *Journal of the National Cancer Institute Monographs* 25: 27–35.

Le Fanu, J., ed. 1994. *Preventionitis: The exaggerated claims of health promotion*. London: Social Affairs Unit.

Lerner, J.S., R.M. Gonzalez, D.A. Small, and B. Fischhoff. 2003. Effects of fear and anger on perceived risks of terrorism: A national field experiment. *Psychological Science* 14: 144–50.

Leiss, W. 1995. 'Down and dirty': The use and abuse of public trust in risk communication. *Risk Analysis* 15, no. 6: 685–92.

———. 1996. Three phases in the evolution of risk communication practice. *Annals of the American Academy of Political and Social Science* 545: 85–94.

Löfstedt, R.E. 1996a. Risk communication: The Barsebäck nuclear plant case. *Energy Policy* 24, no. 8: 689–96.

———. 1996b. Fairness across borders: The Barseback nuclear power plant. *Risk: Health, Safety and Environment* 7, no. 2: 135–44.

———. 1997. Evaluation of two siting strategies: The case of two UK waste tire incinerators. *Risk: Health, Safety and Environment* 8, no. 1: 63–77.

———. 1999. The role of trust in the north Blackforest: An evaluation of a citizen panel process. *Risk: Health, Safety and Environment* 9: 7–30.

———. 2003. Regulatory negotiation: The case of International Paper's Hydrodam Relicensing procedure. *Risk Management: An International Journal* 5, no. 4: 37–53.

———. 2005. *Risk management in post trust societies*. Basingstoke: Palgrave-MacMillan.

Löfstedt, R., and O. Renn. 1997. The Brent Spar controversy: An example of risk communication gone wrong. *Risk Analysis* 17, no. 2: 131–6.

Lupton, D. 1995. *The imperative of health: Public health and the regulated body*. London: Sage.

———. 1999a. *Risk*. London: Routledge.

Lupton, D., ed. 1999b. *Risk and sociocultural theory: New directions and perspectives*. Cambridge: Cambridge University Press.

Maibach, E. 1999. Cancer risk communication: What we need to learn. *Journal of the National Cancer Institute Monographs* 25: 179–81.

McComas, K.A. 2006. Defining moments in risk communication research: 1996–2005. *Journal of Health Communication* 11: 75–91.

Miller, D., and G. Philo. 1995. *Communicating change in the NHS: A review of health communication campaigns, media coverage of health care issues and the process of public opinion and belief information*. Report for the Communication Directorate of the NHS Executive, Leeds.

Moran, M., and B. Wood. 1993. *States, regulation and the medical profession*. Buckingham: Open University Press.

Morgan, G.M., B. Fischhoff, A. Bostom, and C. Atman. 2001. *Risk communication: A mental models approach*. Cambridge: Cambridge University Press.

Moscovici, S. 2000. *Social representations: Explorations in social psychology*. Cambridge: Polity Press.

National Research Council. 1989. *Improving risk communication*. Washington, DC: National Academy Press.

———. 1996. *Understanding risk*. Washington, DC: National Academy Press.

Noelle-Neumann, E. 2000. Foreword. In *Between understanding and trust: The public, science and technology*, ed. M. Dierkes and C. von Grote, xi–xii. Amsterdam: Harwood.

Patrick, K. 1999. Prevention, public health, and interactive health communication. *American Journal of Preventive Medicine* 16, no. 1: 46–7.

Peterson, A., and D. Lupton. 1996. *The new public health: Health and self in the age of risk*. London: Sage.

Peterson, A., and C. Waddell, eds. 1998. *Health matters: A sociology of illness, prevention and care*. Buckingham, UK: Open University Press.

Petrie, K.J., B. Sivertsen, M. Hysing, E. Broadbent, R. Moss-Morris, H.R. Eriksen, and H. Ursin. 2001. Thoroughly modern worries: The relationship of worries about modernity to reported symptoms, health and medical care utilisation. *Journal of Psychosomatic Research* 51: 395–401.

Petrie, K.J., and S. Wessely. 2002. Modern worries, new technology, and medicine: New technologies mean new health complaints. *British Medical Journal* 324: 690–1.

Phillips, L., J. Bridgeman, and M. Ferguson-Smith. 2000. *The BSE inquiry – report: Evidence and supporting papers of the inquiry into the emergence and identification of bovine spongiform encephalopathy (BSE) and variant Creutzfeldt-Jakob disease (vCJD) and the action taken in response to it up to 20th March 1996*. London: Stationery Office. Available at http://www.bseinquiry.gov.uk/report/index.htm.

Pidgeon, N., K. Henwood, and B. Maguire. 1999. Public health communication and the social amplification of risks: Present knowledge and future prospects. In *Risk communication and public health*, ed. P. Bennett and K. Calman. Oxford: Oxford University Press.

Pidgeon, N., R.E. Kasperson, and P. Slovic, eds. 2003. *The social amplification of risk*. Cambridge: Cambridge University Press.

Porter, D. 1999. *Health, civilisation and the state: A history of public health from ancient to modern times*. London: Routledge.

Power, R., S. Jones, G. Kearns, J. Ward, N. Gibson, J. Perera, and J. Smith. 1995. *Coping with illicit drug use*. London: Centre for Research on Drugs and Health Behaviour and Tufnell Press.

Powell, D., and W. Leiss. 1997. *Mad cows and mothers milk: The perils of poor risk communication*. Montreal: McGill-Queen's University Press.

Renn, O. 1990. Risk perception and risk management: A review, part 1: Risk perception. *Risk Abstracts* 7: 1–9.

———. 1998. Three decades of risk research: Accomplishments and new challenges. *Journal of Risk Research* 1: 49–71.

———. 2001. Environmental risks-perception, evaluation and management: epilogue. *Environmental Risks: Perception, Evaluation and Management* 9: 275–99.

Renn, O., and D. Levine. 1991. Credibility and trust in risk communication. In *Communicating risks to the public: International perspectives*, ed. R.E. Kasperson and P.J. Stallen. Amsterdam: Kluwer.

Renn, O., T. Webler, and H. Kastenholz. 1996. Procedural and substantive fairness in landfill siting. *Risk: Health, Safety and Environment* 7, no. 2: 145–68.

Renn, O., T. Webler, and P. Wiedemann, eds. 1995. *Fairness and competence in citizen participation*. Dordrecht: Kluwer.

Renn, O., W. Leiss, H. Kastenholz, and R.E. Lofstedt. 2002. *OECD Guidance Document on Risk Communication*. Paris: OECD.

Repper, J. 1996. Public attitudes towards mental health facilities in the community. *Health and Social Care in the Community* 4, no. 5: 290–9.

Rosenstock, I.M. 1966. Why people use health services. *Millbank Memorial Foundation Quarterly* 44: 94–124.

———. 1974. The health belief model and preventative health behaviour. *Health Education Monographs* 2: 354–6.

Rothman, A.J., and P. Salovey. 1997. Shaping perceptions to motivate behaviour: The role of message framing. *Psychological Bulletin* 121, no. 1: 3–19.

Rutter, D., and L. Quine, eds. 2002. *Changing health behaviour*. Buckingham: Open University Press.

Schuster, M.A., B.D. Stein, L. Jaycox, R.L. Collins, G.N. Marshall, M.N. Elliot, A.J. Zhou, D.E. Kanouse, J.L. Morrison, and S.H. Berry. 2001. A national survey of stress reactions after the September 11, 2001, terrorist attacks. *New England Journal of Medicine* 345, no. 20: 1507–12.

Slovic, P. 1986. Informing and educating the public about risk. *Risk Analysis* 6, no. 4: 403–15.

———. 1987. Risk perception. *Science* 236: 280–5.

———. 1993. Perceived risk, trust, and democracy. *Risk Analysis* 13, no. 6: 675–82.

———. 1997. Trust, emotion, sex, politics, and science: Surveying the risk-assessment battle field. In *Environment, ethics, and behaviour*, ed. M.H. Bazerman, M. Messick, A.E. Tenbrunsel, and K.A. Wade-Benzoni. San Francisco, CA: New Lexington.

———. 1998. Do adolescent smokers know the risks? *Duke Law Journal* 47: 1133–41.

———. 2000a. What does it mean to know a cumulative risk? Adolescents' perceptions of short-term and long-term consequences of smoking. *Journal of Behavioural Decision Making* 13: 259–66.

———. 2000b. Rejoinder: The perils of Viscusi's analysis of smoking perceptions. *Journal of Behavioural Decision Making* 13: 273–6.

———. 2000c. *The perception of risk*. London: Earthscan.

———, ed. 2001. *Smoking: Risk, perception and policy*. London: Sage.

Starr, C. 1969. Social benefit versus technological risk. *Science* 165: 1232–8.

Stern, P.C. 1991. Learning through conflict: A realistic strategy for risk communication. *Policy Sciences* 24: 99–119.

Swann, C., and A. Morgan, eds. 2002. *Social capital for health: Insights from qualitative research*. London: Health Development Agency.

Takahashi, L.M. 1997. Information and attitudes towards mental health care facilities: Implications for addressing the NIMBY syndrome. *Journal of Planning Education and Research* 17: 119–30.

Takahashi, L.M., and M.J. Dear. 1997. The changing dynamics of community opposition to human service facilities. *Journal of the American Planning Association* 63, no. 1: 79–92.

Thornton, J.G., M.P. van den Borne, and A.J. de Bruijn. 1996. Risk communication: The patient's view. *Early Human Development* 47: S13–S17.

Tversky, A., and D. Kahneman. 1974. Judgement under uncertainty: Heuristics and biases. *Science* 185: 1124–31.

UK Strategy Unit. 2002. *Risk: Improving government's capability to handle risk and uncertainty*. London: Cabinet Office, Strategy Unit.

Viscusi, W.K. 1983. *Risk by choice: Regulating health and safety in the workplace*. Cambridge, MA: Harvard University Press.

———. 1990. Do smokers underestimate risks? *Journal of Political Economy* 98: 1253–69.

———. 1991a. Age variations in risk perceptions and smoking decisions. *Review of Economics and Statistics* 73: 577–88.

———. 1991b. *Reforming product liability*. Cambridge, MA: Harvard University Press.

———. 1992a. *Smoking: Making the risky decision*. Oxford: Oxford University Press.

———. 1992b. *Fatal tradeoffs: Public and private responsibilities for risk*. New York: Oxford University Press.

———. 1998. *Rational risk policy*. Oxford: Oxford University Press.

———. 2002. *Smoke-filled rooms: A postmortem on the tobacco deal (studies in law and economics)*. Chicago: University of Chicago Press.

Wåhlberg, A.E.A. 2001. The theoretical features of some current approaches to risk perception. *Journal of Risk Research* 4, no. 3: 237–50.

Warner, F. 1998. Viewpoint. *Journal of Risk Research* 1, no. 1: 13–15.

Weinstein, N.D. 1998. Accuracy of smokers' risk perceptions. *Annals of Behavioural Medicine* 20, no. 2: 135–40.

———. 1999. What does it mean to understand a risk? Evaluating risk comprehension. *Journal of the National Cancer Institute Monographs* 25: 15–20.

Wessely, S., and King's College Gulf War Research Unit. 2001. Ten years on: What do we know about the Gulf War Syndrome? *Clinical Medicine (JRCPL)* 1: 28–37.

White, G.F. 1945. *Human adjustment to floods: A geographical approach to the flood in the United States*. Chicago: University of Chicago Press.

———. 1961. The choice of use in resource management. *Natural Resource Journal* 1: 23–40.

Wildavsky, A., and K. Dake. 1990. Theories of risk perception? Who fears what and why? *Daedalus* 119, no. 4: 41–60.

Wilkinson, R.G. 1996. *Unhealthy societies: The afflictions of inequality*. London: Routledge.

Wynne, B. 1992. Sheep farming after Chernobyl: A case study in communicating scientific information. In *When science meets the public*, ed. B.V. Lewenstein, 43–67. Washington, DC: American Association for the Advancement of Science.

———. 1996. May the sheep safely graze? In *Risk, environment and modernity: Toward a new ecology*, ed. S. Lash, B. Szerszynski, and B. Wynne. London: Sage.

Wynne, B., and K. Dressel. 2001. Cultures of uncertainty-transboundary risks and BSE in Europe. In *Transboundary risk management*, ed. J. Linnerooth-Bayer, R.E. Lofstedt, and G. Sjostedt. London: Earthscan.

Communities of risk research and risk practice: divided by a common language?

Tom Horlick-Jones

Cardiff School of Social Sciences, Cardiff University, Cardiff, UK

This short commentary paper provides a critical, but supportive, discussion of the paper by Löfstedt and 6 (L&6). Three suggestions are made. First, that L&6 make certain unsatisfactory assumptions about the underlying problems faced by risk communication initiatives. Second, that one should not be too surprised about the lack of dissemination of risk communication knowledge across different substantive risk areas, given the diversity of the risk research and risk practice communities. Third, that L&6 present an agenda for risk communication that has perhaps not kept pace with some recent developments, in particular those concerning citizen engagement.

Introduction

Ragnar Löfstedt and Perri 6 (2008; henceforth L&6) have produced a spirited opinion piece on the relative lack of dissemination of risk communication research findings between different substantive areas of risk-related practice. They focus on research and its applications in two areas: regulatory and siting issues associated with environmental and technological risks, and health promotion. They express concern about the limited extent to which findings in each of these domains have found their way into the other. They conclude by arguing for the need for comparative risk communication studies that encompass a number of different areas of application.

There is not sufficient space in this short commentary paper to do complete justice to the range of issues raised by L&6, which include conceptual claims and assumptions, advocacy for a particular vision of risk research, speculation about the reasons underlying the apparent barriers to dissemination between different domains of application, and a bold programmatic statement regarding future research. I have therefore chosen to concentrate on a few matters that I find particularly interesting. These are concerned with three main themes: certain issues about the nature of risk perception; questions about the character of research and the application of its findings; and some recent developments in the use of risk ideas and techniques in policy-making. The title of the paper reflects my view that, despite being ostensibly united by a shared language, it is not surprising that the various communities of risk research and risk practice are in fact divided; for reasons that reflect the diverse nature of both risk and research.

Conceptualising risk

Perhaps the most contentious conceptual claims that L&6 make are that 'social fear is basically a single phenomenon' and 'the fields of risk communication practice all face the same underlying challenges'. I recognise that, on the face of it, these claims have a certain plausibility. It seems reasonable to argue that different risk issues have the capacity to engender a similar sort of emotional arousal. Similarly, issues like the credibility and trustworthiness of agencies seeking to communicate risk information would appear to present similar sorts of challenge across different specific settings. However, there is an instrumental view of risk that seems to underlie L&6's arguments that, as I see it, is simply at variance with the ways in which people commonly make sense of, and act towards, risk issues in the real world.

L&6 mention psychometric work, which, of course, has shown (albeit in 'laboratory' settings) how people's sensibilities towards risk issues reflect a range of subjective associations and connotations often out of kilter with 'objective' assessments. More recent 'interpretative' risk perception studies (Zonabend 1993; Irwin, Dale, and Smith 1996; Horlick-Jones, Sime, and Pidgeon 2003; Horlick-Jones, 2005a; Timotijevic and Barnett 2006) and work in medical sociology (Silverman 1994; Bloor 1995; Maynard 2003) and socio-linguistics (Candlin and Candlin 2002; Myers 2007), has highlighted the central role of meaning, social interaction and context in shaping the sense that people make of given risk issues. My own recent work has suggested a possible means of relating psychometric and interpretative approaches to risk perception studies, by focusing on the constraints and opportunities afforded by the material character of risk issues to sense-making processes in specific social settings (Horlick-Jones 2007). Put quite simply, people's response to risk issues is not instrumentally calculative in any simple sense; rather, it reflects the specificity of the risk issue in question and of the settings in which it is encountered, and the matrix of social expectations, commitments and understandings in which such people find themselves.

It is worth recalling that others have advanced similar instrumental arguments as a basis for developing universal approaches to risk communication; most notably Calman and Royston's (1997) attempt at formulating a 'Richter Scale' for risk. Significantly, a subsequent study of physicians' views of such proposals for a 'standardized language of risk' found considerable scepticism about their potential utility in communicating with patients in practical clinical situations (Edwards et al. 1998). This matter is also addressed in the ESRC/MRC review on *Risk and Health* (Alaszewski and Horlick-Jones 2002, 2003), to which L&6 refer. This drew attention to the potential for misunderstanding and confusion about what is meant by risk in different operational situations. It argued that different types of knowledge may be relevant in order to fully contextualise decision-making processes for the real-world circumstances in question. This is another way of describing the diversity of risk rationalities that emerge from the 'force of circumstances' of different contexts; a matter I have discussed in detail in the pages of this journal (Horlick-Jones 2005a).

The reader will note that the conceptual resources upon which I have drawn in order to present an alternative view to L&6's are not associated with any of the risk theories to which they refer. Indeed, it is a perspective that is agnostic about the capacity of grand theory to fully capture risk phenomena, precisely because such theories purport to provide generic understandings. Therein lies the problem; for risk issues, the devil is truly in the detail.

Research and its uses

In reading L&6's paper, one gains the impression that they regard everyone in the 'risk business' as being somehow involved in the same historical project. For them, this seems to make the lack of integration between risk communication-related scholarship and practice across different substantive fields all the more surprising. In order to analyse this position, we need to consider a number of inter-related issues.

First, seen from an interpretative perspective, all risk communication processes might reasonably be regarded as attempts to impose a technically accredited form of rationality on target audiences whose likes and dislikes, and behaviours, are regarded as 'irrational'. Clearly there is a strongly persuasive, and possibly coercive, element to this activity. I hasten to add, that in making these observations, I have no wish to label all risk communication initiatives as morally dubious. Rather, I wish to highlight the importance of taking into account their political and moral nature when one seeks to understand the potential for dissemination of risk communication from one substantive area to another, or, for example, the recruitment of scholars to support the risk communication work of government agencies and businesses.

Second, although researchers working in areas concerned with different substantive risk issues are all, in some sense, involved in work on 'risk', it does not follow that these communities of practice need necessarily recognise any kinship or mutual interests. Research in each area reflects the specific character of the risk issues in question; its associated practices, institutions and markets; and its funding base and mechanisms. I suspect that most researchers in health-related areas tend to regard themselves as health researchers rather than risk researchers; and the same kind of argument applies for researchers in food safety, climate change, railway engineering, nuclear reactor design, criminology and social care, and a host of other areas that might be categorised as being 'risk-related'. Of course, the increasing use of risk categories as a regulatory and standard-setting basis for professional practice complicates this picture. In this latter sense, virtually all professionals are now, to some extent, concerned with risk. However, whether these changes could be said to have engendered a shared inter-professional culture of risk is surely in doubt (see Horlick-Jones 2005b; Rothstein 2006).

Third, there is an important distinction to be made between different professional settings for risk research. Although there is a great deal of overlap between academic research, consultancy, and applied and policy-specific research, important differences do exist in terms of the relationships between researchers and policy-makers, and the orientation of different researchers towards the topics of research. Academics and researchers in industry and within government departments are subject to different sorts of incentive structure, and they work in settings with different power structures and cultures. In particular, the nature of academic life creates a particular focus on the need to engage in scholarly publication and debate. Of course, a focus on practical matters certainly does not preclude the production of high quality scholarship. However, it is clear that some scholars' interest in risk is more 'academic' than that of others.

In making these distinctions and observations, I have attempted to indicate how one might start to unpack some of the complexity of social relations within and between different communities of risk scholarship and practice. Seen in this light, perhaps the lack of dissemination of research findings between different substantive areas is not quite so surprising.

The policy agenda

Whilst risk communication continues to be an important consideration for public and private sector bodies, I couldn't help but feel that L&6's arguments reflected a slightly old-fashioned research agenda. Here, I had in mind developments within UK government policy over the last decade.

It is true that in the late 1990s a number of government departments became interested in applying risk communication research findings. A cross-departmental working party, the Interdepartmental Liaison Group on Risk Assessment (ILGRA), for which the Health and Safety Executive provided a secretariat, played a central role in facilitating these developments. It did so by commissioning research and promoting the dissemination of knowledge. This commissioned work included a review of risk communication in government (Buley 1996) and a benchmarking exercise (AEA Technology 1998). Subsequent developments included some notable initiatives and publications concerned with the application of risk communication ideas in health promotion (Department of Health 1998; Bennett and Calman 1999). A year or so later, interest in ILGRA circles had moved on to a specific focus within risk communication studies; namely the Social Amplification of Risk Framework, an approach based on an underlying signals transmission metaphor, and one which will be familiar to many readers of this journal. More work was commissioned, which led, *inter alia,* to some serious scholarly publications (e.g., Petts, Horlick-Jones, and Murdock 2001; Pidgeon, Slovic, and Kasperson 2003).

Whilst drawing on these initiatives, subsequent developments concerning risk communication in government became subsumed within a broader framework of risk-related policy development. The Cabinet Office (2002) published a report which recognised that risk management, which it describes as 'getting the right balance between innovation and change on the one hand, and the avoidance of shocks and crises on the other', was 'now central to the business of good government' (2002, 2). In practical terms, the new governance of risk took two, overlapping forms: the adoption of the private sector machinery of corporate governance, with its underlying concern with control and accountability; and a citizen engagement agenda. The latter reflected thinking in international policy networks about citizen engagement, and seems to have also drawn upon a strand of management thinking that advocated a need to be responsive to stakeholder needs. With these changes a new interest in issues of transparency, stakeholder views, and ethical considerations came into regulatory discourse (these developments are discussed in detail in e.g., Horlick-Jones 2005b; Walls et al. 2005; Rothstein 2006; Horlick-Jones et al. 2007).

More recently, UK government policy-makers seem to have come to recognise that shifting behaviour is far more than a matter of educating citizens and expecting them to 'behave sensibly'. Consider, for example, initiatives which seek to address anthropogenic climate change by promoting sustainable consumption. These have taken a more sophisticated form than being simply about 'public education'. Significantly, they include elements of communication, advertising, incentives, and citizen engagement (e.g., HM Government 2005). Such approaches seem much more in keeping with a recognition of the socially-embedded nature of risk, and the need to attend to its specific features. This seems to me a very positive development.

Conclusions

In this short commentary paper, I first suggested that L&6 make certain unsatisfactory assumptions about the underlying problems faced by risk communication initiatives. Second, I discussed some of the diversity of the risk research and risk practice communities, and indicated that perhaps one should not be too surprised about the lack of dissemination across different substantive risk areas. Third, I suggested that L&6 present a slightly old-fashioned risk communication agenda; one that arguably has not kept pace with some recent developments, in particular concerning citizen engagement.

Despite these reservations, one cannot help but feel a certain amount of sympathy for L&6's arguments, which reflect a serious attempt to promote the spread of effective risk communication practice between various risk communities. Perhaps most notably absent from their analysis is a recognition of the potentially pivotal role of certain organisational structures in brokering knowledge, and so facilitating its dissemination. In this respect, as a sideshow to the grand themes of British history, the work of ILGRA may come to be seen as having played an invaluable role in the development of risk communication practice within British state institutions.

References

AEA Technology. 1998. *Risk communication – benchmarking in government*. London: Health & Safety Executive.

Alaszewski, A., and T. Horlick-Jones. 2002. *Risk and health: Review of current research and identification of areas for further research*. Report to Economic & Social Research Council and Medical Research Council, University of Kent and Cardiff University, Canterbury and Cardiff.

———. 2003. How can doctors communicate information about risk more effectively? *British Medical Journal* 327: 728–31.

Bennett, P., and K. Calman, eds. 1999. *Risk communication and public health*. Oxford: Oxford University Press.

Bloor, M. 1995. *The sociology of HIV transmission*. London: Sage.

Buley, P. 1996. *Risk communication by government departments*. London: Report to the Health & Safety Executive.

Cabinet Office. 2002. *Risk: Improving government's capability to handle risk and uncertainty*. London: Strategy Unit Report, Cabinet Office.

Calman, K., and G. Royston. 1997. Risk language and dialects. *British Medical Journal* 315: 939–42.

Candlin, C., and S. Candlin. 2002. Discourse, expertise and the management of risk in healthcare settings. *Research on Language and Social Interaction* 35: 115–37.

Department of Health. 1998. *Communicating about risks to public health: Pointers to good practice*. London: The Stationery Office.

Edwards, A., E. Matthews, R. Pill, and M. Bloor. 1998. Communication about risk: The responses of primary care professionals to standardizing the 'language of risk' and communication tools. *Family Practice* 15, no. 4: 301–7.

HM Government. 2005. *The UK Government sustainable development strategy*. Cm 6467. London: The Stationery Office.

Horlick-Jones, T. 2005a. Informal logics of risk: Contingency and modes of practical reasoning. *Journal of Risk Research* 8, no. 3: 253–72.

———. 2005b. On 'risk work': Professional discourse, accountability and everyday action. *Health, Risk & Society* 7, no. 3: 293–307.

———. 2007. On the signature of new technologies: Sociality, materiality and practical reasoning. In *Risk and the public acceptance of new technologies*, ed. R. Flynn and P. Bellaby, 41–65. Basingstoke, UK: Palgrave Macmillan.

Horlick-Jones, T., J. Sime, and N. Pidgeon. 2003. The social dynamics of environmental risk perception: Implications for risk communication research and practice. In *The social amplification of risk*, ed. N. Pidgeon, P. Slovic, and R. Kasperson, 262–85. Cambridge: Cambridge University Press.

Horlick-Jones, T., J. Walls, G. Rowe, N. Pidgeon, W. Poortinga, G. Murdock, and T. O'Riordan. 2007. *The GM debate: Risk, politics and public engagement*. London: Routledge.

Irwin, A., A. Dale, and D. Smith. 1996. Science and Hell's kitchen: The local understanding of hazard issues. In *Misunderstanding science? The public reconstruction of science and technology*, ed. A. Irwin and B. Wynne, 47–64. Cambridge: Cambridge University Press.

Löfstedt, R., and P. 6. 2008. What environmental and technological risk communication research and health risk research can learn from each other. *Journal of Risk Research* 11, nos. 1/2: 141–67.

Maynard, D. 2003. *Bad news, good news: Conversational order in everyday talk and clinical settings*. Chicago, IL: University of Chicago Press.

Myers, G. 2007. Commonplaces in risk talk: Face threats and forms of interaction. *Journal of Risk Research* 10, no. 3: 285–305.

Petts, J., T. Horlick-Jones, and G. Murdock. 2001. *Social amplification of risk: The media and the public*. Sudbury: HSE Books.

Pidgeon, N., P. Slovic, and R. Kasperson, eds. 2003. *The social amplification of risk*. Cambridge: Cambridge University Press.

Rothstein, H. 2006. The institutional origins of risk: A new agenda for risk research. *Health, Risk & Society* 8, no. 3: 215–21.

Silverman, D. 1994. Describing sexual activities in HIV counselling: The cooperative management of the moral order. *Text* 14, no. 3: 427–53.

Timotijevic, L., and J. Barnett. 2006. Managing the possible health risks of mobile telecommunications: Public understanding of precautionary action and advice. *Health, Risk & Society* 8, no. 2: 143–64.

Walls, J., T. Horlick-Jones, J. Niewöhner, and T. O'Riordan. 2005. The meta-governance of risk: GM crops and mobile telephones. *Journal of Risk Research* 8, nos. 7/8: 635–61.

Zonabend, F. 1993. *The nuclear peninsula*. Cambridge: Cambridge University Press/Editions de la Maison des Sciences de l'Homme.

Meaningful communication among experts and affected citizens on risk: challenge or impossibility?

Anne Bergmans

Department of Sociology, University of Antwerp, Belgium

Experience to date demonstrates that it remains challenging to engage experts and concerned citizens in a meaningful and mutually comprehensive dialogue on complex and technical risk-bearing projects. In search of an explanation we found Niklas Luhmann's interpretation of modern society very useful. Luhmann describes modern society as the aggregate of more or less self-sufficient functional subsystems becoming more and more isolated from each other in a spiral of progressive specialisation. With each system developing its own expectations, language, rationality and ways of observing and interpreting reality, communication between systems becomes progressively problematic; according to Luhmann, even impossible. Contrary to Luhmann, however, we consider communicating human beings (and not communication in itself) the constituting elements of society. From that perspective we see a connection with Ulrich Beck's thesis on modern society as an individualised risk society and his call for 'reflexive science and decision making'. We will use Beck's negotiation model to build communicative bridges between (Luhmann's) social (sub)systems, in particular, by engaging as many concerned parties as possible. Further, we will argue that the Belgian experience with the siting of a radioactive waste repository demonstrates that the creation of an environment in which experts and citizens can enter into dialogue as individuals, rather than as representatives of interests or (scientific) disciplines, can help bridge differences in the rationality and jargon of systems, and result in finding common ground.

Introduction

Discussions between experts, decision-makers and lay people in different fields of environmental and technological risk often lead to conversations of the deaf; even though a careful listener can detect many communalities in the arguments. From our own experience,[1] it often felt as if the participants in the discussions lived in different worlds and lacked the ability to communicate in a meaningful way. What are the consequences of these observations for the relatively widely accepted view that for any kind of policy to be effective and legitimate the affected people should be involved in the decision-making? How meaningful can participation be if communication between various actors involved proves to be so difficult?

We consider that each individual is linked to a broader network of people and resources, which affect his or her competence in taking part in any participatory set up. Could these networks help to explain why some citizens can engage in meaningful communication with some experts, while others keep getting their wires crossed? Surely there must be more to it than a mere difference in personal capacity for making and criticising validity claims in a discourse?

In this paper we will draw on both case study findings and the combined insights of Beck and Luhmann to create a framework that should help understand why constructive dialogue between various actor groups is needed and yet so difficult to realise.

Reconciling Beck and Luhmann: introducing the self-organising society as a means to build communicative bridges between social systems in a functionally differentiated society

Beck and Luhmann describe modern society as increasingly dependent on decisions, burdened with an uncertain future, and devoid of an overall social steering mechanism. Both see this as a consequence of modernisation, which they consider an ever advancing process that leads to a saltatorial emergence of new types of society (e.g., Beck 1992, 1995; Luhmann 1993). But both authors differ distinctively in their explanation of this phenomenon.

The world according to Ulrich Beck

For Beck the process of modernisation is driven by processes of individualisation and technological progress which have reached the limits of their possibilities and have turned our society into a 'risk society'. Changes in society are in his view instigated by (social) conflict.

Modern society=an individualised society

According to Beck, society becomes more and more an aggregation of individuals:

> People are being freed from the certainties and standardized roles of industrial society. Men are no longer automatically fathers, fathers no longer the sole wage earners, and sole wage earners often no longer husbands. In earlier generations social stratification, income, profession, spouse, and political attitude were generally a unified world, today this biographical package is disintegrating into its components. (Beck 1995, 134)

While the existing institutions lose meaning and influence, individuals gain more freedom of choice and self-determination. This however places the individual on an 'endless chain of decisions', eventually turning a freedom of choice into an 'institutionalised compulsion to take decisions' (Laermans 1999, 211–2). Having done away with institutions that give meaning to life, modern man and woman have to decide for themselves and this repeatedly, over and over again. This evolution calls for a new type of community, which

> … no longer rules from the top down, nor can be poured into people through the funnel of the functionalist model of socialization. Instead, it must be waited for, won, fought out, invented and negotiated from below. (Beck 1995, 139)

Laermans therefore points to the fact that a process of individualisation creates a greater need for negotiation: shared meanings, expectations and interpretations of reality will still be possible, but they become more reflexive, permanently renegotiable and only valid for those directly concerned (Laermans 1999, 212–13). Negotiation thus becomes the imperative, the lead model for decision-making in modern society.

Modern society=a risk society

In Beck's view the ever progressing technological development has led not only to an affluent society, but also to a society characterised by intangible and uncontrollable risks constituting a global threat to society. Accidents and threats related to technology form an inherent part of everyday operational management because of the way we have organised our industry, regulatory framework, etc. (Hajer and Schwarz 1997, 12). For Beck, the nuclear accident at Chernobyl in 1986 is the emanation of this phenomenon, the definite proof of a fundamental transition to a new type of society (Beck 1997a). In this society we experience a change in logic from wealth distribution (*a distribution of goods*) to risk distribution (*a distribution of bads*), which Beck sees emerging in Western society since the 1970s (Beck 1992; Hajer and Schwarz 1997). Conflict in society will no longer be about gaining wealth and scarce commodities, but about how to 'limit and distribute away the latent side effects of modernization' (Beck 1992, 19).

The world according to Niklas Luhmann

For Luhmann, society is a system, a 'selective arrangement of related elements' in view of reducing complexity (Laermans 1999, 55). Social systems emerge 'whenever the actions of several persons are meaningfully interrelated', distinguishing them from the environment surrounding them (Luhmann 1982, 70). Because environments offer more possibilities than a system can use, systems aim at reducing complexity. According to Luhmann society is the 'all-embracing social system' (Luhmann 1982, 73), the ultimate social system, incorporating all other social systems, beyond which no social environment exists (Laermans 1997a, 33).

Changes in society, in Luhmann's view, know a salutatory development, and the transition from one society type to another is characterised by radical changes in the means of communication and structural organisation of a society (Laermans 1997b). Modernisation, or more general sociocultural evolution, is in that respect seen as the (ever uncompleted) result of a continuing process of 'functional differentiation' (Luhmann 1982, 77).

Modern society=a functionally differentiated and polycentric society

In trying to reduce complexity, society has become more and more specialised, leading to the development of various systems within society focusing on a particular aspect of our complex world. Consequently a number of bigger subsystems emerged in society, such as the legal system, the economy or the political system, forming sub-structuring entities that organise social life around a particular function or issue (Blom 1997; Dobbelaere 1997; Laermans 1997b). In improving themselves to deal with their particular social questions, society's function systems have grown slowly apart, developing their own rationality, framing of reality and language, and becoming to a large extent autonomous. Autonomous and yet strongly inter-dependent, as many issues and problems of society are so complex that they cannot be solved within one particular function system:

> Precisely because function systems cannot replace one another, they support and burden one another reciprocally. It is their irreplaceability that imposes the continual displacement of problems from one system into another. (Luhmann 1989, 110)

In this way, social systems find themselves in a relation of peers, with no hierarchical structure above them; making for example the political system just one among numerous others. As a result, society today is faced with a decentralised decision structure, without a centre that guides, co-ordinates and controls; leaving little room for a conscious guiding influence from one social system over another (Blom 1997; Laermans 1997b; Luhmann 1989, 49–50).

Modern society=a risk-oriented society

While Beck explicitly assumes that the nature and magnitude of the risks society faces have changed fundamentally as a consequence of technological development, Luhmann merely acknowledges a fundamental change in the way society today deals with risks and perceives them. Modern man's preoccupation with risk, Luhmann attributes to three main causes (Luhmann 1993):

(1) *An increased dependence of society on decision-making*: The autonomy of social function systems means that they independently determine how to respond to challenges from the outside. In addition, there are many different possible responses to challenges, some obvious, some less obvious. The chosen response is thus only one of the possible responses and people are aware that alternative options or responses were or are available (Blom 1997, 260–3). Therefore, society becomes more contingent and dependent on decision-making. People become more aware of the possibility of taking the wrong decisions.

(2) *The dominance of a linear notion of time*: In modern society we tend to think of time in terms of 'before' and 'after', making us aware that the future is open, undefined and unpredictable (Luhmann 1993, 34–8).

(3) *The contrast between decision-makers and those affected*: Faith and coincidence are no longer looked at as driving forces in society. More and more, events and occurrences are understood as related to or consequences of decisions. Since it is impossible to have everybody participate at all times in every decision, there will inevitably be a discrepancy between those who make the decisions and those affected by them (Luhmann 1993, 105). When those affected by decisions experience a lack of control over their own situation, they could feel threatened by other people's decisions. Hence an increased risk-consciousness in today's society.

While Beck's theory of individualisation has its merits, we see more explanatory power in Luhmann's continuing process of functional differentiation to help understand the need to involve a broader range of knowledge bases than 'just' science and politics in the decision making. In the following sections, we therefore would like to build on Luhmann's systems theory first, before re-introducing Beck as a possible way out of Luhmann's relatively pessimistic view of the opportunities for social systems to interact.

How systems frame opportunities for communication and for structuring reality

Contrary to Luhmann,[2] we consider communicating social units (either individuals or collective actors) and not communication in itself, the constitutive elements of society. We interpret communication thus as any form of interaction, verbal or non-verbal, between at least two social units.[3]

Each time a social unit enters into communication with another unit there is 'social system generation', as well as formation of society (Laermans 1999, 232). This does not mean that with each act of communication a new function system arises. The broader systems in society (e.g., science, economics, law, ...) prescribe to a large extent the framework within which communications can occur. Within these overarching social systems new systems can develop, with a temporary or long term character. But these new systems are grafted onto the existing framework, following its prescribed expectations and rationality. In this way social systems frame and limit possibilities for future communication within the system's boundaries. For example, whether people meet in a court of law, in a formal meeting with colleagues, in their sports club, or in a pub, will be determining for the context within which they can enter into dialogue.

Interaction between social systems

Acting relatively autonomously social systems can often unintentionally and unconsciously present other systems with a fait accompli. Think, for example, of unforeseen and unwanted side effects on man and the environment from technological development. In a context of functional differentiation, the system in which these chains of reactions originate, as a rule, cannot adequately evaluate the effect of its decisions on other systems (other social systems as well as external factors like nature or the environment) or on all possible interferences that these might bring about. Again this demonstrates the need to involve a broad knowledge base before, for example, introducing new technologies.

Contrary to Luhmann, we do not consider social systems to be completely closed,[4] as there will always be interfaces and forms of interaction between them. Individual people, for instance, will always relate to various social systems: the politician could at the same time be a lawyer, a wife and mother, growing biological vegetables in her hobby garden, hence being part of different social systems. This way individuals can bring framings and rationalities from one system into another. Non-human actors that co-create communication can also serve as interfaces between function systems. One example could be a book written by a physicist, read and commented on by a biologist, being picked up by an environmentalist group introducing it in the political debate. Such interfaces can stimulate information, framings, worldviews, jargon, etc. to be picked up by other systems and pave the way for meaningful communication between divergent social systems.

It seems self-evident that the more specialised systems become, the more difficult it will be to find communicative bridges. A particular example is the difference in perception of risk between scientists or experts and the general public. In the following section we will put the focus on the notion of risk as a communicative problem and suggest possibilities for overcoming this.

On the notion of risk in a risk-oriented society

Today it seems widely acknowledged that the difficulty in engaging experts and lay people in meaningful communication on the subject of risk (or a particular risk situation) cannot simply be overcome by raising the level of (scientific) knowledge on the side of the public. The actor groups use a different kind of rationality, a different way of dealing with knowledge and interpreting (factual) information. In other words, they adhere to different social systems, and, accordingly, have developed different ways of framing the problem and the notion of risk in general.

Linking risk to decisions

Interpreting risk as a statistical 'fact', open to objectification, and, consequently, treating differences in risk perception as a rationality deficit between 'true' and 'perceived' risk, is not adding value. Luhmann and Beck, however, offer a more useful approach, linking risk to decisions and choices of action. Both define risk as a threat, or potential damage, as a possible consequence of a decision or choice of action (Beck 1995; Luhmann 1993). Risks are always considered avoidable, without a particular decision or choice of action the risk would not have been there.[5]

Modern man's preoccupation with risk

Since every decision entails a certain risk, there is no such thing as risk-free behaviour. The more we realise this, the more risk-conscious we become:

> The more we know, the better we know what we do not know, and the more elaborate our risk awareness becomes. (Luhmann 1993, 28)

An elevated level of knowledge combined with a strong sense of lack of control over a decision thus forms a solid basis for risk-consciousness. It is not because one knows more that one necessarily feels safe; on the contrary.

Without taking position in the debate on whether there actually are more and different kinds of risk today than there were before (as Beck claims), the observation stands that we appear to become more aware of the existence of risks. We are therefore inclined to follow Luhmann's description of today's society as a 'risk-oriented society', rather than Beck's labelling of it as a 'risk society'. Our preoccupation with risk influences how we organise our lives and work; how we maintain relationships; how we manage our finances and look at issues of life and death. Many people today desire almost absolute control over their private lives. Because we believe that we hold our faith in our own hands, we tend to leave as little as possible to coincidence unless by explicit choice (e.g., taking up a risky sport or investing on the stock market). However, this risk-consciousness appears to increase to almost excessive proportions in cases of environmental and technological risks. Why?

The particularity of environmental and technological risks

The lack of control is an important characteristic for situations involving environmental and technological risks. Although there is a strong link to human intervention, it is often not possible to point to one particular decision or a particular culprit. Beck speaks in this regard of 'everyone (being) cause and effect, and thus non-cause', which in a complex industrialised world leads to a state of 'organized irresponsibility' (Beck 1992, 32–3, 50). Hence our sensitivity concerning this kind of threat.

Communicating on environmental and technological risk in a functionally differentiated and risk-oriented society

Summarising, we could say that modernisation has led to an 'over-specialised' society. The autonomous behaviour of social systems in an otherwise strongly interdependent society has also resulted in unintended and unforeseen impacts on other systems. This appears to be the end point of what can be achieved by specialising ever further. In the field of science and technology, for example, autonomous system behaviour in many cases leads to the introduction of new procedures and techniques, while the risk assessments are solely conducted from an expert's (often only from one particular discipline's) point of view. This behaviour does not stem from malicious intent, but is simply a consequence of 'systematism', of being trapped in 'systemic thinking'. So how do we get out of these traps? How do we beat the systems?

Complex and risk-related problems, like nuclear power and waste, climate change or BSE, are often described as needing a multi-disciplinary or multi-actor approach. Beck in this respect points out that statements on risk always contain statements on how people want to live, and that science and politics do not have the monopoly of wisdom on this issue (Beck 1992). Deciding on risks in Beck's view therefore is a matter of overcoming a 'struggle among rationality claims, some competing and some overlapping' (Beck 1992, 59). In terms of our functionally differentiated society, this would imply recognition of the mutual interdependence of systems, in turn leading to cross-system cooperation and communication. According to Luhmann an impossible task, but we prefer not to share his pessimism and suggest to rely on the 'careful listeners' identified in the very first paragraph of this paper to create openings for meaningful communication across different systems.

Beating the systems

Of all briefly discussed interfaces as a basis for interaction between social systems, the individual person appears to be the most promising in terms of creating and grasping opportunities to overcome the logic of the system and to seek alliance with other framings. In that respect, individualisation could be understood as both a consequence and a way out of (too much) functional differentiation.

Individualisation as a consequence of functional differentiation. We see social systems crosscutting relations between individuals. While collective actors and organisations could adhere to one single function system, this could never be the case for an individual. As the example of our vegetable-growing lawyer demonstrates, individuals always relate to various social systems. Today, social relations are determined by a multitude of environments (before, one could argue, people acted and communicated in a more homogenous environment). Assuming that an individual's actions are strongly determined by their context, we have, as a consequence of modernisation, ever more diverse contexts to found our actions on. This is because through functional differentiation the interfaces between the various communications among individuals narrow and their mutual relations become more 'specialist'.

Individualisation as a way out of (too much) functional differentiation. Having various contexts at their disposal to found their actions on, individuals establish networks of relations across the various function systems within which they operate as communicating social units. Individuals thus become the main connecting link between each of these networks.

The main difference between Beck and Luhmann is that Luhmann mainly described what he saw, while Beck mainly presents projections of a future that he sees dawning, focusing thereby on how he would like to see things develop. But it is not so much the existence of systems that poses a problem; it is the fact that we tend to ignore this. Pretending that systems are not there, creates situations in which, for example, politicians believe they can prevent multinational enterprises from closing down local branches, or in which scientists put monopolist claims on rationality. What we are looking for is not a simple abolition of society's function systems, but the creation of opportunities for communication between social systems and for (self-) reflection within these systems. Self-reflection, as understood by Beck (e.g., Beck 1994), is self-confrontation and the ability to question and re-think the own system's logic. By trying to understand systems and their role in relation to each other, we should be able to escape complete entrapment in one particular system's logic and broaden opportunities for meaningful communication and creative solution seeking across systems. Individuals can be driven by the logic of a dominant system (for instance related to their job or education), but they all carry in them many different fragments of the framings in other systems to which they also relate. By stimulating self-reflection, opportunities could be created to look beyond the framing of the dominant system, and to open up to other ways of looking at a particular situation.

A round table model for negotiation

Beck does not offer much tangible material on how to bring reflexive society about. Mostly he refers to a self-organisation of society or incidental and bottom-up organisation of individuals participating in political decisions from outside the system (e.g., Beck 1994, 1997b). The only more concrete idea we found for putting reflexivity into practice, was the suggestion of a 'round table "model for negotiation"' (Beck 1994) as a means to build communicative bridges between different subsystems by bringing different rationality claims together in an open forum, where all are considered equal and none is forced to yield to the techno-scientific rationality. In order for it to be effective, such a model, according to Beck, would have to at least 'break up monopolies and delegate power' and be capable of 'transforming negotiation between experts and decision-makers behind closed doors into a public dialogue' (Beck 1994, 28–30).

To truly make this possible, however, people participating in such a setting must be prepared to reconsider their '*a priories*', to open their minds to other rationales and framings of reality. And, as we mentioned at the beginning of this paper, this appears to be a very difficult thing to do. Difficult, but not impossible, as the case described below indicates.

A case study perspective: discussing options for a low and intermediate level radioactive waste (LILW) repository in Belgium

Following a failed, strong top down, and expert driven approach by the Belgian radioactive waste management agency (ONDRAF-NIRAS) to site a LILW-repository, an attempt was made to discuss different options with concerned parties on the local level. For this purpose, local fora (called 'partnerships') were set up in communities already hosting nuclear activity and willing to discuss the possibility of hosting a repository facility. Three such partnerships were established: the partnerships STOLA and MONA in the neighbouring municipalities of Dessel and Mol (Flanders region), and the partnership PaLoFF in the municipalities of Fleurus and Farciennes (Walloon region). Eventually only Dessel and Mol were candidates as host community for the repository facility. Both municipalities already host some of the LILW in interim storage facilities, which was an important incentive for them to join ONDRAF-NIRAS in its search for a more lasting solution.

Our research team actively participated in the design and establishment of these local partnerships, in particular in the Flanders region, and worked in joint collaboration with the Department of Environmental Sciences and Policy from the University of Liège, for the Walloon region. The adopted research approach can be situated in the tradition of action-research. We did not participate in the negotiations, but facilitated the process of establishing the partnerships, and closely followed their activities and developments, as participatory observers and external guardians of the process. In this respect, we had regular contact, both formal and informal, with both local people participating in the partnerships and representatives from ONDRAF-NIRAS. In what follows we will mainly draw on experience (from 1999 till present) in the partnerships of STOLA and MONA, on analysis of all written communications (minutes of meetings, reports, etc.) from both partnerships over the period fall 1999–summer 2005, and on a series of in-depth interviews with participants (summer–fall 2006).

Jointly developing a proposal for an integrated repository project

The partnerships were created in potential host communities as fora for communication between experts and citizens in developing a LILW-repository project. Their mission was:

> ... to conduct studies and develop an integrated project proposal for the disposal of LILW that can be presented to the federal government. In this respect, the organisation becomes the core for a thorough investigation of the site, the community and (if applicable) also the larger region. This investigation will be carried out through studies covering technical and safety issues, ecological, socio-economic and socio-cultural issues, as well as studies in the field of urban planning. The organisation will serve as a platform for structured project negotiation and local deliberation. The organisation will also ensure communication with and provide information to the local public on all its activities and all relevant studies carried out by ONDRAF-NIRAS. (Partnership bylaws, art.3)

In practice the partnerships studied the possibility of siting a repository facility in their community, both from a technical and safety perspective, as well as in terms of social acceptance. STOLA and MONA developed a proposal for an integrated repository project, which was put to the respective Municipal Councils. In both cases the Council decided to volunteer to become a host community.

The concept of 'integrated repository project' refers to a project that does not only serve the national cause of dealing with the LILW, but also brings added value to the host community:

> ONDRAF-NIRAS is convinced this repository project will not only serve national goals, but that at the same time it can bring added value to the area where it will be located. ... (therefore) The repository project has to be developed in such a way that it creates a positive net impact for the region. A well outlined project creates new prospects: instead of being a burden to the local community, the repository should offer an important potential, whether in terms of economic, socio-cultural or ecological development. (ONDRAF-NIRAS quoted in Anfrie et al. 1998, 2)

In practice, the proposed projects consisted of a repository, and an accompanying local project of a socio-economic or other nature. In order to ensure a win-win situation, both aspects were developed as interdependent and inseparable.

From an *organisational perspective*[6] (see Figure 1) the partnerships were set up as not-for-profit organisations, run by representatives from the local (municipal) councils, local civil society and ONDRAF-NIRAS; the latter providing the necessary funding. Representatives from all founding partner-organisations form the *General Assembly*, supervising the organisation's activity and having the final decision on any advice or (project) proposal put forward by the partnership. This assembly is composed of representatives from the main local interest groups and municipal advisory councils. It is intended to offer a balanced representation of the various interests and points of view relating to a potential repository project in the community. The General Assembly appoints an *Executive Committee* in charge of the day to day management and co-ordination of the organisation. Several *working groups* discussed different aspects of the implantation of a LILW-repository in the community. The working groups are composed of representatives of the founding organisations as well as interested non-organised citizens. At the time, each group was attended by one or two permanent members representing ONDRAF-NIRAS, having a particular expertise regarding the topics discussed. The ONDRAF-NIRAS representatives made the local members acquainted with their plans and views on how the repository should be set up. They entered into discussion about the reasons behind the technical options proposed and about possible alternatives. The working groups considered additional relevant research, evaluated the need for additional studies, and invited various experts to participate in the debate. They reported regularly to the Executive Committee and drafted the various constituent parts of the integrated repository project. All local people participated on a voluntary basis. Each partnership therefore recruited two *project coordinators* to support the organisation.

Particularity of the working groups

As a consequence of their composition, the General Assembly and Executive Committee were the most appropriate places to represent and defend different

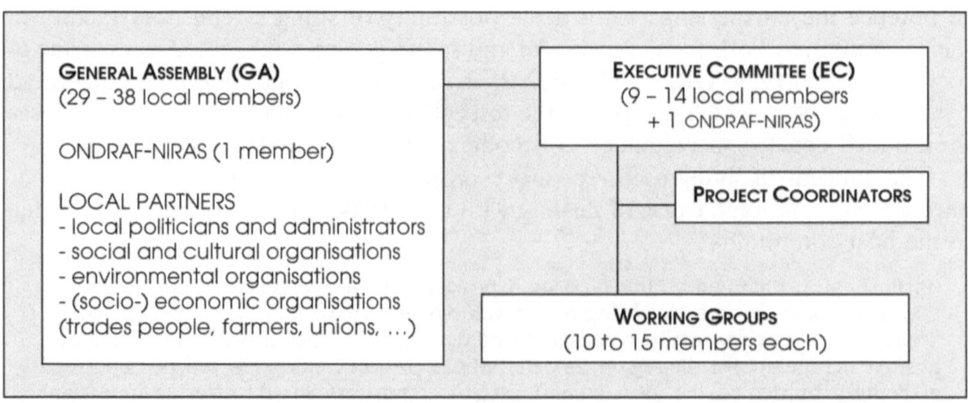

Figure 1. Organisational structure of a local partnership.

interests. Here the project was to be given final approval. The participants in the working groups, however, positioned themselves mainly as interested individuals, as representatives of the community in search of a common good, the good of 'their community'. In a group interview with people participating in STOLA, respondents spontaneously indicated that 'the strength of the working groups had been the joint effort in striving for the common good of the community'.[7]

None of the respondents had the impression that participants had acted in favour of particular group interests. This reference to the personal interest of concerned citizens manifestly returns when talking to participants about their role in the process. When questioned in 2005 about their main reason to participate, the answer 'I considered it my civic duty' featured in first place in MONA and PaLoFF and among the four main reasons in STOLA.[8] Furthermore 13 out of 19 involved citizens, interviewed in 2006, explicitly considered a personal interest and concern as affected citizen the main motivation for their participation.[9] Or as one person put it:

> I think this of great importance to the people in my community, and I wanted the interest of the inhabitants of Mol to be defended in a decent manner. (Local participant in MONA)

So where the General Assembly and the Executive Committee were more about finding allies to support a particular decision, the working groups were mainly about creating networks across different positions and knowledge bases and the identification of common ground. For the purpose of this paper, we will focus on the activities of the working groups dealing with the issue of implantation and design of the repository (further referred to as the working group(s) I&D) in the partnerships STOLA and MONA. One of the many obstacles in developing a common view in this group was the differences in definitions and framings between experts and lay people of concepts like 'safety' and 'retrievability'.

Shifts in framings and roles

All working groups met on a regular basis for about four years before presenting their final conclusions.[10] Over this period of time, they developed a specific dynamic. In particular in the working groups I&D, a clear shift in role appeared after a while. Although all agreed that ONDRAF-NIRAS should retain all responsibility for developing (and eventually realising) the repository project, the situation soon developed to one where the partnerships acted as the patrons of the project, positioning ONDRAF-NIRAS in the role of architect commissioned for its development. This continued until the partnerships put their propositions to the Municipal Council for approval. This shift in roles can be deduced from the way in which the working group gatherings were set in scene and from the vocabulary used in the meetings' minutes, making ample reference to particular roles and responsibilities in the field of building and constructing. About a year onward in the process, the STOLA working group I&D, for example, paused to reflect on its own role and came to the conclusion that:

> STOLA is not a designers group, but takes up the role of a 'patron' commissioning a repository facility. ONDRAF-NIRAS acts as architect, presenting STOLA with one or more proposals and a number of options. The partnership will scrupulously evaluate and critically assess the proposed options. At any time, STOLA can subject these proposals

to external advice from experts of its own choosing. STOLA will not make any technical choices as such, but will decide to agree (or not to agree) with the technical proposals from ONDRAF-NIRAS.[11]

In MONA this evolution was less explicit, but here too, the working group I&D, for example, drafted 'order books' through which ONDRAF-NIRAS was to account for the presented technical choices (Bergmans 2005, 602–4; Bergmans, Van Steenberge, and Verjans 2006, 56). In the beginning the introduction of these concepts by the local partners was largely symbolic, but in the end it proved to be a lasting change in role patterns.

When questioned in the summer of 2005 about the impact of the local participants on the process, many ONDRAF-NIRAS collaborators felt the local people had to a large extent 'bought into' their framing of the problem. According to the waste managers, the local people eventually agreed to most of the initial proposals, after long and fruitful (but from an expert's view sometimes too long and unnecessary) discussions.[12] However, although the position of the local participants might have evolved slightly more in the direction of ONDRAF-NIRAS than the other way around, there has undeniably been an evolution on both sides, as the respondents from ONDRAF-NIRAS also admitted:

> On the technical side, a number of things did change. You no longer start developing a priori your own ideas without contemplating about how the local people will react and what their concerns for this issue are. ... You no longer tell the people 'this is the way it is', but you try and explain things to them, which forces you to better prepare. ... The pressure to present things in a comprehensive matter forces you to get more to the bottom of it, which raises your own understanding of them.
>
> In a first instance we did go there hoping to make the people see through our eyes. So you do try to be persuasive and to 'sell' your stuff. But then you come to develop an attitude of 'let's put all the elements on the table, discuss them and see where that gets us'. You do become more flexible and compliant. (ONDRAF-NIRAS collaborators)[13]

Local participants in the STOLA group discussion indicated they felt they had been listened to, and that they had been able to contribute in a constructive way to the outcome of this 'joint venture'.[14] Furthermore, 94% of the respondents from MONA and STOLA to the summer of 2005 questionnaire shared this appreciation. Up to 75% were convinced they had been able to influence the project at certain points.[15] With the exception of two people, all respondents agreed that it had been worth the while to dedicate all this time and effort to this issue.[16]

Meaningful communication?

As the quote from the ONDRAF-NIRAS collaborators suggests, there has been a change in the communicative style of the experts that participated on a regular basis. Most of them appear to have been able to develop a common framework with the lay people in the group. From attending meetings, reading through the minutes and talking to both the waste management agency's experts and local participants, it became clear that the former over time developed a different kind of language.

> Quite a number of these experts have climbed down from their ivory tower and taken the trouble to explain things in an understandable way.
>
> There were some changes noticeable, both with ONDRAF-NIRAS and the nuclear companies. It must have been something completely new for them: they were the experts

> who had to 'lower their standards' to talk to ordinary citizens, and take into account that there are other people who do have an interest in this. I assume they had to change their ways to come and talk to us. (Local participants in the partnerships)
>
> Preparing for these meetings was a process in itself for us. You had to climb down from your ivory tower, adapt your language and change your way of building an argument.
>
> It was a gradual process of adapting to each other's needs. (Experts from the nuclear research centre, SCK-CEN, who participated in several working group meetings)

Similarly, the local participants became progressively more knowledgeable in technical matters, and gradually managed to understand more 'engineers lingo' than at the outset.

> I often received very pertinent questions, the kind that made you wonder why nobody had asked these questions before. You don't expect that from a lay audience.
>
> To be honest, I was very surprised that so many people went through the trouble of reading our reports. (SCK-CEN experts)

Needless to say that the personal communicative skills of each of the participants and their ability to listen to others played a role in this process. But three other elements also seem to have had an impact.

A first element is *the recognition of the local people that the problem was theirs.* Since they were the ones affected by it, it was as much their problem as it was the experts' and waste managers'. This feeling of being personally affected appears to have stimulated people to go through great lengths in trying to understand difficult technical issues, and convincing the experts to take them seriously.

> I participated because I wanted to understand it, to see from close by how the project was developed. To understand, so that I could pass that on to my family, friends, neighbours. It is something real that is happening in our community, so we all should be aware of it.
>
> I wanted to make sure they knew we were keeping an eye on them. I think it is our duty to protect the interests of future generations.
>
> The time investment was not the problem. What was important, was that people felt concerned and that they really felt they could create and achieve something together. (Local participants in the partnerships)
>
> I was astonished by the enthusiasm and motivation of these people; and by the time and effort they put into it, trying to comprehend these issues. ... The level of involvement was really quite high. Ultimately, I think, this led to a far richer project.
>
> The intensity with which they followed our work ... the effort and engagement of those people, that was truly amazing. ... It was clear that there was great consciousness about the problem among the local people. This made them prepared to make the effort, but also it made them more receptive for 'our message'. I fear that with people who live further away, who feel less personally affected, you can never achieve this. They will be more inclined to take principled positions, leaving no room for dialogue. (SCK-CEN experts)

A second important element in the process of mutual recognition was *personal contact and group dynamics*. After a while, the experts from the agency and others that participated on a fairly regular basis were seen as full members of the group, almost as local people.

> Very important and much appreciated was the role of the ONDRAF-NIRAS representative in our steering committee. He has always taken a very open position. He was there as representative of ONDRAF-NIRAS, but I often felt he was even more a person from MONA. (Former president of MONA)

The intensity and frequency of the meetings encouraged this, as did the field trips and the 'after meetings' drinks that, particularly in STOLA, became an institution in itself. Problems that had emerged during a meeting often found a solution during such 'post-mortem'.

> In Dessel we got a relatively good idea of what people thought and how they felt about our input. This was because the meetings always ended over drinks in the kitchen. Then we sat and talked a bit more and often people came up with questions they had been a bit reluctant to ask in the bigger group before. After a while you did feel trust growing that way. (SCK-CEN expert)

Through these informal contacts in the margin of the working group activities, local people and agency representatives learned to know and appreciate each other better, resulting in mutual understanding of each others' opinions, no matter how conflicting they remained.

> The approach of the waste manager has changed. Opening up the issue and using participatory tools has become normal for them. Before you were snarled at and fobbed off. (Local participant in MONA)

> You can't just tell people to trust you. Trust you have to earn and that is not about trying to convince them. Trust only develops when people can actually comprehend your line of reasoning. (SCK-CEN expert)

One particular example is that of the ONDRAF-NIRAS representative in the MONA working group I&D and one very principled and critical local environmental activist who found out they shared the same, rather unusual passion of playing the bagpipes. It did not resolve (all of) their differences, but to some extent it made them less insurmountable. Or, as one person who participated in a number of these meetings indicated: 'in the end they seem to have become the best of mates and this clearly affected their view on things' (SCK-CEN expert).

A third element playing a significant part, is *the role of knowledgeable local people as 'communicative bridges'* vis-à-vis the ONDRAF-NIRAS and other experts. One expert we interviewed spontaneously referred to this in quite an explicit way:

> It were the knowledgeable and well respected local people that were key. It was they who could take over if discussions got bagged down in endlessly repetitious arguing. … There was this one person in one of the working groups that was very much against the repository at the outset. But in the end he did re-adjust his position to some extent. That was certainly not entirely our doing. What contributed far more, was the fact that his colleagues in the group began explaining things to him and eventually were able to convince him the concept could be safe and our research was sound. (SCK-CEN expert)

In particular in the STOLA working group I&D, this 'bridging' was clearly visible. The members of this working group had quite different backgrounds, but most had some basic technical knowledge, in some cases directly related to the nuclear sector. However, there was also a house wife and a local fruit farmer, neither of them with any prior knowledge concerning nuclear waste. From the start these people played a pivotal role in attaining a meaningful level of communication between experts and citizens. At the end of their first meeting, the chairman of the group suggested members could put their personal thoughts and considerations to paper for discussion.[17] Both the house wife and the fruit farmer took this invitation to heart and presented the group with an extensive list of issues.[18] Looking back, those formed the basis of a concerted problem definition and a relatively clear-cut group vision on the task ahead. From early in the process, the two least technically knowledgeable participants became the yardsticks for passing from one topic to another: 'It is not before Francine says she understands that we continue', indicated several working group members.[19] From occasionally participating in the meetings, talking to participants, and analysing the minutes, it became clear that both Francine and her male counterpart became the unremitting challengers of the technical experts, demanding them over and over again to better explain, to show them the necessity of particular technical details and to comprehensively demonstrate their safety case. On more than one occasion the other local members of the working group played the role of translators for their 'lay' group members. In particular the local people who (had) worked in the nuclear sector played an important role in bridging the experts and the other local members, as they could respond to both 'worlds'. Almost at all times these people took a very critical stance as knowledgeable citizens, rather than as defenders of the industry. Or as one such a member put it:

> I joined STOLA to share my knowledge with others; to help them establish a basic understanding. In that respect I also posed more basic questions to the experts, to make the information they gave us more accessible to the whole group and to make sure this was explicitly noted down in the minutes. It has never been my intention to persuade anybody. All I aspired to, was to help give my fellow citizens enough basic knowledge, so they could judge for themselves. (Local participant in STOLA)

Together the local members thus formed a critical mass that confronted the ONDRAF-NIRAS and other experts with both the inaccessibility of their professional language and their lack of 'local lay knowledge', forcing them to communicate in a more understandable, meaningful and two-way manner.

Unfortunately not all working groups consisted of the same challenging mix of individuals. Still, traces of developments as described above can be found in almost all working groups. In any case, there was a clear difference in communication between the group and the experts they saw frequently (the permanent ONDRAF-NIRAS members and a handful of others), on the one hand, and the various experts that only attended the working group meetings occasionally or for just one consultation, on the other hand. Notwithstanding the proverbial exemption to the rule, the latter generally took a more formal stance and mostly kept to technical jargon:

> It wasn't always easy to follow these technical matters as ordinary citizens. ... and some of these experts, like these two professors who came and talked about environmental impact assessments. Well ... they certainly didn't make it any easier. And mind you, I have a masters degree in applied economics, so there were plenty of people who understood them even less. (Local participant in STOLA)

Conclusion

The need for meaningful interaction between experts and scientists of different disciplines, decision-makers and affected citizens in cases involving environmental and technological risk stems, in our view, from the fact that all these people are related to various systems that affect and are affected by these complex problems and by the outcomes of decisions taken to resolve them. However, engaging people from such different backgrounds, coloured by particular systems' rationalities and speaking different 'languages' is not easy. From our own experience, we believe that the key lies in giving affected citizens more control over their own situation. To accomplish this at least two things are imperative. First, a win-win situation of some kind should be found or created, and a common goal for the interaction defined. Second, the interaction should be structured so that it encourages people to think in terms of confronting formal positions and interests and that settings are created in which concerned parties can meet as individuals with multiple ties to various framings of the issue at hand.

In the case-study we presented, engaging in meaningful communication was not realised at once. It was the result of a long process. First, communalities were sought and a common goal defined for the partnerships. The presence of the LILW in interim storage, made the waste managers and their local interlocutors allies in developing a suitable repository project to deal with the waste on a long term basis. Second, technical experts and lay-people gained mutual understanding through each of the following mechanisms:

(a) meeting on a regular basis;
(b) getting to know each other in formal and informal settings;
(c) being confronted with different frames, rationales and different system's logics; and
(d) learning to bridge those by gradually discussing the issue as interested individuals rather than defenders of a particular interest.

Some people by nature have better capabilities in bridging different systems within a partnership network. Others played this role more as a result of their 'position'. This was particularly the case for the local people that (had) worked in the nuclear sector. Standing with one foot in the nuclear world and with the other in their own living room, these people could see the situation both through the eyes of the nuclear experts and of the affected citizens. In the partnership/working group setting, they were primarily identifying themselves as local citizens. Their alliance with their fellow residents was therefore far stronger than their bond with the nuclear experts. But since they could speak the expert's language and understood their framing of the problem, they were more easily understood and taken seriously by the experts, than were the other local participants, in the first instance. Hence they played an important role in bringing the experts closer to the locals and vice versa.

These observations make us cautiously conclude that within the partnerships the regular participants (citizens and experts alike) managed to some extent to put Beck's reflexivity into practice. As it appears, the partnerships' working groups have been able to 'beat the systems', by creating a new system in which all of the participants could function more as equals. This system developed a rationality and even some jargon of its own. Communicating developments and outcome of discussions to the rest of the partnership and the community at large, introduced openness in the decision-making process. Power was maybe not so much delegated, but definitely shared. Most importantly, through the partnerships, the potential host communities gained greater control over their future, which both forced and allowed them to judge the risks involved in a different way.

Notes

1. Cases concerning asbestos, Seveso-industry, carbon dioxide, mad cows, airports, domestic waste incinerators and nuclear waste.
2. On Luhmann's definition of social systems as being made up of communications, see: Blom (1997), Laermans (1997b), Laermans (1999), Luhmann (1989), Luhmann (1995).
3. Nonetheless we understand human communication to include the interference of non-human actors in the communicative process. Things like books, artefacts, machines, dogs, or – borrowing from Callon – scallops (Callon 1986) all have a communicative part to play in society, but in themselves do not constitute communicating social units.
4. On Luhmann's view of social systems as being closed entities, see: Blom (1997), Luhmann (1989), Luhmann (1995).
5. This in contrast with danger (Luhmann) or disaster (Beck) which refers to a threat or potential damage caused by outside sources (e.g., the forces of nature) and in that respect unavoidable.
6. For some more substantial detail on the functioning of the partnerships, see: Bergmans (2001a, b, 2005), Bergmans, Van Steenberge, and Verjans (2006), Hooft (1999), Hooft et al. (2002), MONA (2005), NEA/FSC (2004), NEA/FSC (2005), STOLA (2004).
7. Group interview with six participants in the STOLA working groups (29 June 2005).
8. Questionnaire sent to all local people that at some time participated in one of the partnerships (summer 2005).
9. In depth interviews with various types of participants in the Belgian partnerships, providing input for the CARL project – see:www.carl-research.org – (June–October 2006). Unless indicated otherwise, all quotes in this case study section are taken from these interviews.
10. The working group I&D in STOLA assembled 43 times, its counterpart in MONA 33 times. Both had an average attendance rate of 69% (Bergmans 2005, 394, 472).
11. STOLA working group Implantation and Design - Minutes meeting 28 February 2001: author's free translation.
12. Group discussions with ONDRAF-NIRAS collaborators (21 June and 5 July 2005).
13. Statements taken during a group discussion on 5 July 2005.
14. Group interview with six participants in the STOLA working groups (29 June 2005).
15. Questionnaire sent to all local people that at some time participated in one of the partnerships (Summer 2005).
16. In all: approximately four years of working group gatherings, three months of preparations and nine months of rounding up. The groups met on average about seven times a year.
17. STOLA working group Implantation and Design - Minutes meeting 1 March 2000.
18. STOLA working group Implantation and Design - Minutes meeting 29 March 2000.
19. Informal contacts with members of the STOLA working group I&D.

References

Anfrie, M.-N., A. Bergmans, M. Mormont, and E. Van Hove. 1998. *Werken met Lokale Partnerschappen. Een nieuwe benadering voor de berging van laagradioactief afval. Deel 1: Basisdocument. (Working with local partnerships. A new approach for the disposal of LILW - Part 1)*. UIA-PSW, FUL-SEED. Report commissioned by ONDRAF - NIRAS.

Beck, U. 1992. *Risk society. Towards a new modernity*. London: Sage.

———. 1994. The reinvention of politics: Towards a theory of reflexive modernization. In *Reflexive modernization. Politics, tradition and aesthetics in the modern social order*, ed. U. Beck, A. Giddens, and S. Lash, 1–55. Cambridge: Polity Press.

———. 1995. *Ecological enlightenment. Essays on the politics of the risk society*. New Jersey: Humanities Press.

———. 1997a. Tsjernobyl en de antropologische shock. In *De wereld als risicomaatschappij. Essays over de ecologische crisis en de politiek van de vooruitgang*, ed. M. Hajer and M. Schwarz, 23–37. Amsterdam: De Balie.

———. 1997b. Wereldrisicomaatschappij en de toekomst van de ecologische politiek. In *De wereld als risicomaatschappij. Essays over de ecologische crisis en de politiek van de vooruitgang*, ed. M. Hajer and M. Schwarz, 83–104. Amsterdam: De Balie.

Bergmans, A. 2001a. Enhancing transparency and public participation through local partnerships. In *VALDOR 2001*, ed. K. Andersson, 390–7. Stockholm, Sweden: Congrex Sweden AB/Kista Snabbtryck.

———. 2001b. *Local partnerships: towards a new approach in nuclear waste management in Belgium, Proceedings, ICEM'01*. Bruges: ASME-International.

———. 2005. Van "de burger als beleidssubject" naar "de burger als partner": de Belgische queeste naar een langetermijnoplossing voor het beheer van het laagradioactief en kortlevend afval. (From "citizens as policy subjects" to "citizens as partners": The Belgian quest for a long term solution for the management of low and intermediate level radioactive waste). PhD dissertation, University of Antwerp, Faculty of Social and Political Sciences, Department of Sociology.

Bergmans, A., A. Van Steenberge, and G. Verjans. 2006. *CARL Country Report - Belgium*. Antwerp: University of Antwerp, Department of Sociology/University of Liège, Group SEED. Available at: http://www.carl-research.org/index.php?pg=10.

Blom, T. 1997. *Complexiteit en Contingentie. Een kritische inleiding tot de sociologie van Niklas Luhmann*. Kampen: Kok Agora.

Callon, M. 1986. Some elements of a sociology of translation: Domestication of the scallops and the fishermen of St Brieuc Bay. In *Power, action and belief: A new sociology of knowledge?*, ed. J. Law, 196–233. London: Routledge & Paul Kegan.

Dobbelaere, K. 1997. Luhmann en de analyse van de godsdienst: enkele denkpistes. In *Sociale Systemen bestaan. Een kennismaking met het werk van Niklas Luhmann*, ed. R. Laermans, 181–96. Leuven: Acco.

Hajer, M., and M. Schwarz, eds. 1997. *De wereld als risicomaatschappij. Essays over de ecologische crisis en de politiek van de vooruitgang*. Amsterdam: De Balie.

Hooft, E. 1999. Working with local partners: ONDRAF/NIRAS's approach to the disposal of short-lived low-level radioactive waste. In Conference Proceedings, *ENS Topseal '99: 'Commitment to the Future Environment'*, 107–11. Antwerp, Belgium: Belgian Nuclear Society.

Hooft, E., A. Bergmans, K. Derveaux, and L. Vanhoof. 2002. Local partnerships: Achieving stakeholder consensus on low-level waste disposal? In Proceedings *Waste Management Conference: HLW, LLW, Mixed, Hazardous Wastes and Environmental Restoration - Working Towards a Cleaner Environment*. Tucson, Arizona.

Laermans, R. 1997a. "De rest is zwijgen": Luhmanns visie op sociale systemen. In *Sociale Systemen bestaan. Een kennismaking met het werk van Niklas Luhmann*, ed. R. Laermans, 19–35. Leuven, Amersfoort: Acco.

———. 1997b. "We kunnen ons geen alternatief voorstellen": Luhmanns analyse van de moderne maatschappij. In *Sociale Systemen bestaan. Een kennismaking met het werk van Niklas Luhmann*, ed. R. Laermans, 37–53. Leuven, Amersfoort: Acco.

———. 1999. *Communicatie zonder mensen. Een systeemtheoretische inleiding in de sociologie*. Amsterdam: Boom.

Luhmann, N. 1982. *The differentiation of society*. New York: Columbia University Press.

———. 1989. *Ecological communication*. Cambridge, MA: Polity Press.

———. 1993. *Risk: A sociological theory*. New York: Walter de Gruyter.

———. 1995. *Social systems*. Stanford: Stanford University Press.

MONA. 2005. *MONA, een weg naar de aanvaardbaarheid van een berging van categorie A-afval in Mol?* Mol: Geïntegreerd eindrapport.

NEA/FSC. 2004. *Belgian Community Visit and Workshop (18–21 November 2003) - Executive Summary and International Perspective*. Paris: OECD Publishing. Available at: http://www.nea.fr/html/rwm/docs/2004/rwm-fsc2004-4.pdf.

———. 2005. *Dealing with Interests, Values and Knowledge in Managing Risk. Report on Belgian Community Visit and Workshop, 18–21 November 2003*. Paris: OECD Publishing. Available at: http://www.nea.fr/html/rwm/fsc.html.

STOLA. 2004. *Het Belgisch laagactief en kortlevend afval: Thuis in Dessel?* Dessel: STOLA.

Risk and safety communication in small enterprises – how to support a lasting change towards work safety priority

Christina Stave, Anders Pousette and Marianne Törner

Occupational and Environmental Medicine, Department of Public Health and Community Medicine, Sahlgrenska Academy at Göteborg University, Göteborg, Sweden

The frequency of occupational accidents in Swedish farming is high. An intervention methodology aimed at influencing safety attitudes and behaviour was previously tested. The aims of the present study were to evaluate long-term effects of the intervention and to assess these results in relation to a randomly selected comparison group.

The intervention was based on regular group discussions, focusing on dialogue and reflection in networks for social support. Eighty-eight farmers and farm workers, divided into nine groups, participated. Each group gathered on seven occasions during one year, assisted by a process consultant, supporting safety communication and the change process. A questionnaire was used to evaluate the effects. The long-term results of the intervention showed that safety activity increased significantly and risk acceptance (fearlessness) and work stress decreased. Measurements were made at T1, T2 and T3, i.e., pre-, post- and long-term, in the intervention group and at T1 and T3 in the comparison group.

The results of the present study showed a long-term effect of the intervention on safety activity and risk acceptance (fearlessness) in relation to the comparison group. However, the comparison group also showed a decline in work stress, which indicates that the stress reduction may have had other causes than the intervention.

Introduction

Background

The question under study was how to communicate risk and safety in small enterprises to influence safety attitudes, cognitions and behaviour in order to encourage and increase safety activity. The focus was on safety, i.e., the desired outcome, rather than on risk. In many cases the cause of an accident is attributed to the person who releases the accident, normally the injured person. In accident investigations, 'the human factor' is often stated as the faulty link. The myth of individual control leads to a search for someone to blame instead of searching for the causes of accidents (Pidgeon 1998). To promote a sustainable change, it is of interest to investigate what made the human act as he/she did, and what factors supported or hindered safe behaviour (Dekker 2002). Gurabardhi, Gutteling, and Zuttchreuter (2005) describe different perspectives on risk communication; the democratic perspective underlines a constructive dialogue among all stakeholders, and empowering the persons whose safety is at stake is considered desirable. The intention in the present intervention was not to modify behaviour but to support both employer and worker in re-evaluating safety attitudes and to assist in the change process towards increased safety priority and safety management.

The most commonly used forms of safety management are (a) traditional safety methods (including the enforcement of laws and state regulations) and (b) quality methods and philosophies of quality in conjunction with safety methods (Willquist 2005). The present intervention focused on the latter, including theories perceiving the relation between enterprise (farmer) and the broader context (Bang Dyhrberg and Langaa Jensen 2004). For these often micro enterprises the vast majority of injuries affect the farmer or his/her family. Therefore the aim was to support a re-evaluation of safety attitudes, not to stress compliance with regulations.

In Sweden, all enterprises with five or more employees should have a safety representative, appointed by the trade union. Systematic work environment management (SWEM), should be applied meaning that the employer should 'investigate, carry out and follow-up activities in such a way that ill-health and accidents at work are prevented and a satisfactory work environment is achieved' (Swedish Work Environment Authority 2001). This process should be integrated in the ordinary work. In one of the final paragraphs in the regulation it is stated that enterprises have to do 'what is needed, depending on size, and occupational hazards'. In practice, this means that the Work Environment Inspectorate normally accepts much simpler systematic work environment management in small enterprises than in large ones. In Swedish agriculture and forestry, 94% of the enterprises have no employees and 99% have fewer than five employees. Therefore very few have a safety representative and SWEM is rarely performed (Antonsson, Birgersdotter, and Bornberger-Dankvardt 2002).

Farmers and farm workers are people who often have to develop, plan and solve their health and safety problems on their own. Their expertise is in production rather than health and safety issues, and improvements in occupational health and safety may therefore not be so easily acquired and incorporated. The farms in Sweden are often managed by the family and assisted by a substitute worker or a small number of employees. Safety communication is not formalised, and the private and occupational spheres are often integrated. The requirement for manageability is therefore high, i.e., there is a need for uncomplicated, cost-effective safety solutions, designed to fit the conditions in farming (Lundqvist 1996).

The focus of safety interventions has often been on technical measures, aiming at controlling specific hazards. Many safety projects in agriculture have been based on the assumption that, to make agriculture safer, it is necessary to educate farmers about dangers and negative outcomes. However, the question has been raised as to whether farmers actually use the educational information about elimination, reduction, or control of physical hazards (Murphy, Kiernan, and Chapman 1996). Murphy (2003) therefore discussed the farm safety risk paradox, that is, the incongruence between farm people's safety knowledge, values, and practices. In order to influence safety behaviour, attitudes towards risks and safety are of interest as are the associations between attitudes and behaviour.

Bornberger-Dankvardt, Ohlson, and Westerholm (2003) claimed that, in small companies such as farming, financial margins are small. Time pressure and high workload lead to little time for reflection about the work environment and health issues. Health and safety management are often perceived as being separate from production.

Learning in small companies takes place within the work, in meetings with others, through a verbal dialogue. In order to proceed from safety knowledge to activity, someone must supply entrepreneurs with this knowledge, face to face, using examples from their working life (Bornberger-Dankvardt, Ohlson, and Westerholm 2003).

Regional safety representatives (RSR) may assist farmers in focusing on physical problems in the working environment. Implementation of a safety management system requires competence in how to support processes of change, and such competence is seldom available. Workers' participation and managers' motivation are important in terms of prioritising safety (Antonsson, Birgersdotter, and Bornberger-Dankvardt 2002)

In Sweden, there have been few long-term safety interventions reported in the agricultural sector directly addressing attitudes and sustained behavioural change. A methodology was therefore developed and first tested in Swedish fishery (Törner and Eklöf 2005), thereafter adjusted and applied to conditions in the agricultural sector (Stave, Törner, and Eklöf 2007).

The intervention

The aim of this intervention study was to analyse effects of a methodology addressing attitudes and behavioural change, in order to increase safety activity. Networks of fellow farmers were created in order to stimulate safety facilitators such as *participation and social support*. The intention was that the network should support the development of a *positive safety culture* (Pidgeon 1998), motivating increased safety activity and counteracting risk acceptance (Mullan 2004). The *process of change* was supported by a process consultant (Schein 1987), stimulating farmers' own initiatives rather than giving expert advice (Blake and Mouton 1989), and supporting the participants' *process and stages of change* (Prochaska and DiClemente 1982) over a one-year period. An open and *non-judgemental social climate* within the group was supported, so that mistakes could be candidly discussed since injury risks are problems that may produce *feelings of stress*, adding to other work stressors. The ambition was to increase risk awareness without increasing work stress (Lazarus and Folkman 1984). This was to be managed through a positive approach *focusing on and stimulating risk manageability* rather than focusing on problems. However, hindrances to safety performance were not neglected, and a structure was provided to facilitate the understanding of how to gain *control over hazards,* finding safety solutions. *Reflection about practice* was encouraged through the use of an incident diary, making analyses of experienced incidents, *following the chain of events* upstream, and identifying different *sources of risks*. By discussing real examples, the *gap between practice and procedures* was revealed, and solutions were sought through *communication with others in the same work situation*. Furthermore, information on risks and their consequences on a more general level was provided to stimulate risk perception, to reduce risk acceptance and hopefully to convince the farmers of the benefits of *prioritising safety*. The time, place and intensity of the intervention were *adjusted to the context* and limitations of work in agriculture. The basic concept was:

- To create *participative arenas* where farmers and farm workers could exchange experience and find *social support* for increased safety activity.
- To encourage new *perspectives, discussions and reflections* on practice.

- To facilitate ways of *increasing perceived risk manageability* and stimulating an *action-oriented* way to manage safety.

The intervention period was from October 2000 to October 2001. Nine networks of approximately ten farmers each were formed. Over the first six-month period each group met once a month for 1.5 to 2 hours. A seventh meeting took place six months later, twelve months after the start of the intervention. Two experienced safety engineers from the agricultural OSH Services and one doctoral student (behavioural science) were assigned as process leaders. The process leaders received a two-day training session from an experienced psychologist and safety researcher for their role.

To illustrate the intervention strategy, a few short examples may be given: the effect of using the incident diary with the structured analysis, following the chain of event upstream, was stated as a change in reasoning, e.g., 'Now I think before I act' or 'I'm more careful now'. One farmer who had been telling his group about his experiences from falling over things, finally after the third incident analyses realised that his problem was that he had to prioritise good order and keeping tools and material away from the floor. One group leader found that there had been a real understanding of the analysis due to an open and thorough discussion. However, general information about rates, types, effects and cost analyses of accident generated little interest. 'It is as though it concerned others but not them', one process leader stated. The socially supportive network was the most appreciated part of the intervention.

The short-term results were reported in Stave, Törner, and Eklöf (2007). Results of the intervention showed significant differences between pre- and post-intervention. Self-reported 'safety activity' increased, while 'work stress', and 'risk acceptance (fearlessness)' decreased. No significant differences were found for 'risk perception' and 'risk manageability'. The contents of these variables are presented below. A qualitative process evaluation was also performed. At baseline the questionnaire was also administered to a representative sample of Swedish farmers and farm workers. However, no follow-up in this non-treatment comparison group was carried out one year later after the intervention.

Aim

The aim of the present study was (1) to evaluate whether any effects of the intervention still remained four years after finishing the intervention and (2) to evaluate the long-term results in the intervention group in relation to a randomly selected comparison group.

Methods

This study had a longitudinal quasi-experimental design with repeated measures from the same individuals in the intervention and comparison group respectively. The present paper focuses on measurement of the long-term effect of the intervention, using a questionnaire survey at baseline (T1), in October 2000, prior to the intervention in both intervention and comparison group, repeated for the intervention group only at the end of the intervention, one year after baseline in October 2001 (T2), and repeated again five years after baseline, in October 2005 for both the interventions group, and the comparison group (T3). The comparison group was measured using the same questionnaire as the intervention group at baseline, T1 and again at T3.

Participants

The intervention group; T1, T2 and T3

Nine farmers were initially assigned to the intervention. They recruited approximately ten group members each, with an emphasis on diversity with reference to production type, gender, age and professional status as employer or employee. Ninety-two participants were thus recruited to the intervention. Four participants dropped out early; an additional four members did not answer the follow-up questionnaire at T2. In total, 84 persons remained at T2. At T3, 74 of the 84 participant answered the same questionnaire. Five of the ten persons who did not answer had stopped farming. The response rate was 88%.

The comparison group; measurements T1 and T3

A random selection of 400 farmers and 99 farm workers was made from the members' register of the LRF (Swedish Federation of Farmers) and Farm-Workers Associations' list of members, in accordance with the proportion between farmers and farm workers in Sweden, which was approximately 4:1. In a second scrutiny, 38 farmers and 15 farm workers were excluded since they did not fit the criterion of being active in farming. The total final selection consisted of 446 farmers and farm workers. A questionnaire was mailed to the respondents in July 2000, and data collection ended in December of the same year (T1). The number of respondents was 315 persons, which represented a response rate of 70%. Differences between farmers and farm workers in all of the variables of the study were tested and found to be non-significant at the 5% level. Consequently, farmers and farm workers were treated as one group, hereafter called farmers. At T3, the comparison group received a second round of questionnaires. At this time point 186 persons answered. Of the non-respondents, 81 had closed down their business or changed employment, leading to a response rate of 79%.

As can be seen in Table 1 the intervention group did not differ significantly from the comparison group in any variable at baseline (T1).

Table 1. T-test for the intervention and comparison group at baseline (T1, before the intervention); based on respondents with valid observations at T3; scale ranged from 1–6.

Variables	Intervention group M (SD)	n	Comparison group M (SD)	n	t (p)
Risk perception	3.98 (0.81)	74	3.84 (0.83)	186	−1.26 (0.210)
Risk manageability	4.51 (0.77)	74	4.43 (0.94)	186	−0.67 (0.501)
Work stress	3.71 (1.16)	74	3.91 (1.47)	185	1.16 (0.247)
Risk acceptance	2.98 (1.12)	74	3.21 (1.15)	185	1.45 (0.150)
Safety activity	4.25 (0.86)	74	4.13 (1.10)	185	−0.84 (0.402)

The demographic information in Table 2 shows that the mean age of the intervention group was somewhat lower than that of the comparison group. Women were generally underrepresented in the comparison group as compared to women in Swedish farming in general (33%); however, as compared to women registered at LRF as leaders or owners of farms (10%), they were fairly representative for this group.

The questionnaire

The questionnaire was designed mainly on the basis of a pilot study among Swedish fishermen (Eklöf and Törner 2002), adjusted to the specific characteristics of agriculture. Questions concerning work stress were based on the pre-study (Törner et al. 2002), where work stress was stated by 65% of a strategic sample of farmers and farm workers as the primary cause of the high frequency of accidents in agriculture. Main stressors put forth were financial worries, time pressure and high workload. Two safety engineers specialised in farming examined the questionnaire before it was used.

Risk perception was explored by the respondents by rating risks associated with 12 different working activities at farms. The 12 activities were selected through an analysis of serious accidents in Swedish agriculture undertaken in a pre-study (Stave and Törner 2000). The items in the questionnaire were presented as 'How high do you estimate the risk to be for you to be injured in connection with X?' The endpoints of the six-point rating scale used were 'very low risk' and 'very high risk'. The 12 items were averaged to produce a variable called 'risk perception'.

Risk manageability was rated for each of the 12 hazardous work tasks. The items were presented as 'Accident risks associated with X can be reduced with technical equipment and changed working methods'. The endpoints of the six-point rating scale were 'not at all correct' and 'entirely correct'. The 12 items were averaged to produce a variable called 'risk manageability'.

Work stress was covered by asking the respondents to rate their level of stress on the basis of the following two statements: 'Stress due to time pressure and high work load is a problem to me', and 'I worry about the future'. The endpoints of the six-point rating scale were 'not at all correct' and 'entirely correct'. The two items were

Table 2. Demographic information about the intervention and comparison group based on respondents at follow-up (T3).

	Intervention group	Comparison group
Respondents at baseline T1	92	315
Respondents at follow-up T3	74	186
Age M (SD), distribution	49.1 (8.7)	53.4 (10.6)
Women (%)	19	5
Married or equivalent (%)	86	83
Children under 18 years (%)	47	34
Years in farming M (SD)	26 (10.7)	27.8 (11.2)
Hours worked/week M (SD)	49.2 (20.6)	39.4 (21.5)
Farm owner (%)	96	86
No. of co-workers on the farm M (SD)	1.0 (1.1)	0.8 (1.4)

averaged to produce a variable called 'work stress'. The variable 'work stress' was validated against the stress variable in the Kjellberg and Iwanowski (1989) stress-energy checklist ($r=0.55$, $p<0.01$).

Risk acceptance (fearlessness) was covered by four items based on previous research (Eklöf and Törner 2002): 'You have to learn to live with the risks at work'; 'A farmer/farm worker should be able to take care of himself'; 'A farmer/farm worker should be prepared to take risks'; and 'A farmer/farm worker should enjoy challenging the forces of nature'. The endpoints of the six-point scale were 'not at all correct' and 'entirely correct'. The four items were averaged to produce a variable called 'risk acceptance'.

Safety activity was covered in this study by three items: 'I try to find methods and equipment to improve safety'; 'We who work on the farm co-operate to improve the safety work'; and 'I and my co-workers often discuss how to make our work safer'. The endpoints of the six-point rating scale were 'not at all correct' and 'entirely correct'. The three items were averaged to produce a variable called 'safety activity'.

Statistical analysis

Results for the intervention group pre- and post-intervention (T1 – T2) were reported in Stave, Törner, and Eklöf (2007). A reanalysis of these results is presented in Table 4. In the present study the analyses were based on participants remaining at T3 ($n=74$).

Measures of change in variables for the intervention group between T2 and T3 were then analysed by a two-sided dependent t-test. Secondly, differences in variables at T1 and T3 were analysed for the comparison group by a two-sided dependent t-test. Finally, changes in production type and changes in work stress in the comparison group were analysed by independent t-test at T1 and T3. Bivariate correlations were also calculated for the comparison group between variables. P-values of <0.05 were considered statistically significant. SPSS 12.0 was used.

Results

There were no significant differences for the intervention group in the variables explored between the end of the intervention (T2) and four years later (T3) as can be seen in Table 3. Thus the effect of the intervention, i.e., increased safety activity and decreased risk acceptance (Stave, Törner, and Eklöf 2007) remained stable four years after the intervention ended.

Table 3. Dependent t-test for the intervention groups after the intervention (T2) and four years later (T3) based on respondents with valid observations at T3, scale ranged from 1–6.

Variables	After the intervention (T2) M (SD)	Follow-up four years later (T3) M (SD)	n	t (p) df=73
Risk perception	3.95 (0.91)	3.82 (0.87)	74	−1.54 (0.129)
Risk manageability	4.44 (0.82)	4.54 (0.82)	74	0.97 (0.333)
Work stress	3.34 (1.32)	3.35 (1.37)	74	0.08 (0.936)
Risk acceptance	2.66 (0.98)	2.64 (0.97)	74	−0.21 (0.835)
Safety activity	4.54 (0.83)	4.53 (0.81)	74	−0.07 (0.945)

As previously reported (Stave, Törner, and Eklöf 2007) and now reanalysed with participants remaining at T3, Table 4 shows that safety activity in the intervention group increased from T1 to T2 and risk acceptance and work stress decreased, while risk perception and risk manageability remained unchanged.

A comparison with a randomly selected group was made in order to evaluate the long-term effect of the intervention. Table 5 presents measures for variables for the comparison group before the intervention (T1) and five years later (T3). The results showed that safety activity, risk acceptance (fearlessness) and risk manageability remained without significant changes but a significant decline in risk perception and work stress from T1 to T3, was found in the comparison group.

Risk perception and work stress were significantly positively correlated ($r=0.37$, $p<0.01$). The 30 farmers (16%) at T3, who had changed their production type had a significantly lower score for work stress than those who had not changed their production ($t=2.4$, $df=181$, $p<0.05$).

Figure 1 illustrates the significant changes pre- and post intervention, with increased safety activity and decreased risk acceptance (fearlessness) and work stress between T1 and T2. It also shows that these effects remained at T3, i.e., four years later. The comparison group did not show any significant changes from T1 to T3 in safety activity, risk acceptance (fearlessness) or risk manageability but a significant decrease in work stress and risk perception. The comparison between groups thus showed that the decreased work stress at T2 in the intervention group may have had

Table 4. Dependent t-tests for the intervention group at baseline (T1) and one year after the intervention (T2), based on respondents with valid observations at T3; scale ranged from 1–6.

Variables	Base-line (T1) M (SD)	After the Intervention (T2) M (SD)	n	t (p) df=73
Risk perception	3.98 (0.81)	3.95 (0.91)	74	−0.35 (0.728)
Risk manageability	4.51 (0.77)	4.44 (0.82)	74	−0.82 (0.414)
Work stress	3.71 (1.16)	3.34 (1.32)	74	−2.91 (0.005)*
Risk acceptance	2.98 (1.12)	2.66 (0.98)	74	−3.05 (0.003)*
Safety activity	4.25 (0.86)	4.54 (0.83)	74	3.14 (0.002)*

Note: * significant (p<.05).

Table 5. Dependent t-test for the comparison group at baseline (T1) and five years later (T3), based on respondents with valid observations at T3; scale ranged from 1–6.

Variables	Before the intervention (T1) M (SD)	Follow-up five years later (T3) M (SD)	n	t (p)
Risk perception	3.84 (0.83)	3.62 (0.87)	186	−3.96 (0.000)*
Risk manageability	4.43 (0.94)	4.32 (0.87)	186	−1.60 (0.112)
Work stress	3.93 (1.45)	3.37 (1.29)	182	−5.73 (0.000)*
Risk acceptance	3.21 (1.15)	3.28 (1.09)	185	1.04 (0.299)
Safety activity	4.13 (1.10)	4.16 (1.04)	185	0.39 (0.693)

Note: * significant (p<.05).

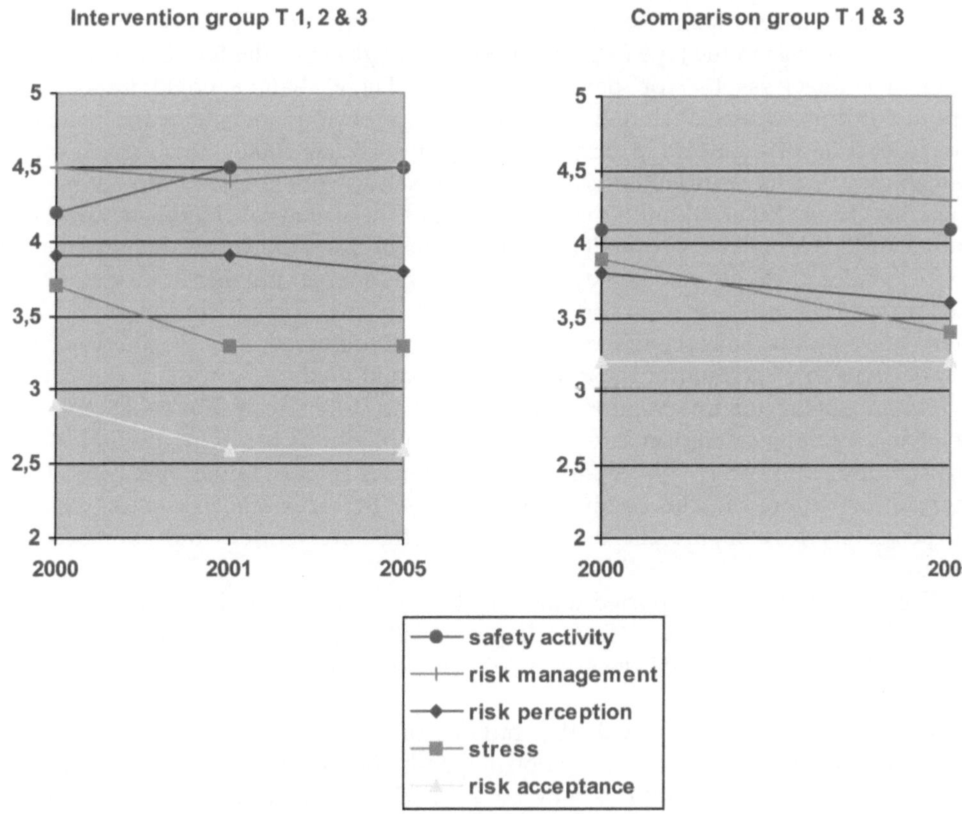

Figure 1. Mean scores for variables in intervention group at T1 (n=92), T2 (n=84) and T3 (n=74) and in comparison group at T1 (n=315) and T3 (n=186) for respondents with valid observations at T3.

other causes than the intervention. The assumption that the increased safety activity and decreased risk acceptance (fearlessness) was a result of the intervention, was however supported.

Discussion

The interesting differences between the intervention group and the comparison group were the significant increase in safety activity and decrease in risk acceptance (fearlessness) in the intervention group between T1 and T2 (Stave, Törner, and Eklöf 2007), which remained at T3, while the comparison group remained stable during the same time period. The changes in safety activity and risk acceptance (fearlessness) may therefore be considered effects of the intervention that remained four years later.

Previously the decrease in work stress in the intervention group was considered a possible effect of the intervention. However, the comparison group also showed a decrease in work stress and additionally a decrease in risk perception. In both the intervention group and comparison group there was a significant correlation between risk perception and work stress. This relation was also found in a previous study that tested a model of the relation between risk perception and safety activity (Stave,

Pousette, and Törner 2006). The reduction in risk perception and stress may have been a result of a change in the type of production in a subgroup of the farmers. The most common change was to stop dairy production. These changes could imply less exposure to animals. Animal handling was found in a previous study to be the cause of a large portion of accidents in farming (Stave and Törner 2000). Less exposure to animals may thus have affected risk perception as well as work stress. We do not know when this change occurred but it may be a result of the structural changes in farming and new EU rules effectuated in 2005, where demand for production was separated from subsidies. It should also be noted that there was a larger number of farmers that had stopped farming or changed employment in the comparison than in the intervention group. Thus the reduction in work stress observed after the intervention may not have been a result of the intervention method used.

Participants in the intervention were expected to differ from farmers in general by having a greater interest in safety, since the sampling was based on an interest in participating. However, no significant differences were found between the intervention group and the comparison group at T1. This similarity strengthens the possibility to compare the development in the intervention and comparison groups.

The reliability of self-reported data may be questioned (Kjellberg and Franzon 2002). This may be a problem when rating a socially desirable factor such as safety activity. However, repeated measures of the same participants should partly outbalance this problem.

Westlander (2003) discussed the pitfalls of using standardised self-report questionnaires repeatedly in intervention research. Besides the targeted conditions to be improved, other conditions may be affected as well, resulting in a response shift. Since the purpose of the intervention was to create a re-evaluation of attitudes leading to more safety activity, this problem was present and complex, and it was addressed by additional qualitative analysis which also confirmed the statistical results (Stave, Törner, and Eklöf 2007). However, even if many questions about mechanisms remain, the ultimate aim of the intervention was to increase safety activity, which was achieved also in a long-term perspective, while no such effect was noted in the comparison group.

The variable 'safety activity' contained three items and could be improved by adding more validated items. The questionnaire was tested in two prior studies but should be further developed and evaluated in future research.

The present study showed long-term effects on safety activity and risk acceptance (fearlessness). How can this success be explained? The methodology used in this intervention differed from previously reported safety interventions in Swedish farming (Lundqvist and Gustafsson 1992) by primarily targeting both attitudes, behavioural and cultural change. Habits and attitudes take time to change (Prochaska and DiClemente 1982). Therefore a long-term methodology with repeated meetings during one year may have been beneficial for the stable change in risk acceptance (fearlessness) and safety activity. The process of change does not only take place at meetings but also between meetings, when participants reflect and discuss with colleagues and family and encounter new but similar situations.

In the intervention, safety communication took place as a dialogue between fellow farmers, which may have provided a means to better identify with the problems and solutions, making them the owner of the problem as well as of the solutions. Expert advice or communicating safety in an educational manner may create defensiveness rather than an openness to change (Stave, Törner, and Eklöf 2007). Since many farmers have to attend to their safety problems by themselves, a network could provide beneficial support. Non-judgemental risk and safety communication that takes place in a trustful atmosphere between fellow farmers is likely to have had a positive effect. The network was also the basis for developing a positive safety culture, rejecting risk acceptance (fearlessness) and a normalisation of risks. Safety culture was defined by Pidgeon (1998) as a set of assumptions, and their associated practices, which permit beliefs about danger and safety to be constructed. Our aspiration was that safety-conscious and active farmers could act as early adopters, according to the theories of diffusion of innovation (Rogers 1995), paving way for other farmers.

An incident diary was used in the intervention, analysing experienced accidents upstream in a structured manner. Analysis and discussions of real examples of incidents may be of help in finding patterns of behaviour or weak points in planning or giving priority to safety. It is of importance to see and overcome the barriers to safety, but the focus of communication should be the desired outcome, which is safety, rather than dwelling on the risks as such.

Finally we conclude that the present study lent support for long-term effects on safety activity and risk acceptance (fearlessness). Thus the intervention methodology, using discussions and reflection over a longer time period in networks of small entrepreneurs, seems to have positively influenced the participants towards a lasting change in safety priority. This may inspire future interventionists that aim at developing safety communication in continuous processes of change.

Acknowledgements

Financial support was given by Region Västra Götaland. The authors also wish to express their appreciation of the excellent administrative support of Pernilla Håkansson.

References

Antonsson, A.-B., L. Birgersdotter, and S. Bornberger-Dankvardt. 2002. *Small enterprises in Sweden: Health and safety and the significance of intermediaries in preventive health and safety*. Work and Health, 2. Stockholm, Sweden: National Institute for Working Life.

Bang Dyhrberg, M., and P. Langaa Jensen. 2004. Organisations in context: Proposal for a new theoretical approach in prescriptive accident research. *Safety Science* 42: 961–77.

Blake, R.R., and J.S. Mouton. 1989. *Consultation: A handbook for individual and organizational development*. Reading: Addison-Wesley.

Bornberger-Dankvardt, S., C.-G. Ohlson, and P. Westerholm. 2003. *Arbetsmiljö- och hälsoarbete i småföretag – försök till helhetsbild* [*Work life in transition*]. Stockholm: National Institute for Working Life.

Dekker, S.W.A. 2002. Reconstructing human contributions to accidents: The new view on error and performance. *Journal of Safety Research* 33: 371–85.

Eklöf, M., and M. Törner. 2002. Perception and control of occupational injury risks in fishery – a pilot study. *Work and Stress* 16, no. 1: 58–69.

Gurabardhi, Z., J.M. Gutteling, and M. Kuttchreuter. 2005. An empirical analysis of communication flow, strategy and stakeholders' participation in the risk communication literature 1988–2000. *Journal of Risk Research* 8, no. 6: 499–511.

Kjellberg, A., and H. Franzon. 2002. A task questionnaire for participatory job development. Paper presented at the Annual Congress of the Nordic Ergonomics Society (NES). Humans in a Complex Environment, October 1–3, in Kolmården, Sweden.

Kjellberg, A., and S. Iwanowski. 1989. *Stress/energi-formuläret: utveckling av en metod för skattning av sinnestämning i arbetet.* (in Swedish). Report No 1989: 26. Stockholm, Sweden: National Institute for Working Life.

Lazarus, R.S., and S. Folkman. 1984. *Stress, appraisal and coping*. New York: Springer.

Lundqvist, P. 1996. Agriculture - a hazardous work environment. In *Fifteen years of occupational accident research in Sweden*, ed. E. Menckel and B. Kullinger, 138–44. Stockholm: Council for Working Life Research.

Lundqvist, P., and B. Gustafsson. 1992. Accidents and accident prevention in agriculture - A review of selected studies. *International Journal of Industrial Ergonomics* 10: 311–19.

Mullan, J. 2004. Investigating factors that influence individual safety behaviour at work. *Journal of Safety Research* 35: 275–85.

Murphy, D.J., N.E. Kiernan, and L.J. Chapman. 1996. An occupational health and safety intervention research agenda for production agriculture; does safety education work? *American Journal of Industrial Medicine* 29: 392–6.

Murphy, D.J. 2003. *Looking beneath the surface of agricultural safety and health.* Michigan, IL: The American Society of Agricultural Engineers.

Pidgeon, N. 1998. Safety culture: Key theoretical issues. *Work and Stress* 12, no. 3: 202–16.

Prochaska, J.O., and C.C. DiClemente. 1982. Transtheoretical therapy: Toward a more integrative model of change. *Psychotherapy: Theory Research and Practice* 19: 276–88.

Rogers, E.M. 1995. *Diffusion of innovation.* 5th edn. New York: Free Press.

Schein, E.H. 1987. *Process consultation, volume II, lessons for managers and consultants.* Massachusetts: Addison-Wesley Publishing Company, Inc.

Stave, C., and M. Törner. 2000. *Analys av allvarliga olyckor i lantbruket* (in Swedish), proceedings of the 32nd Annual Conference of the Nordic Ergonomics Society, Trondheim, Norway.

Stave, C., M. Törner, and M. Eklöf. 2007. An intervention method for occupational safety in farming; evaluation of the effect and process. *Applied Ergonomics* in press.

Stave, C., A. Pousette, and M. Törner. 2006. A model of the relation between risk perception and self-rated safety activity. *Occupational Ergonomics* 6: 35–45.

Swedish Work Environment Authority. 2001. Provisions on systematic work environment management. *AFS* 1.

Törner, M., C. Stave, M. Eklöf, M. Cagner, B. Nilsson, and R. Örtengren. 2002. *Program in agriculture for prevention of injuries, based on participation and network for social support* (in Swedish), Arbetslivsrapport, 14. Stockholm, Sweden: National Institute for Working Life.

Törner, M., and M. Eklöf. 2005. Participatory analysis of accidents and incidents as a tool for increased activity in safety work among fishermen. A pilot intervention study. *Work and Stress* 19, no. 4: 360–9.

Westlander, G. 2003. The use of standardized self-reporting pre and post measurement in intervention studies. *Arbete Människa Miljö & Nordisk Ergonomi* 1: 3–36.

Willquist, P. 2005. Industrial safety: Principles for safety management based on empirical studies. Eng. Dr. Thesis, Department of Product and Production Development, Production Systems, Chalmers University of Technology, Göteborg.

Audiovisual risk communication unravelled: effects on gut feelings and cognitive processes

Vivianne H.M. Visschers[a], Ree M. Meertens[a,b], Wim F. Passchier[c] and Nanne K. de Vries[b]

[a]Department of Health Education and Promotion, Nutrition and Toxicology Research Institute Maastricht (NUTRIM), Maastricht University, Maastricht, The Netherlands; [b]Care and Public Health Research Institute (Caphri), Maastricht, The Netherlands; [c]Department of Health Risk Analysis and Toxicology, Nutrition and Toxicology Research Institute Maastricht (NUTRIM), Maastricht University, Maastricht, The Netherlands

Audiovisual material is increasingly applied in risk communication (e.g., information films on the Internet) to affect the public's risk perception. This study investigated how the sound/footage and text of videos can influence two processes of risk perception: a primary (associative) and a secondary (deliberate) evaluative process. Our main hypothesis was that the sound/footage of a video would particularly influence the primary evaluative process and its text would affect the secondary evaluative process. This was investigated using a two (text: yes/no) by two (sound/footage: yes/no) design. We applied an indirect test to measure the videos' effects on the primary evaluative process (the Extrinsic Affective Simon Task) and a direct test (questionnaire) to assess the effects on the secondary evaluative process. These two tests were applied immediately after the respondents had seen one of the videos and two weeks later. Text appeared to affect the self-reported risk perception (questionnaire) only at the first measurement. Sound/footage influenced risk perception as measured by the EAST merely at the second measurement. The results are discussed in light of the two risk perception processes.

The development of new communication technologies, such as the Internet and digital television, stimulates the use of audiovisual means in risk communication (e.g., information films on television and on Internet sites). Only little research has focused on how moving pictures (i.e., footage) and sound can affect how people evaluate risks (cf. Meijnders, Midden, and Wilke 2001). Pfau and colleagues (2000) suggested that audiovisual material probably influences someone's general evaluation or impression of an object. Thus, audiovisual material may have more effect on general feelings about a risk than text material only. In this article, it is explored how audiovisual material can affect risk perception.

Like other researchers, we distinguish a global, associative and affective process of risk perception from a deliberate process (see Chaiken, Liberman, and Eagly 1989; Loewenstein et al. 2001; Slovic et al. 2004). The former process is associative and experiential, leading to a 'gut feeling', which we call the primary evaluative process (cf. Zajonc 1980). It forms an instantaneous reaction to a risk. The latter is analytic and elaborative, which we name the secondary evaluative process (Chaiken, Liberman, and Eagly 1989; Strack and Deutsch 2004). This process takes place at a later stage, for example, after being explicitly asked for an opinion. The primary evaluative process is evoked by minimal stimulus input. The secondary evaluative process is only activated when there is opportunity (e.g., time) and motivation to do so (e.g., a strong negative association that stimulates further deliberate processing) (see Chaiken, Liberman, and Eagly 1989; Fazio and Olson 2003). Then this process may overrule the primary evaluative process. It seems worthwhile to develop risk communication methods that pay attention to both processes. For reasons explained later, we examined the effects of various forms of audiovisual risk communication on the two processes of risk perception in the present study.

Two different tests are needed to measure the two risk perception processes, namely a direct and an indirect test. So far, risk communication research has mainly used direct measures, such as questionnaires, to assess risk perception (e.g., Hendrickx, Vlek, and Oppewal 1989; MacGregor, Slovic, and Morgan 1994). In our view, these studies mostly concentrated on the secondary evaluative process as respondents can think relatively deliberately about their responses when they fill in a questionnaire. Such a direct test is less suitable to reveal the primary evaluative process in isolation because the conscious character of a direct test contradicts the spontaneity of the latter process. Instead, this process requires a global and immediate evaluation. We used an indirect measure to assess the primary evaluative process since this test does not allow conscious control and reveals the instantaneous association with an object (cf. Siegrist, Keller, and Cousin 2006). Putting time pressure on people may also reveal the primary evaluative process. Previous studies have shown that people then rely on the affect heuristic (Finucane et al. 2000), which resembles our primary evaluative process. This has been assessed by means of a questionnaire, which is a direct test.

We chose the Extrinsic Affective Simon Task (EAST) as an indirect test in this study (de Houwer 2003) and adapted it to measure risk perception. This computer task has been developed to measure people's implicit attitude towards an object by measuring the association strength between this object (e.g., 'flower') and two contradicting evaluations (i.e., positive vs. negative). The association strength is assessed by the reaction times and the number of errors that respondents make while categorizing variations of the object on their extrinsic valence (e.g., colour), and evaluations on their intrinsic valence. It is expected that the intrinsic valence of the object interferes with the categorization of the object's extrinsic valence. So far, the EAST has been shown to differentiate vegetarians and non-vegetarians on their implicit preference for meat, which appeared to relate to their explicit food preference (de Houwer and de Bruycker 2007). Furthermore, an EAST with pictures in stead of words appeared to result in faster response times than the words-EAST about the same object, when assessing people's implicit attitude of victims of an harassment and that of spiders among spider-fearful people (Callan and Ellard 2003; Huijding and de Jong 2005, respectively).

We already suggested that the EAST is valuable for assessing risk perception (Visschers et al. 2007). In this previous study, the EAST included dangerous-harmless evaluations to assess risk perception, rather than positive-negative evaluations as in the original EAST. We also investigated whether an EAST with dangerous-harmless evaluations would be capable of assessing people's association strength. The respondents appeared to associate high-voltage power lines more strongly to danger than to harmlessness. In the present study, we therefore hypothesized that the (possibly risky) activity we chose as subject, nanotechnology, would generally be associated more strongly with danger than with harmlessness (*Hypothesis 1*). Confirmation of this hypothesis would show that an indirect measure like the EAST may shed some light on how people perceive a risk (although non-confirmation of this hypothesis would not mean that the EAST does not measure the primary evaluation of a risk). In addition, we investigated in an explorative sense whether the two measurements (EAST and questionnaire) were related. The procedure of the EAST is described in the Materials section.

Audiovisual material or video (including footage, text and sound) is often classified as a vivid information source because of its multimedia characteristics (Chaiken and Eagly 1983). Video information can provide its audience with a concrete visualization due to the combination of images and corresponding text, whereas readers of only text have to visualize the information themselves. Consequently, people may think that events presented in video format are more likely to happen in reality than events reported only in text information (cf. Sherman et al. 1985). Furthermore, video information may affect the primary evaluative process because of its vividness (cf. Loewenstein et al. 2001) and because it relies on images and impressions, which are all primary evaluative elements (cf. Finucane, Peters, and Slovic 2003; Pfau et al. 2000). Thus, audiovisual material may have more effect on the primary evaluative process than text information.

Text information is expected to mainly affect the secondary evaluative process because it relies on arguments, which need to be processed deliberately. Visschers et al. (2007) indeed showed that text information influenced risk perception (of high voltage power lines) as measured by a questionnaire but not as measured by the EAST. In the present study, we unravelled the effects of text and sound/footage in video material on the two risk perception processes. We hypothesized that the different elements of audiovisual material would influence the two risk perception processes differently. This hypothesis consisted of two parts: sound/footage videos mainly affected the primary evaluative process (*Hypothesis 2a*) and text videos particularly influenced the secondary evaluative process (*Hypothesis 2b*).

Video information often includes both text and images, which are stored in multiple codes in memory (symbolic and analogous, respectively, see Clark and Paivio 1991). It may thus be easier to process and recall a message that includes both text and images, than a message presented in one code, such as text (Mousavi, Low, and Sweller 1995). Hence, we expected that the effect of both sound/footage and text in one video on risk perception would persist over time while the effect of a video that only includes text or sound/footage would decrease (*Hypothesis 3*).

Method

Respondents

Ninety respondents participated in the first measurement of the study. All respondents were students at Maastricht University who were recruited in their first year to join a respondent pool that could be called up for studies in our laboratory. At the second measurement, four respondents did not show up. The remaining sample of 86 respondents was used in all analyses. Its mean age was 21.02 years (SD=2.66), 75 were women and 11 were men. The respondents received €5 for each measurement, €10 in total. They were randomly assigned to one of four video conditions and one of two EAST versions.

Materials

Informed consent

Before starting the experiment, the respondents read and signed an informed consent form in which it was stated that they could stop the experiment at any time. The respondents also agreed that their anonymous data were used for analyses and that they would not inform other students about the content of the experiment.

Risk

The video addressed the risks associated with nanotechnology. Nanotechnology is a relatively new technique, in which materials are manipulated at atomic, molecular and macromolecular level to develop and produce new structures, systems and applications (The Royal Society & The Royal Academy of Engineering 2004). Nanotechnology is used in various fields, among which are medicine, pharmacology, defence, and information and communication technology (van Est, Malsch, and Rip 2004). It may lead to beneficial applications, such as more specific administration of chemotherapy, increasingly smaller computers and sensor networks that can detect radioactivity and explosives. Opponents of this technology fear that terrorists might misuse it, that it can lead to super-intelligent systems no longer understood by humans and that nanoparticles in the atmosphere become the 'asbestos of the future'.

Video risk communications

We used four different videos, which varied on the presence of footage and music, and of text (subtitles). This resulted in four videos: a text and sound/footage video, a text video without sound/footage, a sound/footage video without text, and a control video (no text and no sound/footage). A collage of nine nanotechnology pictures that would appear in the EAST was shown during the first seven seconds of all videos so that these would be recognized in the EAST. The videos lasted between 2.5 and 4.1 minutes; except for the control video, which lasted 14 seconds. The text and sound/footage video presented the possible negative consequences of nanotechnology, such as nanoparticles in the human body that cross the blood-brain barrier and damage the brain, and terrorist attacks. The footage fitted the subtitles of the video (text) and it included threatening music. The sound/footage video without text consisted of the same music and footage as the previous video, but without subtitles. The text video without sound/footage included the text, formatted as videotext, and no music or footage. The control video consisted of the collage that appeared for seven seconds after which the title 'Nanotechnology' was presented for seven seconds and the video ended. This video did not provide any information about nanotechnology; it simply included sufficient material to get sensible EAST data.

EAST

The EAST was programmed in E-Prime (Schneider, Eschman, and Zuccolotto 2002). We adapted the original EAST (de Houwer 2003) to measure risk perception by changing the evaluations in dangerous-harmless. Also, we used pictures instead of words to express the risk and evaluations, as these could 'describe' nanotechnology better than words.

The procedure of the EAST was as follows. The instructions on the computer screen told the respondents that this was a reaction time task and that they should try to respond as fast as possible without making mistakes. The EAST consisted of three types of blocks. The first type of block was a practice block in which participants learnt to respond to the intrinsic valence (i.e., meaning) of the evaluation pictures (dangerous-harmless). They had to press the right key (P-key) when a dangerous picture (warning sign) was presented and had to use the left key (Q-key) for a harmless picture (information sign). Pictures were considered dangerous when they implied a threat to life or health. We chose five warnings, e.g., a poison and an electric shock warning. Harmless pictures should communicate safety. These were five information signs, e.g., a toilet and a parking sign. All were black-and-white pictures. The inter-trial interval, presentation time of the pictures and feedback interval were similar to the original EAST (de Houwer 2003). The second type of block was a coloured pictures practice block, in which the respondents learnt to classify nanotechnology pictures according to their extrinsic characteristics (i.e., colours: blue-left and green-right). The intrinsic valence of the pictures should be ignored. We chose 10 different nanochips, nanotubes and nanostructures, which were converted in a grey scale and covered under a blue or green colour film, similar to Callan and Ellard (2003). This resulted in 20 coloured nanotechnology pictures. Two versions of the EAST were developed to switch the colour-key combination and control for hand preference. Then, the third type of block followed: four test blocks. In each block, five dangerous, five harmless, 10 blue and 10 green pictures were randomly presented. Reaction times and response accuracy to the coloured pictures in the test blocks were recorded.

We expected that respondents who perceived nanotechnology as risky would react faster and make fewer mistakes on the coloured nanotechnology pictures that had the same response as the dangerous pictures (warning signs) than on coloured pictures that had the same response as the harmless pictures (information signs). Conversely, respondents who perceived nanotechnology as safe would react faster and make fewer mistakes on the coloured pictures that were associated with harmless than on coloured pictures that were associated with danger. The same EAST was used in the second measurement.

Questionnaire

The questionnaire at Measurement 1 included items about demographic characteristics (gender, age and current main course) and two items to measure the video's effects, namely the extent to which the respondents thought the video to be clear (*clarity*) and the extent to which they considered the video to be vivid (*vividness*). The questionnaires at Measurement 1 and 2 had the same risk perception and affect items. Ten items measured risk perception, namely perceived health damage, perceived environmental damage, health consequences, environmental consequences, probability estimation, order of seven risks in which the risk of nanotechnology should be placed (*order of risks*, ranging from 'the risk of drowning in a flood' to 'the risk of lung cancer caused by smoking'), six quotations of which the respondents selected the one that described their opinion about nanotechnology (*order of quotations*, ranging from '...not at all concerned...' to '...a very serious risk...'), concern, vulnerability and acceptability, all items related to nanotechnology.

At both measurements, the risk perception items perceived health damage, perceived environmental damage, health consequences, environmental consequences, probability estimation, order of quotations, concern and vulnerability had high internal reliabilities, Cronbach's α's$>.88$, $K=8$, $N=86$. We calculated the mean sum scores for these items in both measurements to form a risk perception scale. The items order of risks and acceptance had low item-test correlations. These were excluded from the risk perception scale at both measurements and were analysed separately.

Three items measured affect: a feeling thermometer about nanotechnology (ranging from negative (0) to positive (100)), positive feelings about nanotechnology, and negative feelings about nanotechnology. Except when mentioned differently, all items were measured on seven-point Likert scales; higher numbers indicated higher levels of the items measured.

Procedure

Respondents were invited to the laboratory of the Faculty of Health Sciences, Maastricht University, with a maximum of six respondents at a time. The experiment was presented as 'Information processing of visual material'. After entering the lab, the respondents were instructed about the experiment and the order in which they should do the different tasks. Respondents took place in separate cubicles in which they filled in the informed consent form. Next, they put on the headphone that was next to the computer screen. After watching the video, the computer automatically switched to the EAST and then the respondents completed the questionnaire. This was the end of Measurement 1. Respondents got paid and made an appointment for the second measurement. Each respondent received a unique number to link their data of Measurement 1 to those of Measurement 2.

Measurement 2 took place approximately two weeks later. The same procedure was used as before, except that the participants did not watch a video. After completion of this measurement, the respondents left the cubicle, were 'sworn to secrecy', thanked and paid for participation. Finally, they received written information about the experiment and nanotechnology.

Results

Manipulation check

The items clarity and vividness (Measurement 1) functioned as a manipulation check of the videos' effects. We conducted a Multivariate Analysis of Variance (MANOVA) with these two items as dependent variables and text (text videos vs. no-text videos) and sound/footage (sound/footage videos vs. no-sound/footage videos) as independent variables. Text and sound/footage had significant multivariate main effects on clarity and vividness, $Fs>8.07$, $ps<.001$. The univariate tests showed an effect of text on the item clarity, $F(1, 82)=79.27$, $p=.0001$. The text videos (text video without sound/footage, and text and sound/footage video) were perceived as more clear ($M=5.17$, $SD=1.46$) than the no-text videos (sound/footage video without text and control video, $M=2.33$, $SD=1.52$). The univariate tests also revealed an effect of sound/footage on vividness, $F(1, 82)=14.36$, $p=.0001$. The sound/footage videos were judged to be more vivid ($M=3.48$, $SD=1.70$) than the videos without sound/footage ($M=2.33$, $SD=1.15$). Thus, risk communications with moving pictures and sound were perceived as more vivid, while risk communications with text were perceived as more clear.

In the remaining section, the results are presented per hypothesis. First, we report the EAST results and then the questionnaire results that are relevant for each hypothesis.

Hypothesis 1

EAST

We used the mean reaction times and accuracy of the first response to the coloured pictures in the test blocks for the analyses. Reaction times that were lower than 300ms or higher than 3000ms were recoded as 300ms and 3000ms, respectively. Only the reaction times of correct first responses were used (de Houwer 2003; Greenwald, McGhee, and Schwartz 1998). The reaction times were skewed to the right and not normally distributed at both measurements, Ds $>.15$, $ps<.0001$. Therefore, we conducted a log-transformation on the reaction times. This somewhat improved the distributions at both measurements, Ds$<.13$, ps$>.001$.

The difference scores of the log-transformed reaction times and percentage of errors between the pictures associated with danger and the pictures associated with harmlessness were used to test whether nanotechnology would be more strongly associated with danger than with harmlessness in the EAST (Hypothesis 1). This would be the case if the respondents reacted faster to and/or made fewer errors on the association nanotechnology-danger than on the association nanotechnology-harmlessness.

The one-sample t-tests using the log-transformed reaction times and the percentage of errors showed that the difference scores of the log-transformed reaction times ($\Delta M_{\log}=-.01$, $SD_{\log}=.04$) and of the percentage of errors ($\Delta M=-2.38$, $SD=6.43$) were significantly negative at Measurement 1, ts<-2.64, ps$<.01$ (see Figure 1).[1] These findings suggest that the respondents associated nanotechnology more strongly with danger than with harmlessness.

We did the same analyses for the results of Measurement 2. The difference score of the log-transformed reactions times was significantly negative ($\Delta M_{\log}=-.01$, $SD_{\log}=.04$), $t(85)=-3.10$, $p=.003$ (see Figure 1). The respondents reacted faster to the pictures that were associated with danger than those associated with harmlessness. The difference score of the percentage of errors was not significantly different from zero, $t(85)=1.14$, $p=.26$ (see Figure 1). Thus, there is also some evidence that the respondents associated nanotechnology with danger at the second measurement. The findings of the first and second measurement combined lead to the conclusion that respondents associated nanotechnology more strongly with danger than with harmlessness.

Questionnaire

The risk perception scale showed that the respondents scored nanotechnology around the midpoint (4) of the seven-point Likert scale at Measurement 1 ($M=3.62$, $SD=1.01$). The respondents compared the risk of nanotechnology most often with the risk 'dying of cancer due to regular eating of scorched meat' (25.6%) in the order of risks item, which was the second lowest risk in the list. They scored the acceptability of nanotechnology around the average of the scale ($M=4.17$,

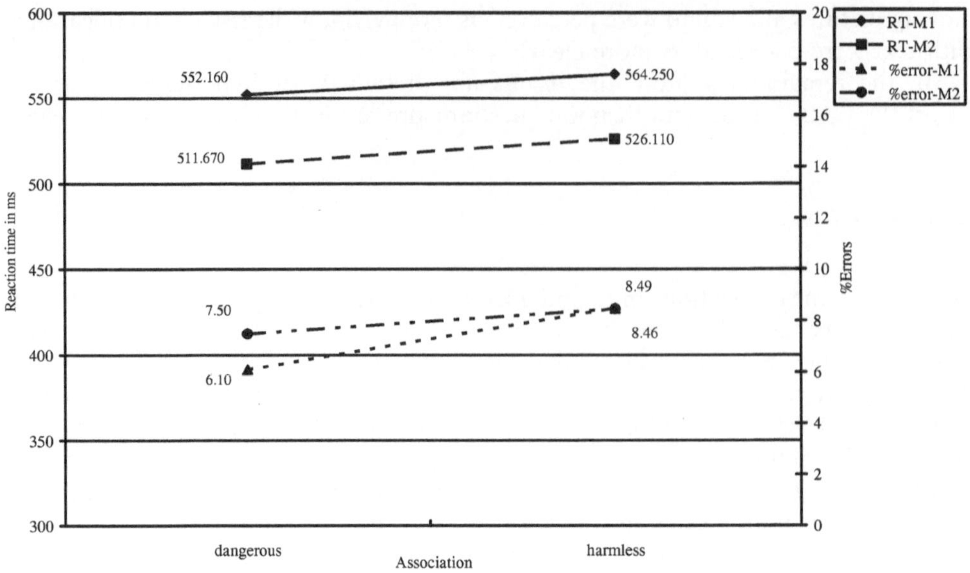

Figure 1. Mean untransformed reaction times in ms (left y-axis) and percentage of errors (right y-axis) per association in EAST, at Measurement 1 and Measurement 2. Note: For ease of understanding, we report the mean *untransformed* reaction times and corresponding standard deviations.

SD=1.40). The mean score on the feeling thermometer was around the neutral label of the scale from 0 to 100 (*M*=48.95, *SD*=16.51). The respondents were not extremely negative or positive about nanotechnology (3.41<*M*s<4.31, 1.26<*SD*s<1.32). So, the questionnaire results suggested that respondents had neutral feelings about nanotechnology and did not perceive it as risky.

In conclusion, the EAST data indicated that respondents associated nanotechnology more strongly with danger than with harmlessness. However, the questionnaire data suggested that respondents had a neutral perception of nanotechnology. Thus, the EAST data supported Hypothesis 3, whereas the questionnaire data did not. This may imply that a questionnaire and the EAST reveal different aspects of risk perception and that the EAST may be a complementary measure to a questionnaire.

Hypothesis 2a

EAST

Next, we tested Hypothesis 2a, namely that sound/footage videos would particularly affect the primary evaluative process as measured by the EAST. Therefore, we analysed a 2 (text: text videos vs. no-text videos) × 2 (sound/footage: sound/footage videos vs. no-sound/footage videos) × 2 (association: harmlessness vs. danger) Analysis of Variance (ANOVA) with repeated measures for the log-transformed reaction times. This analysis was repeated for the percentage of errors.

The results of both analyses showed that the interactions that would show an effect of text and/or sound/footage on association were not significant for both outcomes of the EAST, *F*s<2.30, *p*s>.13. This suggests that the stronger association between nanotechnology and danger than between nanotechnology and harmlessness was not influenced by the type of video the respondents had seen. Thus, Hypothesis 2a was not supported at Measurement 1.

We performed the same ANOVAs with repeated measurements on the EAST data of Measurement 2. The interaction between sound/footage and association was significant, $F(1, 82)=4.96$, $p=.03$. Paired samples t-tests (Bonferroni correction: $\alpha=.025$) showed that the sound/footage videos may have led to faster responses to the pictures associated with danger ($M_{\log}=2.70$, $SD_{\log}=.07$) than to the pictures associated with harmlessness ($M_{\log}=2.72$, $SD_{\log}=.07$), $t(41)=-3.96$, $p<.0001$ (see Figure 2). The no-sound/footage videos did not show this difference, $t(43)=.68$, $p=.50$.[2] The main effect of text and the other main and interaction effects were not significant, $Fs<1.88$, $ps>.19$. The ANOVA with repeated measures for percentage of errors showed no significant interaction effects, $Fs<3.19$, $ps>.08$.

Thus, after a delay period, only the sound/footage videos may have caused the respondents to associate nanotechnology more strongly with danger than with harmlessness as measured by the reaction times. These results are in accordance with Hypothesis 2a; the primary evaluative process was particularly affected by sound/footage videos, but only after a delay period.

Hypothesis 2b

Questionnaire

Hypothesis 2b related to the effects of text and sound/footage in videos on the secondary evaluative process, as measured by the questionnaire. At the first measurement, the MANOVA with the risk perception scale and the items order of

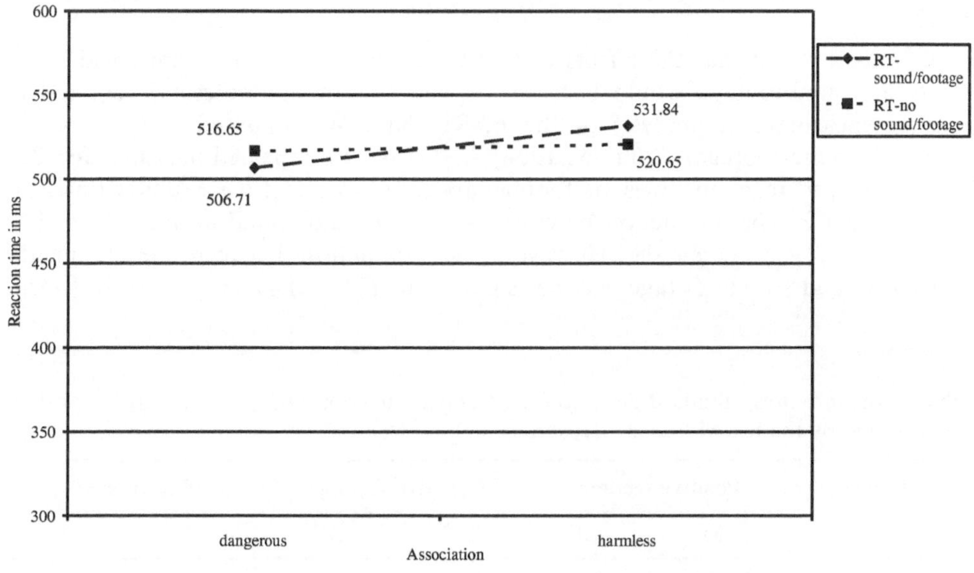

Figure 2. Mean untransformed reaction times in ms per association in EAST for sound/footage videos and no-sound/footage videos, at Measurement 2. Note: For ease of understanding, we report the mean *untransformed* reaction times and corresponding standard deviations.

risks, acceptance, feeling thermometer, positive feelings and negative feelings as dependent variables, and text and sound/footage as independent variables showed a significant multivariate main effect for text, $F(6, 77)=3.21$, $p=.01$, but not for sound/footage, nor for the interaction between text and sound/footage, $Fs<1.22$, $ps>.31$. The univariate tests showed that the respondents who had seen a text video evaluated nanotechnology less positively, more negatively, and reported less acceptance than the respondents of the no-text videos (see Table 1).

In short, one risk perception item (acceptance) and two affect items (positive and negative feelings) seemed to be influenced by the text in the videos at Measurement 1, which resulted in a more negative perception. These findings supported Hypothesis 2b; text videos appeared to influence the secondary evaluative process as measured by the questionnaire. However, risk perception as such (i.e., the risk perception scale) was not affected by text.

Next, we analysed the results of the second measurement. We conducted a MANOVA with the risk perception scale, the remaining risk perception items and the affect items as dependent variables, and text and sound/footage as independent variables. There were no significant multivariate effects for text, sound/footage or text and sound/footage on these items, $Fs<1.38$, $ps>.26$. Thus, at Measurement 2, the text and sound/footage variations did not influence any of the risk perception or affect items as measured by the questionnaire.

In sum, as expected, the text of the videos appeared to influence some of the questionnaire items immediately after watching the video. After a delay period, the effect of text video variation on the questionnaire items seemed to have declined.

Hypothesis 3

EAST

Hypothesis 3 stated that the effects of text would deteriorate over time, while the effects of sound/footage would last. An explicit test of Hypothesis 3 required a repeated measurement analysis of the EAST data. We analysed a 2 (measurement) × 2 (sound/footage) × 2 (association) ANOVA with repeated measures for the log-transformed reaction times, as footage appeared to affect the reaction times at Measurement 2. The interaction between association and sound/footage showed a trend, $F(1, 82)=3.11$, $p=.08$. However, the interaction between measurement, association and sound/footage was not significant, $F(1, 82)=1.71$, $p=.20$. In short,

Table 1. Mean scores, standard deviations and univariate effects of questionnaire items that were affected by the text videos at Measurement 1.

Type of video	Positive feelings		Negative feelings		Acceptance	
	M	SD	M	SD	M	SD
Text videos	3.07	1.09	4.76	1.01	3.81	1.45
No-text videos	3.73	1.45	3.89	1.33	4.53	1.27
Univariate effect	$F(1, 82)=5.37$, $p=.02$		$F(1, 82)=11.44$, $p=.001$		$F(1, 82)=6.01$, $p=.02$	

the effect of sound/footage on the association strength with danger was only present in Measurement 2, it was not an overall effect and did not seem to change over time.

Questionnaire

The finding that text had an effect on risk perception as measured by the questionnaire at Measurement 1 but not at Measurement 2 seemed to support Hypothesis 3. We conducted 2 (measurement) × 2 (text) ANOVAs with repeated measures on the items positive feelings, negative feelings, and acceptance, as these items were affected by text at the first measurement. Measurement appeared to have an effect on the positive feelings and negative feelings, Fs>6.11, ps<.02. Nanotechnology evoked more positive feelings (M=4.05, SD=1.26) and less negative feelings (M=3.85, SD=1.26) at Measurement 2 than at Measurement 1 (M=3.41, SD=1.32; M=4.31, SD=1.26, respectively). For negative feelings, the main effect of measurement was qualified by an interaction between measurement and text, $F(1, 37)=4.29$, $p=.05$. The respondents who saw a text video appeared to evaluate nanotechnology less negatively at Measurement 2 (M=3.77, SD=1.27) than at Measurement 1 (M=4.59, SD=1.14), $t(21)=3.37$, $p=.003$ (Bonferroni-correction: $\alpha=.025$). This was not found for the respondents who saw a video without text, $t(18)=.36$, $p=.73$. The scores on the item acceptance were not influenced by measurement or the interaction between measurement and text, Fs<1.39, ps>.24.

Thus, it appeared that respondents evaluated nanotechnology somewhat less negative over time and the text videos increased this effect on one item. We conclude that the questionnaire data did not provide clear support for Hypothesis 3.

Relation between EAST and questionnaire

In an explorative sense we wanted to find out whether the results of the EAST and of the questionnaire were related. Hence, we correlated the difference score of the mean log-transformed reaction times and the difference score of the percentage of errors of the EAST to the risk perception scale, the remaining risk perception items, and the affect items of the questionnaire, for the two measurements separately. At Measurement 1, we only found that the difference score of the percentage of errors of the EAST correlated significantly with the questionnaire item order of risks, $r=-.30$, $p=.01$. There were no other significant correlations between the EAST and the questionnaire items at Measurement 1 and 2, $-.01<rs<.16$, ps>.14. This may imply that the indirect and direct measures tap two separate processes of risk perception.

Discussion

Overall, we may conclude that the respondents spontaneously associated nanotechnology more strongly with danger than with harmlessness, as was shown by the EAST results (Hypothesis 1). The questionnaire results suggested a relatively neutral opinion about nanotechnology. Therefore, we believe that the EAST can be a valuable complement to a direct test in risk perception research. This suggestion may be supported by the finding that the EAST and the questionnaire data almost did not correlate. This may imply that the EAST reveals another aspect of risk perception, such as the primary evaluative process.

Our second and main hypothesis was that text would particularly influence the secondary evaluative process and sound/footage would mainly affect the primary evaluative process. Texts appeared to affect the acceptance of nanotechnology and the positive and negative feelings as tested by the questionnaire at Measurement 1. Thus, the text of the video seemed to have influenced the secondary evaluative process only directly after the intervention. The prediction that sound and footage would influence the primary evaluative process was only supported at the second measurement. At this measurement, the participants who had seen a sound/footage video associated nanotechnology more strongly with danger than with harmlessness; whereas those who had seen a video without sound/footage did not. This may imply that the sound/footage of videos can only affect risk perception through the primary evaluative process after a delay period. We further discuss the effect of time below.

Hypothesis 3 concerned the long-term effects of video risk communication. We proposed that the effect of sound/footage video would be more persistent than a text-only video. Nanotechnology was evaluated more negatively in the questionnaire at Measurement 1 than at Measurement 2. At the same time, the effect of sound/footage videos on the primary evaluative process was only present after a delay. However, an analysis that included both measurements did not show clear effects over time. Thus, there is no conclusive proof of Hypothesis 3 that the effects of sound/footage would last longer than the effects of text.

We interpret our findings as a support for the idea that text is suitable to affect the secondary evaluative process, while the combination of footage and music is fit to affect the primary evaluative process. However, one may also infer from the present findings that the text and sound/footage variation only influenced the measurement that required the same type of processing (Clark and Paivio 1991). The text videos may have influenced the questionnaire results because both require verbal processing, whereas the sound/footage videos may have affected the EAST results, because both involve visual processing. Nevertheless, the sound/footage videos only seemed to influence the EAST results after two weeks, so a match in type of processing cannot fully account for this effect.

We had expected that the sound/footage videos would influence the primary evaluative process directly after watching the video and not only after a delay period. This may imply that it takes longer to establish the effect of sound/footage information on the primary evaluative process than the effect of text information on the secondary evaluative process. The sound/footage videos seemed to require a sort of incubation time.

A so-called sleeper effect may be able to explain the delayed effect of sound/footage videos (Kumkale and Albarracín 2004). The sleeper effect is described as an initial suppression of a message's persuasive effect due to a discounting cue (e.g., a distrusted source) because the recipients feel that the message influences them incorrectly. At a later stage, the discounting cue is forgotten or decayed so that the message can persuade its recipients. Several studies have shown that the sleeper effect primarily occurs when the discounting cue is presented at the end of the persuasive message. Then, the effect of the discounting cue declines more rapidly than the effect of the message's content, and compared to when the cue is presented at the beginning of the message (Kumkale and Albarracín 2004; Pratkanis et al. 1988). This is called differential decay: the effects of the content and of the discounting cue decline at different rates.

Our findings suggest that the effect of the sound/footage videos on the EAST results may have been suppressed at the first measurement, perhaps because respondents experienced the end of the footage material, about nanotechnology and terrorism (e.g., the airplanes flying into the World Trade Centre at 9/11), as manipulative. This perceived manipulation may have operated as a discounting cue. As this was presented at the end of the video, the discounting cue declined more rapidly than the content of the sound/footage videos due to the sleeper effect and differential decay. Consequently, the persuasive effect of the sound/footage videos only appeared at Measurement 2.

The EAST is a new method to assess risk perception. We chose harmless and dangerous as evaluative valences in this test because these seemed more related to risk perception than positive and negative of the original EAST. These latter two evaluations have affective connotations, whereas harmless-dangerous may be denotations. It might be that the respondents translated the denotations of the risk perception valences into affective connotations. Thus, we are not sure whether the EAST measured risk perception or general affect.

Another limitation of this study may be that the videos varied in length, which may have influenced our findings. It was however difficult to control this factor; we would have had to make the content of the text videos longer with more information than the footage provided, or had to slow down the pace of the video text, which the respondents could have experienced as irritating and could have influenced our findings unexpectedly.

In conclusion, the findings of this study suggest that a video's text particularly affects cognitive processes related to risk perception, while its sound/footage particularly affects people's gut feeling about a risk. Furthermore, this study exemplifies once more that an indirect measure of risk perception, such as the EAST, may be a valuable complement to other risk perception measures, for example questionnaires. The results of this study indicate that people not only rely on cognitive processes when interpreting a risk but also on their gut feeling. It would be interesting to explore further how risk communication can influence these two processes immediately and in a long-lasting way, because both play a role in forming a risk perception and appear to react differently to risk messages.

Notes

1. Because the log-transformed reaction times were still not normally distributed, we also analyzed these data using the non-parametric Wilcoxon Signed Ranks Test. This test showed similar results: the respondents reacted faster to the pictures related with danger than those related with harmlessness at both measurements, $zs < -2.46$, $ps < .01$.
2. The results of the Wilcoxon Signed Ranks Tests revealed similar results: the respondents who had seen a sound/footage video, responded faster to the pictures associated with danger than those associated with harmlessness at Measurement 2, $z = -3.45$, $p = .001$. While the respondents who had seen a no sound/footage video responded just as fast to the pictures associated with danger as to the pictures associated with harmlessness at this measurement, $z = -.05$, $p = .96$.

References

Callan, M.J., and J.H. Ellard. 2003. The effects of justice outcome on implicit victim derogation. Paper presented at the Justice Preconference of the 4th Annual Meeting of the Society for Personality and Social Psychology, February, in Universal City, Los Angeles, CA.

Chaiken, S., and A.H. Eagly. 1983. Communication modality as a determinant of persuasion: The role of communicator salience. *Journal of Personality and Social Psychology* 45: 251–6.

Chaiken, S., A. Liberman, and A.H. Eagly. 1989. Heuristic and systematic information processing within and beyond the persuasion context. In *Unintended thought*, ed. J.S. Uleman and J.A. Bargh, 212–52. New York: Guilford.

Clark, J.M., and A. Paivio. 1991. Dual coding theory and education. *Educational Psychology Review* 3: 149–210.

de Houwer, J. 2003. The Extrinsic Affective Simon Task. *Experimental Psychology* 50: 77–85.

de Houwer, J., and E. de Bruycker. 2007. Implicit attitudes toward meat and vegetables in vegetarians and nonvegetarians. *International Journal of Psychology* 42: 158–65.

Fazio, R.H., and M.A. Olson. 2003. Implicit measures in social cognition research: Their meaning and use. *Annual Review of Psychology* 54: 297–327.

Finucane, M.L., A. Alhakami, P. Slovic, and S.M. Johnson. 2000. The affect heuristic in judgments of risks and benefits. *Journal of Behavioral Decision Making* 13: 1–17.

Finucane, M.L., E. Peters, and P. Slovic. 2003. Judgment and decision making: The dance of affect and reason. In *Emerging perspectives on judgment and decision research*, ed. S.L. Schneider and J. Shanteau, 327–64. Cambridge: Cambridge University Press.

Greenwald, A.G., D.E. McGhee, and J.L.K. Schwartz. 1998. Measuring individual differences in implicit cognition: The implicit association test. *Journal of Personality and Social Psychology* 74: 1464–80.

Hendrickx, L., C. Vlek, and H. Oppewal. 1989. Relative importance of scenario information and frequency information in the judgment of risk. *Acta Psychologica* 72: 41–63.

Huijding, J., and P.J. de Jong. 2005. A pictorial version of the Extrinsic Affective Simon Task: Sensitivity to generally affective and phobia-relevant stimuli in high and low spider fearful individuals. *Experimental Psychology* 52: 289–95.

Kumkale, G.T., and D. Albarracín. 2004. The sleeper effect in persuasion: A meta-analytical review. *Psychological Bulletin* 130: 143–72.

Loewenstein, G.F., E.U. Weber, C.K. Hsee, and N. Welch. 2001. Risk as feelings. *Psychological Bulletin* 127: 267–86.

MacGregor, D.G., P. Slovic, and M.G. Morgan. 1994. Perception of risks from electromagnetic fields: A psychometric evaluation of a risk-communication approach. *Risk Analysis* 14: 815–28.

Meijnders, A.L., C.J.H. Midden, and H.A.M. Wilke. 2001. Role of negative emotion in communication about CO_2 risks. *Risk Analysis* 21: 955–66.

Mousavi, S.Y., R. Low, and J. Sweller. 1995. Reducing cognitive load by mixing auditory and visual presentation modes. *Journal of Educational Psychology* 87: 319–34.

Pfau, M., R.L. Holbert, S.J. Zubric, N.H. Pasha, and W.K. Lin. 2000. Role and influence of communication modality in the process of resistance to persuasion. *Media Psychology* 2: 1–33.

Pratkanis, A.R., A.G. Greenwald, M.R. Leippe, and M.H. Baumgardner. 1988. In search of reliable persuasion effects: III. The sleeper effect is dead. Long live the sleeper effect. *Journal of Personality and Social Psychology* 54: 203–18.

Schneider, W., A. Eschman, and A. Zuccolotto. 2002. E-Prime [1.1] Pittsburgh: Psychology Software Tools Inc.

Sherman, S.J., R.B. Cialdini, D.F. Schwartzman, and K.D. Reynolds. 1985. Imagining can heighten or lower the perceived likelihood of contracting a disease: The mediating effect of ease of imagery. *Personality and Social Psychology Bulletin* 11: 118–27.

Siegrist, M., C. Keller, and M.E. Cousin. 2006. Implicit attitudes toward nuclear power and mobile phone base stations: Support for the affect heuristic. *Risk Analysis* 26: 1021–9.

Slovic, P., M.L. Finucane, E. Peters, and D.G. MacGregor. 2004. Risk as analysis and risk as feelings: Some thoughts about affect, reason, risk, and rationality. *Risk Analysis* 24: 311–22.

Strack, F., and R. Deutsch. 2004. Reflective and impulsive determinants of social behavior. *Personality and Social Psychology Review* 8: 220–47.

The Royal Society & The Royal Academy of Engineering. 2004. Nanoscience and nanotechnologies: Opportunities and uncertainties. www.nanotec.org.uk/finalReport.htm.

van Est, R., I. Malsch, and A. Rip. 2004. *Om het kleine te waarderen… Een schets van nanotechnologie: publiek debat, toepassingsgebieden en maatschappelijke aandachtspunten [To value the small things… A draft of nanotechnology: public debate, fields of application and social points of interest]*. (Working document 93). The Hague: Rathenau Instituut.

Visschers, V.H.M., R.M. Meertens, W.F. Passchier, and N.K. de Vries. 2007. An associative approach to risk perception: Measuring the effects of risk communications directly and indirectly. *Journal of Risk Research* 10: 371–83.

Zajonc, R.B. 1980. Feeling and thinking: Preferences need no inferences. *American Psychologist* 35: 151–75.

The illusion of economic objectivity: linking local risks of credibility loss to global risks of climate change

Annette Henning

Solar Energy Research Center (SERC), Dalarna University, Sweden

This article analyses economic thinking as a conceptual construct and discusses the link it provides between local risks of losing credibility and global risks of resource depletion and climate change. Due to the high-ranking value given to economic thinking in many Swedish contexts, economic arguments are perfect ways to win an argument or negotiation, to prevent risks of losing credibility, or to discredit others. As illustrated in this article, legitimacy-creating acts of communication tend to precede or follow upon a decision to install a solar heating system. An immediate and massive use of solar energy for heating purposes would mitigate the risks of global warming and depletion of energy resources. Unfortunately however, the conspicuous use of economic arguments accompanying solar heating installations tends to be interpreted by policy-makers as a preference for economic motives rather than a sign of questioned consumption. The illusion that economy is an objective entity rising above all other social issues seems to be confirmed, and political measures based on economic thinking can continue to block adequate steps towards the mitigation of climate changing emissions. The cultural representation of economic sovereignty is continuously reproduced through communication. It is a circle; a vicious circle if you like.

Introduction

In this article, 'economic objectivity' is analysed as a cultural construct, continuously reproduced through power-related communication. The article argues that the more one's choice of consumption is being questioned, or the more one needs to discredit the choice of the other, the more important it becomes to emphasise economy as a main motive. Furthermore, it is argued that an immediate and massive use of solar energy for heating purposes would help to mitigate the risks of both energy resource depletion and global warming. Unfortunately, the hegemonic belief in economy as a reified and objective entity tends to prevent house owners, politicians, researchers and others from taking initiatives in this direction.

Risk researchers have often been occupied with the impact on people of so-called 'modern risks' and with how scientific information concerning human health, nutrition, deceases, chemicals and other risk issues best should be conveyed to the general public (Boholm 2003; Caplan 2000). The concern here is rather the opposite. Swedish authorities need to be informed about the risks involved in taking certain culture-specific modes of thought for granted, modes of thought of which they, themselves, are an integral part. Thus, rather than dealing with communication about risk, the article concerns the link between economy discourse in everyday life and two kinds of risks: the local risks of credibility loss and the global risks of climate change.

The data on which this analysis is based were collected during more than four years in the late 1990s. This social anthropological research project had the aim of studying attempts to introduce solar heating in Swedish contexts (Henning 2000). The main methods used were qualitative interviews and participant observation. In total, 100 interviews were made during the project. In this article, empirical data are used from three of the case-studies in particular. Most probably, however, the results are also highly relevant for a much larger area than the particular contexts and social actors described in this article.

Solar thermal systems and economic thinking

By the mid-1990s, solar collectors in Sweden were artefacts still not fully integrated into people's everyday lives or perceived as a self-evident necessity (Figure 1). For this and other reasons the solar collector had an ambiguous position that made it more liable to be questioned than a new car or refitting a kitchen, for example (see more about this in Henning 2000). One could not be sure that others would respect one's choice of purchase. It was therefore especially important to emphasise those parts of one's motives that were perceived as the most legitimate ones.

In Sweden, even today, it is not enough to have the money needed to buy a solar heating system. Often, one also has to be able to prove to others that the heating system would not cost them anything. Preferably, buying this system should even be a way of saving money. This conspicuous need to prove to others the profitability (or non-profitability) of solar energy plants, as well as the particular ways in which this has to be done, primarily illustrates a wish to legitimise and create acceptance for a decision on consumption that is not self-evident in the surrounding society. Because of this situation, negotiations and other communication preceding or following upon plans to install solar thermal systems are suitable examples for illustrating the cognitive structures imbedded in concepts like 'useful actions' and 'pure economy'.

Solar thermal systems and climate change

The cultural representation of economy as a most highly valued, objective and emotionally 'pure' entity is hegemonic and mainly taken for granted in this culture area. Therefore, an excessive use of economic arguments tends to be interpreted by politicians as a confirmation of their assumption that solar thermal systems would become more popular if only they were less expensive. Thus, rather than focusing on the main problems following upon weak legitimacy, such as lack of accessibility and goodwill, the Swedish government predominantly uses subsidies and research funding to accomplish price reduction.

This is particularly unfortunate, since a wider use of solar energy would decrease the use of oil, gas, coal and bio-fuel for heating, thereby reducing climate change emissions from buildings. According to Roaf, Crichton, and Nicol (2005), buildings are responsible for producing over half of all climate change emissions in the world. Furthermore, an extensive and rapid conversion to solar energy could replace a large part of valuable energy-carriers like bio-fuel and electricity. If our global resources of less polluting fuels like bio-pellets are to suffice, we clearly need to reduce demand for energy.

Figure 1. Photo of Swedish house with solar collectors on the roof.

Primarily, solar thermal systems are dependant on a clear sky rather than a warm outdoor temperature. As buildings in Sweden need to be heated during cold days with clear sky in spring and autumn, these heating systems are particularly suited for Swedish conditions. The building sector accounts for 40% of the total energy consumption, and heating purposes accounts for the major part of that (Swedish Energy Agency 2001). A solar heating system replaces about 25% of the energy used for heating a single-family house (Lorenz in Henning 2007). However, so far, solar thermal systems replace merely 33 GWh energy in Sweden per year (Swedish Energy Agency 2007).

Actions perceived as ordered on a ranked scale

Gullestad (1992, 80f) has said that 'Norwegians often play not out in the open but under the cover of doing something practical and reasonable'. In her study of Norwegian households, she describes how projects of home redecoration are most often explained as being for functional or financial reasons rather than for the creative or aesthetic ones people are also very obviously motivated by.

The results from interviews and participant observation, on which I base this article, are congruent with these statements. The concepts of 'usefulness' and 'basic need' tend to be emphasised when priorities are explicitly described to others, while various feelings of pleasure or interest tend to be played down. For example, I interviewed and studied meetings held by 20 men and women in the small Swedish village Österå. Among these villagers, usefulness and basic need were the most common reasons given for giving, for instance, a new kitchen interior or extension to the house a higher priority than a partial change of heating system.

A further conclusion one might draw from these and other similar results from the project, is that, in Swedish contexts, cultural representations of actions are, in fact, ordered on a ranked scale (Figure 2). On this cognitive scale of actions, those perceived as practical and useful are more highly valued, primarily in the sense of being more credible, than actions seen as motivated by the opposite, that is by feelings, enjoyable interests, or ideas that go beyond the practicalities of everyday life. This is not to say that such motives do not have at least as much impact on household priorities.

Of all useful actions, those perceived as motivated by economic reasons are, for most interviewees, the most high-ranking ones. These actions are the ones often referred to as 'pure economy' (*krass ekonomi*). An action which is based on the idea of 'pure economy' involves a judgement of price and a notion of earning or saving

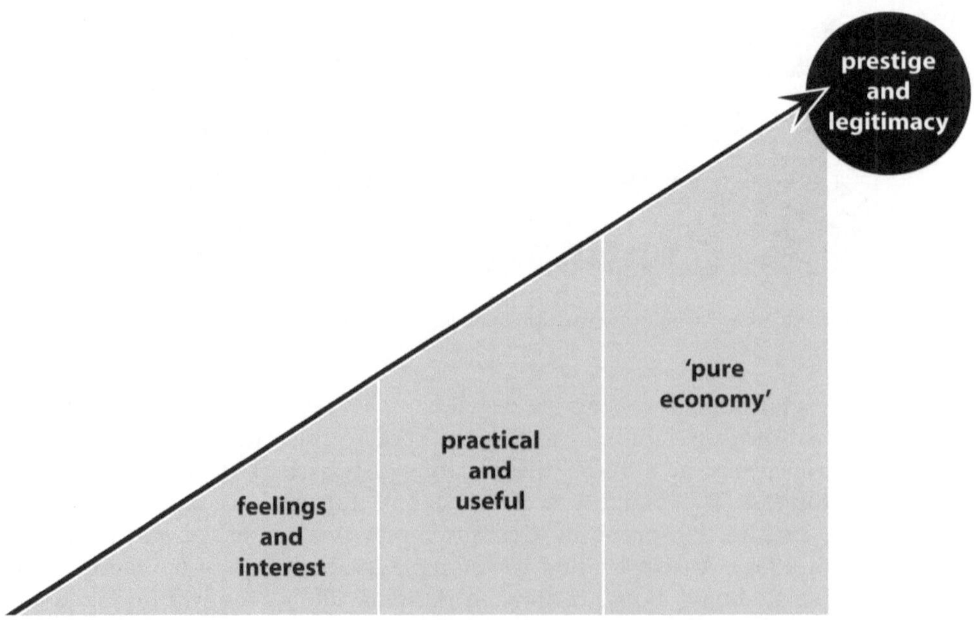

Figure 2. Some motives are more capable than others of giving prestige and legitimacy.

money. It is understood as uncontaminated by sentiment and ideas that are not basic and down to earth. It is also seen as free from any necessary association with specific actions or artefacts. Throughout the entire field-study, my results strongly indicated that actions which are, in this way, culturally constructed as free from emotion, are not only most highly valued, but have a hegemonic position within a large area.

Preventing credibility loss - case study 1

In the following case study, economic thinking and argumentation is used to prevent credibility loss. Furthermore, the social pressure expressed through a shared idea of a value-scale of actions set up limits which finally forced a municipal energy company to change their strategy.

The outcome of the plans for a large-scale solar heating plant was surprising to most of those who knew about these plans. Two years earlier, the municipal energy company had decided to replace the heat-pumps in Torsång village with solar heating. There turned out to be a problem, however, as this area was classified as an area of national interest. A long-term process of negotiations and attempts to reach a compromise was started between the directors and employees at the energy company and the antiquarians at the Department for Culture Environment.

Finally, two years later, it seemed as if the energy company had won the battle. After all this time of endless negotiations and attempts to find a compromise, after nine suggested and carefully planned locations and three applications for building permits, a permit to start the construction process was finally granted. It was in that situation of apparent success the energy company decided to withdraw their application. Their reason for doing so was, in their own words, an additional cost for archaeological diggings. Now, how could this final cost, minor compared with what had already been spent on this project, prevent the company from going through with the project? In my opinion, the true answer is that the company could not take the risk of losing credibility in front of other energy companies. Throughout the entire project period, the company men had fought against a taken-for-granted assumption among energy companies of solar thermal technologies as unprofitable per se. They had persistently contended that this plant would be different; this one would make a profit. Apart from the practical aim of providing the village with heat and the unspoken aim of improving their image and prestige, the energy company had had two main goals with this project. One was to produce environmentally benign energy, the other to produce it cheaply. They would still have been able to produce environmentally benign energy, although not cheaply enough to gain a good reputation. As their understanding of legitimate actions were intimately connected to the shared perception of actions as ordered on a ranked scale, the company simply could not take the risk of losing credibility by agreeing to pay the small final costs for archaeological diggings.

The economic self and emotional other

On the ranked scale of differently valued actions, the interviewees are most inclined to label their own actions economic or useful (and to consider them non-emotional), while saving the lower ranking actions for others. Thus, while solar collector owners themselves usually accentuate the practical, useful or economic sides of their motives, their purchasing decisions may be described as less sincere when commented upon by someone else.

This tendency is particularly obvious in relation to environmental or technical reasons for buying solar collectors. While environmental motives may be referred to by others as philosophy and idealism, technical motives may be referred to as sheer play. A slight laugh often accentuates the perception of such motives as rather ridiculous. A female engineer told me, for instance, that her only interest in technology was that it should work: 'But men can be interested in technology for its own sake. For them, it doesn't necessarily have to be useful', she laughed. 'It is enough if it is interesting'.

Giving an activity an air of 'pure economy', and thus freeing them from emotions, provides men and women with one of the best ways of giving themselves credibility and legitimacy, while making another person's activity seem emotionally loaded can render it less credible. Contrasting one's own actions to the lower-ranking actions of others is also a way of clarifying the legitimacy of one's own activity or decision. The following quotation is typical for how many of the villagers in Österå would explain to me how they had reached their decision to install a solar heating system or not to do it:

> It is a question of priority what you do in your spare time and what you decide to invest in. Right now we're building this extension to our house. But that solar heating system would have come up high on my list of priorities, so to speak, if only I could have earned money on it. Time and money. That's what it's all about. That is, not for those who have a burning commitment for environmental issues or so. The solar collector rushes right up to the top of their list anyway. My reasons, however, were purely economical. I wasn't interested for mere interest's sake. I wanted to see if it could be practically and economically feasible to install a solar collector.

Discrediting others - case study 2

This following example illustrates in more detail how the value-scale of actions may be used to discredit others. In one of my case studies, I followed the attempts made by pre-school teachers to include the installation of a solar heating system in their activities with the children. In this final meeting, three men from the municipal building office enhanced their position by repeatedly referring to their own professionalism and practical knowledge, while discrediting the aims and knowledge of the female teachers and the woman from the child-care department by referring to their feelings.

The pre-school teachers had some environmental aims with the solar heating system as they wanted to be able to show parents and neighbours that the child-care centre cared about the future environment and to show the power of one's own actions. Primarily though, their aims were of pedagogic nature. They wanted to include the children in the practical construction process. They made clear that they wanted the solar collector to be part of their activities with the children. They did not want to stand with them on the other side of the yard while it was being installed. Neither did they want a model, but a full system.

One of the men pointed at a technical drawing of a solar heating system and used technical terms to explain why it would be incredibly difficult for a pre-school teacher to take part in the construction of a plant like this with its 'by-pass groups and tempering valves'. The woman from the child-care department was not at all convinced that it would be impossible to take an active part in the building and installation process, and referred to those who do this on their own houses. 'I am not sure how much you have to know', she said, 'but people do buy self-build collectors and put them together themselves with some instruction and supervision. Perhaps we could get some more help than people normally get, and perhaps it would be possible to build the collector on the ground before it is put on the roof?' One of the men from the building office responded by asking whether they couldn't just construct a solar shower for the summer. She answered that this would be to miss one of the main points of it all. The point was not to produce another model with the children.

Finally after a long meeting where the woman from the child-care department tried to draw the men into a discussion on her terms by asking for their advice, and the men repeatedly explained the impossibility of it all, the head of the building office found it was time to come to an agreement. In formulating the final conclusions, he played down the professional aims of the pre-school teachers by contrasting them to the professionalism of the others: 'Now, we have dropped the self-build philosophy and chosen a professional philosophy'. He also emphasised the importance of their (low-ranking) feelings: '... but we will still let the personnel have information so that they can feel involved the whole time anyway'. This meant that the practical training through which the teachers had hoped to learn how to put together a solar collector was replaced by a 30-minute lecture. It was also agreed that the company that would deliver the ready-made full-scale solar collector would also help the personnel to 'achieve their little model'.

The right frame of thoughts

Due to the ambiguous position of solar collectors in Sweden, usefulness, and its highest form, the emotionally pure economy, are normally emphasised when considering a solar heating installation. In order to strive for this highest form of legitimacy, however, certain implicit rules have to be followed.

When someone in Sweden is contemplating the purchase of a solar heating system for his or her own house, many of the family's different wants and needs are taken into account. Often, the house itself works as a frame for one's thoughts when making priorities between different inputs of work or money. The choice is therefore often a choice between a solar heating installation, a new garage, barn or fence, or some repair or other work done on or inside the house. When, on the other hand, the decision to purchase or not to purchase a solar heating system is talked about with others, or when the price of such a system is actively thought about, it is the idea of 'pure actions' (cleansed from emotions) that is in focus.

The original idea of frame analysis, as proposed by Miller (1992, 100ff), is that the frame helps us recognise the picture inside it as art. The frame is not there to draw attention to itself, but to strengthen one's attention and direct it towards the work of art. However, the frame can also consist of artefacts like furniture, which means that the opinion we form of a person we meet at home for the first time is as much affected by the home environment as by the conversation we have among the furniture. The frame may be an art gallery, a theatre, or a church, settings which give us clues as to what sort of activity or behaviour we are involved in. Even a more abstract category like the word art may, according to Miller (1992), function as a frame that gives us clues to appropriate attitudes and value judgements. We can therefore carry the reasoning one step further and see a threatening loss of credibility among friends, neighbours or colleagues as constituting a 'frame of thoughts' that not only directs thinking towards price estimations, but also directs the idea of how such price estimations should be made.

Comparing the right goods

So how do we know how to estimate a price in a culturally correct way? First of all, the concept of 'pure economy' (*krass ekonomi*) means that the price of a solar heating system is deemed high or low by a comparison with other artefacts or materials that could be used to produce heat. This way of estimating a price and describing this as a main motive to others is used by all kinds of potential or non-potential customers, also those who explicitly talk of other motives for a purchase too and those who have no intention of buying a solar system. By making a 'correct' comparison they can show others that they think and act in a respected and accepted way, independent of their interests or lack of interests in solar collectors. The culture-specific knowledge of how these comparisons should be made is, of course, mainly implicit and taken for granted. The world is filled with an almost infinite amount of totally heterogeneous and separate artefacts, but this is not how we humans experience it. From the moment we are born we classify, grade and assort the immense amount of artefacts and all sorts of other impressions we are constantly inundated with. It is part of our human nature to try and create an understandable universe for us, and classifying is one way to do this. By the cognitive order we thus create for ourselves, we also create room for further thoughts (Kopytoff 1990; Douglas and Isherwood 1979; Douglas 1988). 'What we usually refer to as structure', Kopytoff says, 'lies between the heterogeneity of too much splitting and the homogeneity of too much lumping' (1990, 70). This means that certain artefacts get grouped together and classified as a particular category of things. These may be things perceived as having similar value. One of them is seen as just as good as another within that cognitive category, or 'value class' (cf. Kopytoff 1990; Sahlins 1976, 214f).

In social anthropological literature, value classes used for the exchange of goods are usually discussed as 'spheres of exchange', a concept first used by Bohannan in his study of the Tiv people in the 1950s (Bohannan 1959; Guyer 1995; Hylland-Eriksen 1995). This concept is used to describe a phenomenon where goods from one such sphere of exchange cannot be exchanged with goods from another. Objects which belong to different spheres are cognitively almost completely separated from one another in the context of exchange, and it is therefore very difficult to compare them.

Kopytoff (1990) has pointed at the acknowledged importance of spheres of exchange in every society, and that, in this realm of exchange values, the natural world of singular things must be arranged into several manageable value classes. 'Different things must be selected and made cognitively similar when put together within each category and dissimilar when put into different categories' (1990, 70). Kopytoff, however, emphasised at the same time that 'Westerners are most apt to perceive such value classes in uncommercialized and unmonetized economies'.

My argument, however, is that we do find that artefacts in Sweden (an undoubtedly commercialised monetary economy) are categorised into separate value classes, although in a slightly different form than anthropologists have normally used the concept. Within such a value class the objects are compared for the purpose of buying one of these many comparable things, and not for the purpose of exchanging one for another. They are not comparable in a context of barter, but in a context of purchase and price estimation.

Why a solar heating system is more expensive than a car and most expensive for those who use cheap wood

In a context of price estimation we mainly find only very similar objects grouped together and compared, such as cars of all models and makes or computers of different brands. A comparison between goods is now confined to this one specific value class, and it becomes impossible to compare a car, for instance, with any other kind of artefact.

As prices only mean something in relation to the goods they are connected to, an estimation of whether a price is high or low can only be made by comparing the price of one commodity with the price of another commodity in the same value class. This means that money cannot simply be seen as a relational element between goods, as Simmel has suggested (1978, 128 in Miller 1992, 71). As far as I can see, it is the goods-with-prices that are tied together cognitively in a value class, thereby making the prices comparable. And this, I contend, is the very essence of 'pure economy', of how culturally correct price estimations work. This is also the main reason why 40,000 SEK may seem expensive for a solar heating system, while three times the amount for a new kitchen or car may not.

For some reason I have not, as yet, figured out, Swedish solar heating systems belong to a broadly defined value class, including not only heating systems, but also energy-carriers with their respective present and also expected future prices. It is therefore fully possible to compare the present price for the artefact solar collector with expected future costs for energy-carriers like oil or electricity. The opposite is also common, where the present costs for an old oil-burner in use are compared to the price of a new solar collector. This means that solar heating becomes cheaper for those who switch from an expensive energy-carrier: 'I use cheap wood now, so purely economically I will not save as much as all the others who use electric heating and suchlike'. So far, I have not come across a price estimation made from scratch where the price for a new solar heating system is, for example, compared to a new oil-boiler plus the cost of oil during a certain period of time.

How to save money by spending it

In order to legitimise a solar collector purchase, tempted buyers may not only use price comparisons according to 'pure economy' to show the low price of the collector. They may also try to show that they even save money on the purchase.

Miller (1996) says that, even though the North London consumers he studied spent a lot of money on food and various consumer goods, they perceived this spending themselves as saving. My material from Österå village is congruent with Miller's experience, and I would add to his discussion the suggestion that the reason why this strange way of saving money works is that certain goods and their prices are cognitively connected to certain other goods with prices. For example; people can create a sense of saving by telling others that they have installed a cheaper solar collector than another more expensive type, which they could have bought (but of course never would). Naturally, 'saving money' in this sense does not actually give people any more money in their pockets, but it does give their purchasing decision more credibility.

Myers (2007) has shown how shared and taken-for-granted perspectives and commonplace, generally known arguments may be used in communication to deal with things that threaten the ways one wishes to be seen by others. In two ways, in particular, the Österå villagers used monetary terms to achieve a culturally more accepted way of presenting their motives for installing a solar heating system. Some of them managed to do part of the construction and installation-job themselves during weekends, or used their social contacts to get hold of cheaper components. In describing this situation, they would try to translate their input into the wage for a hypothetical salaried work. Others talked about their wish to 'save time'. This was especially true for those who planned to install a solar heating system in order to reduce the work with firewood. 'Saving time' is a motive usually seen as practical and useful. But by using monetary concepts like 'saving', 'worth', 'fortune', 'money', and 'gain', the solar heating project could also achieve some of the higher value and credibility of the really pure economy:

> A solar collector would save me time. Saving a few working-days per year would be worth a fortune. You may estimate those working days in money if you wish, but the important thing is that to me it would be a real gain.

Conclusion

An important aim of this article was to analyse the cognitive structure of economic thinking in a specific cultural context. Another aim was to describe how the belief in economy as an objective and neutral instrument is constantly reproduced through communication. Due to the high-ranking value given to economy in Swedish contexts, making use of the cultural representation of a socially uncontaminated economy is a perfect way to win an argument or negotiation, or to prevent the risk of losing credibility. In turn, the hegemonic position of economic thinking is created and recreated through these acts of communication. I am far from the first to assert that economic objectivity is nothing more than a culturally formed representation of the world. In 1976, Sahlins (166f) proclaimed that 'Materialism and utilitarianism not only have been the tools of certain social scientists, but constitute the very way in which people in the industrialised countries generally perceive their society and their life. In the native perception, economy is an arena for objective, rational, and need oriented actions'. Also, several other authors have emphasised the many problems with the present state of things where economy and economists are given too much power and influence (Aune 1997; Carrier and Miller 1999; Shove et al. 1998; Wilhite 2000; Wilhite et al. 2000).

Even so, to analyse cognitive characteristics of contextualised economic thinking in the sense this has been done in the present article seems to be a less prominent approach within economic anthropology (Guyer 1995; Parry and Bloch 1991). Dumont's interpretation of classical structuralism may be closest to this approach. According to Dumont, classical structuralism was developed as a technique to analyse the formal organisation of ideas. Dumont added a notion of value and a concept of hierarchy, however. He was of the opinion that meaning arises from making conceptual distinctions that, in turn, are ranked and hence always contain an element of value (Graeber 2005, 449).

In order to understand how people in Sweden tend to act upon the risk of losing social credibility, we need to be able to see how motives for actions are being ranked on a scale of more or less valuable actions. On this cognitive and culture-specific value-scale of actions we find activities motivated by emotions and interest at the bottom, activities seen as practical and useful rather high up on the scale, and the most prestigious activities motivated by 'pure economy' at the very top of the list. As a proper use of this cognitive scale helps to create social legitimacy, the necessity of making active use of it increases, for example, the more a purchasing decision is being questioned.

The value-scale of actions can easily be used as an instrument of power. We saw this happening during the day-care centre negotiations. The two main professional groups who participated in these clearly differed in how they perceived and wished to handle solar collectors and the risks of climate change emissions. Nevertheless, both sides took the value-scale of actions for granted, using it in their struggle to control the outcome of this specific social battle and to prevent the risk of losing it. Consequently, the men in one of the professional groups consistently discredited the women in the other professional group by referring to their emotions, idealism and need of help, and by contrasting these less valued motives for action with what they considered was their own professionalism and economic and more down-to-earth thinking.

The essence of the most valued motive for action, 'pure economy', is to be able to prove that one thinks and communicates in the right way. To communicate in a way that is understood as natural and correct, one needs to use the right frame of thoughts and be able to compare the right kind of goods. In social anthropological literature, value classes used for the exchange of goods are usually discussed as 'spheres of exchange'. My argument is that artefacts in Sweden are also cognitively classified as belonging to separate value classes. However, these artefacts are not comparable in a context of barter, but in a context of purchase and price estimation. I also suggest that it is the goods-with-prices that are tied together cognitively in a value class, thereby making the prices comparable. This reasoning means, for example, that the existence of a broad value class encompassing solar collectors provides us with an explanation to why these artefacts are perceived as being more expensive than a car, why they are seen as most expensive for those who use cheap wood, and how it is possible to save money by spending it.

The analysis presented in this article is based on ethnographic field-research, as proposed by several researchers of risk communication. Phillimore (2006, 2) has pointed out that 'risk communication might also be seen as part of the fabric of everyday life'. Boholm (2006) and Reith (2004) have argued that the concept of risk should be seen as a comparative term, a temporal construct, and an epistemological category. And the authors of the anthology *Risk Revisited* (Caplan 2000, 25) propose a further use of ethnographic methods which considers risk in particular times and places, and through the voices of particular informants. However, Boholm (2003) has also warned us to focus exclusively on relativism, on the socially constructed risk, as this could make us unable to account for the juxtapositions and amalgamations of 'objective' and 'subjective' perspectives on risk. Also Caplan (2000) gives an overview of a number of reviewers who argue that environmental concerns are more 'real' than Douglas and Wildavsky appear to suggest in their work on risk.

In my research on the use and implementation of solar thermal systems, I am repeatedly reminded that the belief in the sovereignty of economy prevents potential buyers from installing solar heating and policy-makers from finding ways to make this technology more popular and accessible. The idea that a low price will solve everything rises like a brick-wall, effectively preventing all other actions. It is a circle, a vicious circle if you like, in which local ways of perceiving and handling social risks have implications for our ability to mitigate the escalating, global and, in a sense, more 'objective' risks of climate change and depletion of resources.

Acknowledgement

The photo in Figure 1 was taken by my colleague Kent Börjesson, and Molly B&B (mollybob.se) helped me with the graphic form in Figure 2.

References

Aune, M. 1997. *'Nöktern eller nytende'. Energiforbruk og hverdagsliv i norska husholdninger*. Trondheim: Center for teknologi og samfunn, Norges teknisk-naturvitenskapelige universitet, Trondheim.

Bohannan, P. 1959. The impact of money on an African subsistence economy. *Journal of Economic History* 19: 491–503.

Boholm, Å. 2003. The cultural nature of risk: Can there be an anthropology of uncertainty? Ethnos 682: 159–78.

———. 2006. *The public meeting as a theatre of dissent: Risk and hazard in land use and environmental planning*. Paper presented at RISKCOM 2006, New perspectives on risk communication: Uncertainty in a complex society, August 21–September 2, in Göteborg.

Caplan, P. 2000. Introduction: Risk revisited, In *Risk revisited*, ed. P. Caplan, 1–29. London: Pluto Press.

Carrier, J.G., and D. Miller. 1999. From private virtue to public vice. In *Anthropological theory of today*, ed. H. Moore, 24–47. Oxford: Polity Press.

Douglas, M. 1988. *Purity and danger. An analysis of the concepts of pollution and taboo*. London: Ark Paperbacks.

Douglas, M., and B. Isherwood. 1979. *The world of goods. Towards an anthropology of consumption*. London: Allen Lane.

Gell, A. 1990. Newcomers to the world of goods: Consumption among the Muria Gonds. In *The social life of things. Commodities in cultural perspective*, ed. A. Appadurai. Cambridge: Cambridge University Press.

Graeber, D. 2005. Value: Anthropological theories of value. In *A handbook of economic anthropology*, ed. J.G. Carrier. Cheltenham: Edward Elgar.

Gullestad, M. 1992. *The art of social relations. Essays on culture, social action and everyday life in modern Norway*. Kristiansand: Scandinavian University Press.

Guyer, J. 1995. Introduction: The currency interface and its dynamics. In *Money matters. Instability, values and social payments in the modern history of West African communities*, ed. J. Guyer. Portsmouth: Heinemann.

Henning, A. 2000. *Ambiguous artefacts. Solar collectors in Swedish contexts. On processes of cultural modification*. Stockholm Studies in Social Anthropology, 44. Stockholm: Almqvist & Wiksell International.

———. 2007. *Värmesystemen i vardagen. Några småhusägares erfarenhet av att byta värmesystem.* Falun: Solar Energy Research Center SERC, Högskolan Dalarna.

Hylland-Eriksen, T. 1995. *Small places, large issues. An introduction to social and cultural anthropology.* London: Pluto Press.

Kopytoff, I. 1990. The cultural biography of things. Commoditization as process. In *The social life of things. Commodities in a cultural perspective*, ed. A. Appadurai. Cambridge: Cambridge University Press.

Miller, D. 1992. *Material culture and mass consumption.* Oxford: Blackwell.

———. 1996 The tribal and peasant house in North London. Paper presented at the Barcelona EASA Conference, July 12–15, in Barcelona.

Myers, G. 2007. Commonplaces in risk talk: Face threats and forms of interaction. *Journal of Risk Research* 10, no. 3: 285–305.

Parry, J., and M. Bloch. 1991. *Money and the morality of exchange.* Cambridge: Cambridge University Press.

Phillimore, P. 2006 'I can remember very different times'. Remembering and communicating about risk in industrial Ludwigshafen. Paper presented at RISKCOM 2006 'New perspectives on risk communication: Uncertainty in a complex society', August 21–September 2, in Göteborg.

Reith, G. 2004. Uncertain times. The notion of 'risk' and the development of modernity. *Time & Society* 13, nos. 2/3: 383–402.

Roaf, S., D. Chrichton, and F. Nicol. 2005. *Adapting buildings and cities for climate change.* Amsterdam: Elsevier, Architectural Press.

Sahlins, M. 1976. *Culture and practical reason.* Chicago: The University of Chicago Press.

Shove, E., L. Lutzenhiser, S. Guy, B. Hackett, and H. Wilhite. 1998. Energy and social systems. In *Human choice and climate change. Vol. 2: Resources and technology*, ed. S. Rayner and E.L. Malone. Ohio: Batelle Press.

Simmel, G. 1978. *The philosophy of money.* London: Routledge and Kegan Paul.

Swedish Energy Agency. 2001. *Energimyndigheten 2001.* Eskilstuna: Energimyndigheten.

———. 2007. *Årlig rapportering av erfarenheter från det statliga bidraget till investeringar i solvärme (SFS 2000:287). Verksamheten 2006.* Eskilstuna: Energimyndigheten.

Wilhite, H. 2000. Energy research on consumption and lifestyles: International experiences and new directions. In *Energi og livsstil*, ed. P. Gundelach and S. Huehn, 61–72. Copenhagen: Sociologisk Institut.

Wilhite, H., E. Shove, L. Lutzenhiser, and W. Kempton. 2000. The legacy of twenty years of energy demand management: We know more about individual behaviour but next to nothing about demand. In *Society, behaviour, and climate change mitigation*, ed. E. Jochem, J. Sathaye, and D. Bouille, 109–26. Dordrecht: Kluwer Academic Publishers.

Risk management in Swedish forestry – Policy formation and fulfilment of goals

Kristina Blennow

Southern Swedish Forest Research Centre, Swedish University of Agricultural Sciences, Alnarp, Sweden

The formation of a risk management policy in Swedish forestry and its consequences for fulfilment of goals was analysed. The risk of wind damage was used as a model where an apparent gap between stated accepted risk and extent of risk-reducing measures taken among south Swedish non-industrial private forest owners was used as a starting point. The results of an enquiry, and personal experiences from the debate after an extensive wind damage event in January 2005, were used and complemented study of literature. It was concluded that risks have not been particularly actively managed in the Swedish forestry culture. This was explained by notions of seeing (i) risks such as wind damage as a natural hazard rather than a technological risk that can be modulated, and (ii) forestry as an enterprise free of valuation in which value aspects of risk were neglected. A narrow agenda in Swedish forest research influenced the risk management policy, where forest consultants played an important role as a link between science and practice and between the public and the private. Neither the legislator nor the forestry consultants declared how they weighted risk which reduced the fulfilment of goals. At the national level relaxation of regulations after 1993 has not yet resulted in markedly diversified privately owned forests and spreading of risk as intended, and the management of the risk of wind damage before January 2005 was in many cases not adapted to the risk the individual forest owner was prepared to take. The results were discussed with respect to possibilities for improved fulfilment of goals in Swedish forestry by facilitating more active risk management.

Introduction

Over one night in January 2005, timber and wood corresponding to almost the annually harvested volume for Sweden blew down in a wind storm (Swedish Forest Agency 2006). The growing stock of several private forest holdings was almost completely damaged and objects of the Swedish cultural heritage, buildings, electricity supply, communications etc. were substantially damaged, mainly from fallen trees. Ten casualties from the salvation work and more than 140 working accidents have been reported (Swedish Forest Agency 2006).

Forestry is of major importance to the Swedish economy. Storms do occur from time to time and there is not much an individual can do but learn to live with them. In contrast, the risk of damage to forests from exposure to strong wind is highly dependent on the state of the forest and thus on the forest management. For example, in production-oriented forestry with a high volume of standing forest, such as in Sweden, there will be more and larger trees that could blow down (e.g., Schelhaas, Nabuurs, and Schuck 2003). The stability of the forest, furthermore, depends on the tree species. In Swedish production forestry the conifers Norway spruce and Scots pine are extensively used (Swedish Forest Agency 2005). In particular Norway spruce is less wind firm than many other tree species in Sweden, including deciduous tree species (e.g., Peltola et al. 2000; Nielsen 2003; Valinger et al. 2006). A silvicultural system in which harvesting is made by clear felling creates forest edges. Along these the trees will be exposed to stronger wind and the edges will initiate the formation of turbulence that could damage trees further downwind (e.g., Alexander 1964; Sommerville 1980; Lohmander and Helles 1987; Morse, Gardiner, and Marshall 2002; Gardiner et al. 2005).

Historically, more forest was damaged by wind in southern than in northern Sweden (Blennow and Olofsson 2004; Nilsson et al. 2004). In southern Sweden (Götaland region), 77% of the productive forest land was owned by 139,608 non-industrial private forest (NIPF) owners in the year 2000 (Swedish Forest Agency 2005). According to an enquiry among south-Swedish NIPF owners made in 1999, only 33% of the respondents had taken action to reduce the risk of wind damage (Blennow and Sallnäs 2002). According to another enquiry, only about 40% of the NIPF owners affected by the storm in January 2005 were insured against loss of income because of wind damage (Swedish Forest Agency 2006). Yet, in 1999, south-Swedish NIPF owners on average had ranked the risk of wind damage as one of the worst risks to their forestry from a financial point of view (Blennow and Sallnäs 2002). Furthermore, on average they ranked the risk of wind damage as one of the risks they were prepared to invest the most effort and money in to reduce. Although lack of information about risk-reducing measures is one possible explanation why few active risk-reducing measures had been taken, only 13% of the respondents in the enquiry in 1999 stated to be unknowing of whether they took action to reduce the risk of wind damage (Blennow and Sallnäs 2002). Thus, it appears as if in the south-Swedish forestry culture risk-reducing measures were not taken to the extent that matched the risk the forest owners were prepared to take. This was further supported by almost one third of the private forest owners responding to an enquiry about one year after the storm in January 2005 who had felt fairly large or large desolation and grief because of the consequences of the storm (Grimby in Swedish Forest Agency 2006).

The objective of this study was to analyse the formation of a risk management policy in Swedish forestry and its consequences for the fulfilment of goals. The hypothesis was that risk is not actively managed. Wind damage risk was used as a model where an apparent gap between stated accepted risk and extent of risk-reducing measures taken in south Swedish non-industrial private forestry was used as a starting point. In reviewing the historical development of Swedish forestry three arenas of importance were identified; the academic arena, the legislative arena, and the private forest owner arena. The forestry consultants serve as a link between science and practice and between the public and the private. The industrial forestry perspective was not analysed. The results were discussed with respect to possibilities for improved fulfilment of goals in Swedish forestry.

Methods

The formation of a risk management policy in Swedish forestry was analysed based on literature study. Additionally, a sub-set of the answers to an enquiry on risk management in forestry together with personal experiences by the author from the debate after an extensive wind damage event in Sweden on 8 January 2005 was used. The enquiry was sent in April 2004 (before the storm) to 3300 Swedish NIPF owners of estates covering an acreage corresponding to a taxation value of at least 11,000 EUR in 2003, respectively. The sub-set of the enquiry refers to those questions relating specifically to the risk of wind damage and only to those sent to 650 forest owners sampled evenly by the National Statistics Office of Sweden over the counties Kronobergs län, Jönköpings län, and Kalmar län in southern Sweden. Sixty percent of the forest owners in the sub-sample responded. The three counties cover a large proportion of the area in which the forest was severely damaged by the storm on 8 January 2005 (Swedish Forest Agency 2006). In these counties, furthermore, the threshold minimum taxation value of 11,000 EUR approximately corresponded to an estate of 5 ha productive forest land (Swedish Forest Agency 2004). A contingency table X^2-test was used to test for difference between groups of responses. Preliminary results have been presented in reports from the project 'Stormanalys', a joint project between the Swedish Forest Agency and the Swedish University of Agricultural Sciences for analysing consequences of the wind damage event over southern Sweden in January 2005, and possibilities for the future (Blennow 2006; Blennow and Eriksson 2006). The rest of the results from the enquiry will be presented elsewhere.

Background

Risk theory

Decision-making in forestry involves a multitude of factors and different actors have different goals and preferences. Long rotation periods at a century-scale in Swedish forestry make the time horizon distant which increases uncertainty. Although simple heuristics sometimes leads to fulfilment of goals (cf. Gigerenzer and Todd 1999), the prospect of achieving set goals could increase if deliberate choices between alternatives were made where risk was part of the valuation.

To actively manage a risk, the decision-maker needs to be aware of the risk and the stake needs to be high enough (Figure 1). Sometimes the risk may be modulated, sometimes it may not. Previous studies (e.g., Rogers 1997; Axelrod, McDaniels, and Slovic 1999) have shown that natural hazards, such as storms, in general are seen as having less impact on ecosystems and on species, and being far less avoidable than technological hazards. To reduce epistemic risk the decision-maker could actively search for information. In this way the decision-maker could see new alternatives and his or her knowledge on how to modulate risk could improve (Gärdenfors and Sahlin 1982; Sahlin and Persson 1994). To refrain from considering possibilities that could lead to new alternatives is in itself a risk (Gärdenfors and Sahlin 1982). Finucane (2005) defined risk as a combination of analysis and emotion that motivates individuals and groups to take on a certain lifestyle. If and how a risk is actively managed then depends on the culture to which the lifestyle or livelihood belongs. Within a system, experts and decision-makers at higher levels are crucial for how risk is valued (Clarke 1993). Trust has, furthermore, been identified as an important component for the successful communication of risk (Slovic 2000; Persson 2004).

Active risk management is most easily performed in relation to well defined goals and where the goal priority is set, and more difficult to make where there is value

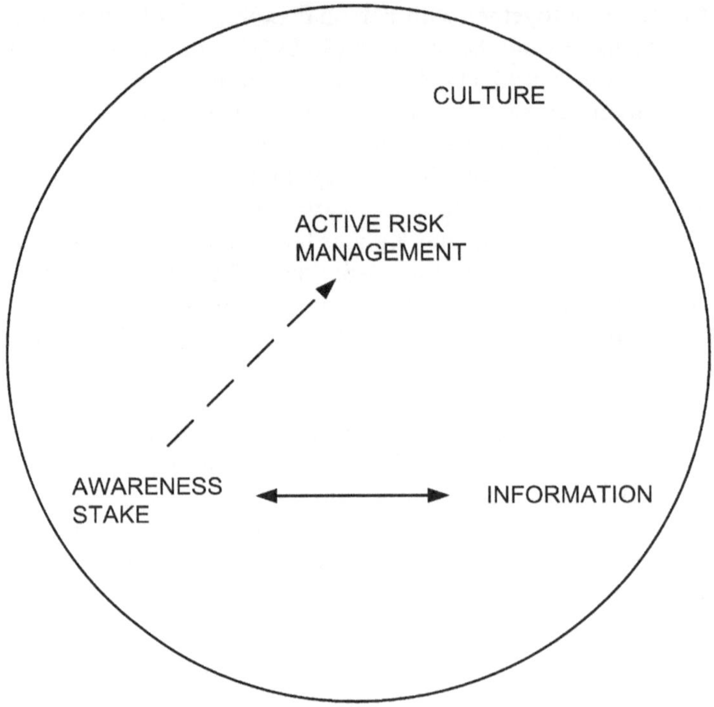

Figure 1. Important components for active risk management.

uncertainty. Since values change over time, maintaining a large freedom of action is a way to reduce risk from unstable values (cf. Andersson, Sallnäs, and Carlsson 2006). Active risk management is, however, not necessarily a matter of minimising risk. Rather, it is about making decisions in accordance with the level of risk that is accepted and where one is aware of the consequences the decision could have.

Formal calculations of risk in terms of the probability of a specified outcome could support decision-making (e.g., Blennow and Sallnäs 2005). But in such quantification important aspects of risk are unaccounted for, such as differences in how the outcome is valued, its dreadfulness, the affect it results in, trust, cultural setting, and aversion against involuntariness (e.g., Gärdenfors and Sahlin 1993; Slovic 2000; Slovic et al. 2002; Hansson 2004a). For example, wind damage from the storm in January 2005, which was the most extensive on record to Swedish forests (Holmgren 2005), only corresponded to some 2% of the total Swedish forest growing stock of forest. Clearly, if risk is weighted collectively from the perspective of the economical welfare of the Swedish nation, the risk of wind damage could be regarded as small, at least as long as secondary damage is unaccounted for. At the same time, the forests of many private estates were severely damaged after the storm. Thus, for individual forest owners the stake may have been higher. Consequently, if the risk of wind damage is weighted individually, the risk of wind damage could be higher (cf. Hansson 2004b). This was illustrated by calculations made recently for three alternative forest management strategies for southern Sweden where the result differed markedly between alternatives depending on the time horizon applied and whether or not the risk of wind damage was taken into account (Nilsson and Sallnäs 2006).

Swedish forestry

A national perspective

In the early Swedish agrarian society, the economic management of the forest resource for multiple purposes was an integrated part of agriculture. During the 1830s and 1840s the systematic cultivation of forest on state-owned and public land was introduced to supply the military and other public interests with raw material. The silvicultural system included clear-felling and measures for rejuvenation. The forest officers made forestry independent and separate from agriculture (Berggren, Björk, and Eliasson 2004). Along with industrialisation, the demand for forest products increased.

The idea of the forest as a national resource for sustained wood production motivated the formulation of a Swedish forestry legislation. The Swedish Forestry Act was constructed as an educative law (Törnqvist 1995) and was first passed in 1903. The Act applied to all private forest land and was held at a low level of specification since more detailed regulation was considered to have interfered too much with the right of possession and to require deeper knowledge in silviculture than was available at the time (Enander 2000). The law required regional forest agencies be established to (i) maintain obedience to the law, (ii) support non-industrial forestry through consultancy, and (iii) provide education to private forest owners. The persuasive rather than commanding working practice that developed made the forest owners have trust in the personnel and in the enterprise (Stjernquist 2001). The regional forest agencies were merged with the Swedish Forest Agency in 2006.

According to Törnqvist (1995) Swedish forestry was treated, early on, as an enterprise free of valuation that was guided by scientific results. In this way there would essentially be one right way of conducting forestry. Törnqvist described the project as *the forestry learning process* where the forest owners were persuaded to fulfil societal goals by building and supporting their senses of duty for forest management. Also the special interest organisation *The Forest Society* established in 1913, and the forest-owner associations, the first one established during the 1930s, worked for the same goal. *The Royal College of Forestry*, established in 1915, was to provide competent personnel and conduct research on forestry. The joint efforts resulted in a folk movement where forestry was made every man's duty (Stjernquist 1973).

Törnqvist (1995) characterised the learning process as a translation of (i) a super-individual scientific language, (ii) to a normative language, and finally to (iii) a practical language (Figure 2). Two common perspectives in Swedish forest science were identified; a dominating forest growth perspective and an economic perspective. In the growth-oriented perspective it is sought to maximise forest growth by optimally making use of the growing capacity of the soil. In the economic perspective it is sought to increase welfare by advocating growth of the national economy and high profitability at an enterprise level. The profitability and economy of private forest enterprises have, however, generally not been matters of priority for Swedish forest research.

As knowledge in forestry improved over time and as the collective goals were increasingly valued more highly in national politics, the Swedish forestry legislation was specified towards the adoption of a silvicultural system of even-aged forest stands where harvesting was made by clear-felling (Enander 2000). Since the 1950s

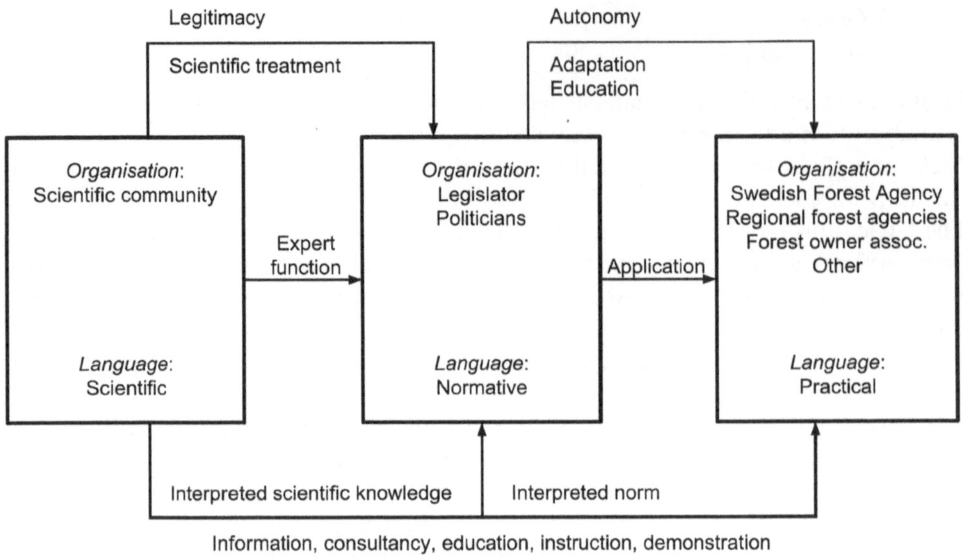

Figure 2. Actors, their functions and languages used in 'The forestry learning process'. Modified from Törnqvist 1995.

this silvicultural system almost completely dominates Swedish forestry (Ekelund and Hamilton 2001).

During the period 1960–1990 the law required that the NIPF owners deliver to the forest industry the largest possible amount of timber and wood at a cost as low as possible (Stjernquist 2001). To reach this goal non-industrial private forestry was to be made more efficient for which industrial forestry served as a model. The demand from the forest industry was mainly for conifer timber and wood and research and development were oriented towards forest production based on conifer species in a clear-felling system (Ekelund and Hamilton 2001). The personnel at the regional forest agencies were directed to pay more attention to the task of maintaining obedience to the law, yet the word advice was used to describe the service provided. In the preparatory work to the Forestry Act of 1979 it was stated that the legal duties of the NIPF owners were not to be hampered by consideration of profitability of private forestry enterprises (in Stjernqvist 2001). The Forestry Act of 1979 authorised the most detailed and production-oriented forest policy during the twentieth century. Harvesting by clear-felling except possibly leaving shelterwood or seed trees, and even harvesting of low-producing private forests was enforced by law (Swedish Forest Agency 1983). During the 1980s the conversion of former pasture and slow-growing deciduous forest to coniferous forest was subsidised (Ekelund and Hamilton 2001).

By the revision of the Forestry Act in 1993 a shift was seen and for the first time regulation was substantially relaxed whereby the legal possibilities were improved for the forest owners to influence the forest management and their right to possession was strengthened. This was motivated by political liberalisation and by the introduction of a second national forestry goal. For the first time two goals were valued equally high in the Act; that of sustained production of timber and wood, and that of maintained biodiversity. To meet the biodiversity goal for Swedish forest ecosystems, deciduous tree species are generally preferred over coniferous species (e.g., Jonsell, Weslien, and Ehnström 1998) but how to reach the goal is still not very well understood. As in 1903, the legislator chose a low level of regulation to spread the risk through diversification of forestry. Again it became more important that the forest owner be educated in forestry (Government Bill 1992/93) to facilitate finding ways to reach the biodiversity goal. Education was also motivated by the fact that the forest owners as a group had become more diversified. In the Bill was expressed that consultancy with respect to the profitability of individual forest enterprises was primarily to be a matter for the forest owner associations but could be provided by regional forest agencies at a cost. Although the responsible minister welcomed alternative harvesting methods, only to a very minor extent these have been practiced until today (Enander 2003) and the build-up of the volume of standing forest has been successful and has continued also after 1993 (Swedish Forest Agency 2006).

A private forest-owner perspective

In the agrarian society, the forest provided raw material for a multitude of uses. As the demand for timber and wood increased alongside industrialisation, also the possibilities for income from the selling of timber and wood increased. The forest, however, maintained a utility value in addition to the trade value. Today, many NIPF owners work outside their estate and the proportion of estate-residents had reduced to 66% in 2000 (Swedish Forest Agency 2005). Also the fraction of forest owners conducting agriculture had reduced. The proportion of female NIPF owners had increased to 38% in 2000 (Swedish Forest Agency 2005).

Based on interviews with Swedish forest owners, Hugosson and Ingemarsson (2004) identified four main, yet unranked, motifs for non-industrial private forestry, each motif consisting of a set of forestry objectives (Table 1). In an enquiry among Swedish NIPF owners and forest officers in 2003 (Mattsson, Boman, and Kindstrand 2003), 77% of the NIPF owner respondents regarded the production of timber as very important or fairly important. Although the majority of the respondents valued timber production highly, not all forest owners were motivated by income from forestry. An average of 12% of the total household income was stated to come from forestry (Mattsson, Boman, and Kindstrand 2003). According to Törnqvist (1995), less than 10% of the forest owners depended on income from their forestry.

Table 1. Swedish NIPF owner forestry objectives classified into motifs for forestry based on interviews with 16 forest owners in 2003 (Hugosson and Ingemarsson 2004).

Utilities	Conservation	Amenities	Economical efficiency
Game Production	Nature Conservation	Emotional Tie	Yield of Capital
Berries Production	Cultural Conservation	Forestry Tradition	Liquidity Reserve
Mushrooms Production	Water Conservation	Challenge of Silviculture	Annual Income
Forest Grazing Production	Soil Conservation	Aesthetics	Tax Planning

The forest grower culture was characterised by Törnqvist (1995) as a rural type of a self-employer livelihood. At the heart of the livelihood lies a high priority of independence. Other highly valued aspects are: (i) The estate as a social space; (ii) The legal type of ownership where the economy of the enterprise is not separate from that of the household; (iii) The ambition to be independent towards capital interests such as banks and credit agencies; (iv) That the household economy is adapted towards resistance to changes in the state of the market; (v) That no difference is made between work and leisure; (vi) That the division of working tasks is made from the point of view of the enterprise, where income from work outside the farm is meant to support the economy of the enterprise; (vii) That competence and qualification is gained from socialisation; (viii) That the qualifications include the ability to carry out practical working operations as well as a more abstract way of thinking that reproduces the ideology of the family-owned estate, interest in conservation, and the estate as an overall life-time family achievement; and (ix) That the estate will be passed on to the next generation. Apparently, forests provide a multitude of goods and services to the owners, all of which are potentially at stake.

Results and discussion

Risk management and forest science

At the beginning of the twentieth century a prominent Swedish forest officer formulated an approach to the prospect of managing the risk of extensive wind damage:

> Any new knowledge about how to protect the forest from storm damage the forest man has likely not been able to collect at this destruction, especially since it probably should be regarded as a natural phenomenon of such an abnormal kind that nothing for the future could be done. (Hollgren 1903, 39) (Author's translation)

The forest officer was referring to damage caused by a storm in December 1902 that affected approximately the same area as did the storm in January 2005 and that was similar to that storm from a meteorological point of view (Alexandersson 2005). Although strong enough wind will also damage the most wind-firm tree, neither the storm in December 1902 nor the storm in January 2005 damaged all trees in the affected area (Holmgren 2005; Swedish Forest Agency 2006). Still the defeatist approach expressed by Hollgren in 1903 has remained. For example, in forestry terminology an uncertain event of negative outcome is termed calamity. This term makes one think of something that is out of control and something one is not responsible for. It inspires few countermeasures to be taken. Judging from arguments raised by many in the debate after the wind damage event in January 2005, this defeatist approach was still widespread.

When organisations and experts are driven to maintain high production, this could lead to omission of low-probability events with marked negative outcome (Clarke 1993), such as the risk of extensive wind damage in forestry. Furthermore, when things develop according to the plan, uncertainty is easily forgotten. Indeed the production effort in Swedish forestry was successful, with the standing volume of forest in Sweden almost doubled in less than a hundred years (Swedish Forest Agency 2006), which likely diverted interest from risks. Ways to modulate the risk of wind damage were described internationally already early during the twentieth century (e.g., Fritsche 1933), but only 24 articles including either of the Swedish words for wind or storm out of a total of 2626 entries were published between 1939 and 11 September 2006, in Swedish forestry magazines covered by the Forest Science Database (CAB International 2006). Indeed, in standard calculations of expected cash flow and net present value of forest land, no consideration is made of risk. Furthermore, a widespread welfare economical perspective in forest science likely helped make collective risk weighing the default among forest academics. According to Clarke (1993) specialisation among experts and decision-makers at higher levels through their background in a common narrow scientific discipline favours normative guidelines of what is considered as truth (cf. Figure 2). The presence of a defeatist approach to the risk of wind damage and the success of a forest growth-oriented forestry motivated by welfare economic interests where the risk of wind damage could be regarded as low, help explain why the risk of wind damage was not a major focus of Swedish forest research.

Risk management and Swedish forestry legislation

The rationale behind the Swedish Forestry Act was to manage the risk of future timber shortage. In the preparatory work before the revision of the Forestry Act in 1993 (Government Bill 1992/93), reducing the risk of damage to production forests from wind, insect attacks, and fungi was seen as important as long as significant values of nature would not be lost. This indicates that the regulator regarded these as technological risks rather than natural hazards. Regulation of the management of the risk of wind damage was, however, not authorised while the management to reduce the risk of damage from insect attack was. The latter was based on experience of extensive insect damage following wind damage to Swedish forests at the end of the 1960s. No motivation was given for not regulating the management to reduce the risk of the primary cause of damage, namely wind damage. Although no formal calculations appear to have been made, it is likely that the risk of wind damage was only seen as a threat to the economy of private forestry enterprises and not to the national forestry goals. This would mean that the legislator weighted risk collectively. How extensive wind damage affects the second major goal of the Forestry Act, that of preserved biodiversity, remains still today largely unknown (Niklasson et al. 2006). At the same time individual risk weighing also appears, however, to have been practised by the legislator since consultancy to prevent working accidents among self-active forest owners was recommended to be supported by the state (Government Bill 1992/93).

According to Törnqvist (1995) the value aspects in forestry have not been adequately observed in Swedish forestry. This appears true also for the value aspects of risk. The risk-weighing principle being not declared by the legislator has likely helped maintain forest management at a risk level that was accepted under the forestry policy formulated in the Forestry Act of 1979, before the revision in 1993. Before the revision, the collective goal of sustained production of timber and wood was given higher priority by the legislator than after the revision.

Although not explicitly formulated as such, there are several examples of how epistemic risk was managed in the Swedish forestry legislation. For example, (i) the level of specification in the legislation was initially held low and was increased successively as knowledge improved, (ii) the working practice of the regional boards of forestry whereby the intentions of the law were to be met by education of forest owners, (iii) the relaxation of the forestry legislation aiming at the spreading of risk when the goal of preserved biodiversity was given higher priority (Government Bill 1992/93), and (iv) the regulation stating that only methods be used for rejuvenation of forest that empirically or through research results and tests in practice have proven to give satisfactory results within an acceptable time span (§4 in Swedish Forest Agency 2003).

The focus on one type of forest management to reach the production goal resulted in a successful build-up of the Swedish volume of standing conifer forest during the twentieth century. Since values generally are known to change over time the effort appears risky. Indeed, preferences in society shifted and the goal of preserved biodiversity was given high priority. Today, the knowledge on alternative cutting methods and the knowledge on silviculture of non-conifer tree species is less compared to that for the sustained production of conifer timber and wood. This gives rise to epistemic risk in relation to alternative forest management and hampers intended spreading of risk through diversification of forests.

Although there has been an increase in the standing volume of deciduous tree species in Sweden since the relaxation of the Forestry Act in 1993 (Swedish Forest Agency 2005), the privately owned forests have not become markedly diversified. This is likely partly explained by negligence of declaring what risk weighing principle was used by the legislator and the presence of epistemic risk.

Risk management by forestry consultants

Several organisations, both private and public, offer forestry consultancy services to forest owners. In so doing they serve as a link between science and practice and between the public and the private.

Nine forest officers working as consultants for *The Forest Society* in southern Sweden were asked in 2004 to rank a given set of risks in south Swedish forestry from a financial point of view (Skov 2006). Their ranking was similar to the ranking of the same risks made by NIPF owners in 1999 (Blennow and Sallnäs 2002). This was the case both when the forest officers were asked to rank them according to how they believed NIPF owners would have ranked them and according to their own assessment of risk. Thus it appears as if the consultants had a good picture of how NIPF owners assessed different risks from a financial point of view and in a relative sense. In terms of goals for forestry, however, forest officers generally over-estimated how NIPF owners valued timber production according to the results of an enquiry in 2003 among both forest officers at the *Swedish Forest Agency*, forest owner associations, and *The Forest Society,* and among NIPF owners (Mattsson, Boman, and Kindstrand 2003). In general, furthermore, they over-estimated how NIPF owners valued the forest as part of the family tradition. In contrast, they under-estimated how NIPF owners valued the forest for recreation and outdoor activity, and as an environment for biodiversity. Putting high value on timber production is in agreement with the two main perspectives among forestry academics; promoting forest growth and economical welfare of the nation. Thus, advice given by consultants was likely too much oriented towards favouring timber production and valuing the forest as part of the family tradition, but too little oriented towards recreation aspects and putting high value on the forest as an environment for biodiversity. This could likely partly be explained by a recent diversification of forest owners including an increase in the fraction of non estate-resident NIPF owners which was not adequately observed by the forest officers. More importantly, however, it was likely an effect of the notion of seeing forestry as an enterprise free of valuation.

In the advice given by consultants employed at regional forest agencies, the risk of wind damage was systematically considered, for example only in exceptional cases was commercial thinning in old spruce forest recommended during the period 1997–2005 (Swedish Forest Agency 2006). Commercial thinning late during the rotation period would have increased the probability of wind damage (e.g., Persson 1975; Gardiner et al. 1997). Indeed, according to statements in discussions with consultants at regional forest agencies and at the *Swedish Forest Agency*, the advice policy was in accordance with the detailed regulation authorised by the Forestry Act before its relaxation in 1993 (Swedish Forest Agency 1983). Thus the advice policy that aimed at fulfilment of one collective goal of sustained production of timber and wood appears to have remained. Furthermore, according to interviews with forestry consultants information about the risk level associated with alternative management strategies was seldom discussed (Swedish Forest Agency 2006). In developing an advice policy after the relaxation of the legislation in 1993, the value aspects of risk appear to have been neglected. If the level of risk associated with the proposed management action was not clarified, it was not possible for the forest owner to decide on a type of forest management corresponding to an accepted level of risk. Thus if the recommendations were followed, which they largely were (Swedish Forest Agency 2006), the risk level would be strongly influenced by the consultants' management policy rather than being a risk level accepted by the individual NIPF owner within the substantially wider action space defined by the legislation.

Risk management among NIPF owners

To reduce the risk of reduced income from forestry, several companies offer insurance covering the risk of wind damage. Before the storm damage in January 2005, the insurance companies had under-estimated the risk and so the cost of insurance was low, less than 1 EUR per ha and year in the area hit by the storm (Wahlqvist, personal communication 2006). Still only about 40% of the forest owners were insured at the time of the storm (Swedish Forest Agency 2006). To some extent, this could be explained by income from forestry not being that important to the private household economy (Törnqvist 1995; Mattsson, Boman, and Kindstrand 2003). Indeed, after the storm 63% of the responding forest owners in an enquiry did not express any anxiety at all with respect to financial consequences of the storm (Swedish Forest Agency 2006). For example, money from insurance would not bring back the highly valued social space that the undamaged forest constituted (cf. Törnqvist 1995). The ambition of independency which is at the heart of the forest owner livelihood could also help explain why few forest owners were insured. This ambition could be linked to an expectation of being able to take care of wind damaged forest by him- or herself, which in many cases was not possible after the extensive damage in January 2005. Even if NIPF owners in general were aware of the risk of wind damage already before the damage event in January 2005, most likely not only the insurance companies but also many NIPF owners had under-estimated the risk.

Only about one third of the NIPF owner respondents took action to reduce the risk of wind damage before the storm in January 2005 according to three separate

Table 2. Fraction of respondents for different answers on the question 'Do you take action to reduce the risk of wind damage?' posed in enquiries in 1999, 2004, and 2005 among south-Swedish NIPF owners in the area affected by a storm in January 2005.

Year	Answer: Yes (%)	n	Reference
1999	33	149	Blennow and Sallnäs 2002
2004	29	361	
2005*	33	721	Ingemarsson et al. 2006
2005**	56	698	Ingemarsson et al. 2006

Notes: * refers to management before extensive wind damage in January 2005; ** refers to planned management after extensive wind damage in January 2005.

enquiries (Table 2). There was only slight difference between the results of the three enquiries in spite of somewhat different geographical sampling of NIPF owners and expected variation over time. When NIPF owners again were asked the same question but with respect to their planned future management of the risk of wind damage after recently having experienced extensive wind damage in 2005, the fraction stating that he or she planned to take risk-reducing action was somewhat higher (Table 2). In the enquiry sent out in 2004 (less than one year before the storm), the forest owners were asked to hypothetically state what tree species to grow if their spruce forest was severely damaged by wind (Table 3). A majority of the respondents answered that they would plant spruce again.

Affect arising from experience of an event is known to influence assessed risk (e.g., Gärdenfors and Sahlin 1993; Slovic et al. 2002). Although experience of extensive wind damage in January 2005 did give rise to both astonishment and affect (Swedish Forest Agency 2006) the influence it will have on the future risk management by NIPF owners appears to be limited (Tables 2 and 3). This is likely explained by several reasons. First, if NIPF owners saw wind damage as a natural hazard there would be no options for forest management. Rogers (1997) found that learning from experience of an event seen as a natural hazard was limited since it was seen as an 'Act of God'. A second reason is related to trust. According to the responses to the enquiry made in 2004, 73% of the respondents sought advice from one or more consultant. Among those who had sought advice, 80% trusted the consultants to be competent in risk management. Sixty-four percent of the respondents stated that he or she had discussed matters of risk with their consultant.

Table 3. Fraction of respondents for different answers to the question 'What would you do if your spruce forest was severely damaged by wind?' posed in an enquiry among south-Swedish NIPF owners in 2004 (before extensive wind damage in January 2005) (n = 380).

Answer	Fraction (%)
Plant Norway spruce again	83
Convert to deciduous tree species	14
Convert to other conifer tree species	4
Other alternative	12

Table 4. Fraction of responses to the question 'From whom do you seek advice in forest management?' in an enquiry to NIPF owners in 2004 (n = 278).

Consulting organisation	Fraction (%)
Regional forest agency	40
Forest owner association	54
The Forest Society	1
Other	24

The NIPF owners had sought advice from consultants at different organisations (Table 4). Although some had sought advice from more than one consultant, less than half of the respondents stated that he or she had been given advice in relation to the risk of wind damage (Table 5). This fraction was independent of whether advice had been sought from a public consultant (*Regional Forest Agency*) or a consultant at a *Forest owner association*, and not statistically significantly larger for the category *Other* ($p=0.28$) in which both organisations and private persons were represented. That little explicit advice had been given in relation to the risk of wind damage helps explain why only few NIPF owners had taken action to modulate risk. The reason behind the consultants sharing the same way to deal with risk irrespective of employer is likely explained by them or their organisations sharing common values. These, in turn, likely originate from common motivations and a common background in terms of their education rooted in the Swedish academic forestry tradition.

The fraction of NIPF owners who had got explicit advice with respect to the risk of damage by insects was not significantly ($p=0.13$) different from that of the risk of wind damage (Table 5). The risk of insect attack was regulated in the forestry legislation and thus likely was assessed a bigger risk to the collective goals in which case the fraction that got advice could be considered low from the legislator's point of view.

Although advice by consultants was generally followed (Swedish Forest Agency 2006) and did include some consideration to the risk of wind damage, less than half of the NIPF owners had got explicit advice with respect to the risk of wind damage. Why a supposed collective risk weighing regulated through the previous forestry legislation was generally accepted by the NIPF owners is likely explained by their trust in the Swedish forestry culture, including the consultants. Many of the forest owners who stayed at the family estate during the development of Swedish forestry during the twentieth century likely socialised with the culture that emerged from the

Table 5. The fraction of NIPF owner respondents stating that he or she had got advice with respect to the risks of wind damage and damage by insects according to an enquiry in 2004. Answers relating to *The Forest Society* were omitted due to few observations.

Consulting organisation	n	Wind damage (%)	Damage by insects (%)
Regional forest agency	85	42	53
Forest owner association	105	42	56
Other	48	54	56
Total	238	45	55

forest management policy that was implemented. But today a large proportion of the NIPF owners do not live at their estates. Often these forest owners are dependent on external forest management services. This dependency also helps explain the large trust the forestry consultants enjoy (cf. Siegrist and Cvetkovich 2000).

Trust in the forestry culture, furthermore, likely made 'to do like everyone else does' appear like an easy way to make reasonable decisions (cf. Gigerenzer and Todd 1999). Several of the objectives of the forest owner livelihood (cf. Törnqvist 1995) are difficult to quantify and thereby easily left unaccounted for. A wish to be part of the forestry culture and the value of the forest as a social space exemplifies some of the less well defined goals a forest owner may have. Indeed the large fraction of NIPF owners who had felt desolation and grief after the extensive wind damage event in 2005 (Grimby in Swedish Forest Agency 2006) is difficult to explain unless these losses are also taken into account since only a minor fraction of the NIPF owners depended on income from forestry (cf. Törnqvist 1995). If cultural aspects including less well defined goals are part of the analysis of forest owner risk management, their actions, or rather lack of deliberate risk-reducing management, is easier to understand. But at the same time, it reveals a lack of active risk management within the forestry culture.

Possibilities for the future

Even if active management aims at increasing the fulfilment of goals, it may not necessarily be a simple task to activate the management of risk through information campaigns. The success of such campaigns will to a large extent depend on the trust the sender enjoys. Trust is a complex matter and if the sender is not trusted, such a campaign would be more successful if launched by someone else (cf. Löfstedt 2005). Although forestry consultants were trusted before the recent wind damage event in 2005, the situation may have changed after the extensive damage in January 2005. Judging from an enquiry administrated by the Swedish Forest Agency (Ingemarsson et al. 2006) it appears fairly intact.

In terms of information to facilitate active risk management (Figure 1), that the risk of wind damage can be modulated would be an important message to put through. Another important message would be that valuation is an important part of risk management at any level. For example, awareness of the 'ideological mechanism or mind-set that leads experts and decision makers to neglect information that contradicts a conviction' (Clarke 1993, 300) would be useful. In the Swedish academic forestry culture, one conviction appears to be that collective risk weighing is the right type of risk weighing in Swedish forestry. The most important factor to guard against such mind-sets is that the system is open to review from outside (cf. Brinck 2005). One should not neglect, however, that more active risk management could slow down decision-making and lead to loss of a sense of security associated with doing 'the way one always has' or like 'everybody else does'. At the same time a structured rationale for the management of uncertainty and risk where the risk weighing principle is declared could make it easier to deal with risk and could help reach set goals at any level within Swedish forestry.

Conclusions

The formation of a risk management policy in Swedish forestry was described and analysed. The analysis showed that risk in general was not particularly actively managed in the Swedish forestry culture. Thus the hypothesis that risk is not actively managed in the Swedish forestry culture could not be refuted. Apparently there was a widespread awareness of the wind damage hazard among actors in Swedish forestry already before the extensive wind damage event in 2005 but fewer seem to have appreciated that the risk of wind damage can be modulated. Furthermore, a notion of seeing forestry as an enterprise free of valuation appears to have contributed to negligence of the value aspects of risk. For example, an at least previously narrow research agenda strongly influenced the Swedish risk management policy in forestry where the forest consultants play an important role as a link between science and practice and between the public and the private. Neither the legislator nor the forestry consultants declared how they weighted risk. Negligence of value aspects of risk appears to have reduced the fulfilment of goals both at a national level and among NIPF owners. At the national level relaxation of regulation has not yet resulted in markedly diversified privately owned forests and spreading of risk as intended, and the management of the risk of wind damage before January 2005 was in many cases not adapted to the risk the individual forest owner was prepared to take.

Acknowledgements

Professor Ola Sallnäs, Southern Swedish Forest Research Centre (SSFRC) of the Swedish University of Agricultural Sciences (SLU), Magnus Fridh of the Swedish Forest Agency, and other colleagues in the project 'Stormanalys' are acknowledged for fruitful discussions. 'Stormanalys' was a joint project between the Swedish Forest Agency and SLU on analysing the causes and consequences of the wind damage event in southern Sweden on 8 January 2005, and suggesting possibilities for the future (Swedish Forest Agency 2006). Associate professor Mattias Boman, Professors Leif Mattson and Ola Sallnäs, MSc Desirée Johansson and MSc Mikaela Petersson all at SSFRC are acknowledged for contributing to the planning and pursuing the enquiry among NIPF owners in 2004. The work was financed by the SLU, the Swedish Forest Agency, the insurance company Länsförsäkringar and Mistra through the research programme SUFOR.

References

Alexander, R.R. 1964. Minimizing windfall around clear cuttings in spruce-fir forests. *Forest Science* 10: 130–42.

Alexandersson, H. 2005. Jämförelse av stormarna 1902, 1969 och 2005. *Väder och Vatten* 2: 11.

Andersson, M., O. Sallnäs, and M. Carlsson. 2006. A landscape perspective on differentiated management for production of timber and nature conservation values. *Forest Policy and Economics I* 9: 153–61.

Axelrod, L.J., T. McDaniels, and P. Slovic. 1999. Perceptions of ecological risk from natural hazards. *Journal of Risk Research* 2: 31–53.

Berggren, L., F. Björk, and P. Eliasson. 2004. Från erfarenhet till vetenskap. Naturresurser och framtidstro 1840–1940. In *Miljö och hållbar utveckling – samhällsvetenskapliga perspektiv från en lundahorisont*, ed. P. Wickenberg, A. Nilsson, and M. Stenetoth Sillén, 25–44. Lund: Studentlitteratur.

Blennow, K. 2006. Riskmedvetenhet och riskhantering. In *Stormen 2005 – en skoglig analys*. Meddelande No. 1, 119–137. Jönköping: Swedish Forest Agency.

Blennow, K., and H. Eriksson. 2006. *Riskhantering i skogsbruket*. Report 14. Jönköping: Swedish Forest Agency.

Blennow, K., and E. Olofsson. 2004. Kan man undvika stormskador? In *Osäkerhet och aktiv riskhantering – aspekter på osäkerhet och risk i sydsvenskt skogsbruk*, ed. K. Blennow, 38–43. Stockholm: SUFOR.

Blennow, K., and O. Sallnäs. 2002. Risk perception among non-industrial private forest owners. *Scandinavian Journal of Forest Research* 17: 472–9.

———. 2005. Decision support for active risk management in sustainable forestry. *Journal of Sustainable Forestry* 21: 199–210.

Brinck, I. 2005. Om riskkommunikation: kartor, klyftor och mål. In *Risk och det levande mänskliga*, ed. I. Brinck, S. Halldén, A.-S. Maurin, and J. Persson, 45–78. Nora: Nya Doxa.

CAB International. 2006. Forest Science Database.

Clarke, L. 1993. The disqualification heuristic: When do organizations misperceive risk? *Research in Social Problems and Public Policy* 5: 289–312.

Ekelund, H., and G. Hamilton. 2001. *Skogspolitisk historia*. Rapport 8A, 45–78. Jönköping: Swedish Forest Agency.

Eliasson, P. 2002. Skog, makt och människor. En miljöhistoria om svensk skog 1800–1875. *Skogs- och lantbrukshistoriska meddelanden* 25. [Supplement to Kungliga Skogs - och Lantbruksakademiens Tidskrift, Stockholm].

Enander, K.-G. 2000. *Skogsvårdslagen 1903 – dess förhistoria och några huvuddrag i utvecklingen,* Rapporter 46. Umeå: Dept. of Silviculture, Swedish University of Agricultural Sciences.

———. 2001. *Skogsbrukssätt och skogspolitik 1900–1950,* Rapporter 48. Umeå: Dept. of Silviculture, Swedish University of Agricultural Sciences.

———. 2003. *Skogsbrukssätt och skogspolitik 1950–2000,* Rapporter 54. Umeå: Dept. of Silviculture, Swedish University of Agricultural Sciences.

Finucane, M. 2005. En berättelse tar form: risk, narrativa värden och global bioteknikdebatt. In *Risk och det levande mänskliga*, ed. I. Brinck, S. Halldén, A.-S. Maurin, and J. Persson, 79–101. Nora: Nya Doxa.

Fritsche, K. 1933. Sturmgefahr und Anpassung. *Tharandter Forstliches Jahrbuch* 84: 1–94.

Gärdenfors, P., and N.-E. Sahlin. 1982. Unreliable probabilities, risk taking, and decision making. *Synthese* 53: 361–86.

———. 1988. Introduction: Bayesian decision theory – foundations and problems. In *Decision, probability, and utility*, ed. P. Gärdenfors and N.-E. Sahlin, 1–15. Cambridge: Cambridge University Press.

Gärdenfors, P., and N.-E. Sahlin. 1993. Varför skall vi minnas våra beslut? In *Huvudinnehåll*, ed. ÅE. Andersson and N.-E. Sahlin, 189–205. Nora: Nya Doxa.

Gardiner, B.A., G.R. Stacey, R.E. Belcher, and C.J. Wood. 1997. Field and wind tunnel assessments of the implications of respacing and thinning for tree stability. *Forestry* 70: 233–52.

Gardiner, B.A., B. Marshall, A. Achim, R. Belcher, and C. Wood. 2005. The stability of different silvicultural systems: A wind tunnel investigation. *Forestry* 78: 471–84.

Gigerenzer, G., and P.M. Todd. 1999. *Simple heuristics that make us smart*. Oxford: Oxford University Press.

Government Bill. 1992/93. *Regeringens skogspolitiska proposition till riksdagen 226*. Stockholm: Swedish Government Press.

Hansson, S.O. 2004a. Fallacies of risk. *Journal of Risk Research* 7: 353–60.

———. 2004b. Weighing risks and benefits. *Topoi* 23: 145–52.

Hollgren, C.A. 1903. Juldagsnattens storm i Halland. *Skogsvårdsföreningens Tidskrift* 1: 38–9.

Hugosson, M., and F. Ingemarsson. 2004. Objectives and motivations of small-scale forest owners; theoretical modelling and qualitative assessment. *Silva Fennica* 38: 217–31.

Ingemarsson, F., J. Malmhäll, B. Merkell, S. Nasic, and S.A. Svensson. 2006. *Hur drabbades enskilda skogsägare av stormen Gudrun? – Resultat av en enkätundersökning,* Rapport 13. Jönköping: Swedish Forest Agency.

Jonsell, M., J. Weslien, and B. Ehnström. 1998. Substrate requirements of red-listed saproxylic invertebrates in Sweden. *Biodiversity and Conservation* 7: 749–64.

Löfstedt, R. 2005. *Risk management in post trust societies*. Basingstoke: Palgrave MacMillan.

Lohmander, P., and F. Helles. 1987. Windthrow probability as a function of stand characteristics and shelter. *Scandinavian Journal of Forest Research* 2: 227–38.

Mattsson, L., M. Boman, and C. Kindstrand. 2003. *Privatägd skog: värden, visioner och forskningsbehov*. Brattåsstiftelsen: SUFOR.

Maurin, A.-S. 2005. (För) enkla lösningar i en komplicerad värld. In *Risk och det levande mänskliga*, ed. I. Brinck, S. Halldén, A.-S. Maurin, and J. Persson, 151–80. Nora: Nya Doxa.

Morse, A.P., B.A. Gardiner, and B.J. Marshall. 2002. Mechanisms controlling turbulence development across a forest edge. *Boundary-Layer Meteorology* 103: 227–51.

Nielsen, C.N. 2003. *Adaptive physiology and tree management*. Copenhagen: Royal Veterinary and Agricultural University.

Niklasson, M., B. Nordén, T. Appelqvist, and T. Edman. 2006. Stora stormfällningar och biologisk mångfald, en literaturöversikt med tyngpunkt på död ved. In *Miljökonsekvenser för biologisk mångfald. Underlagsrapport inom projekt Stormanalys*, ed. M. Andersson, T. Appelqvist, T. Edman, H. Liedholm, M. Niklasson, B. Nordén, and J. Paulsson. Rapport 11. Jönköping, Swedish Forest Agency.

Nilsson, C., I. Stjernquist, L. Bärring, P. Schlyter, A.M. Jönsson, and H. Samuelsson. 2004. Recorded storm damage in Swedish forests 1901–2000. *Forest Ecology and Management* 199: 165–73.

Nilsson, U., and O. Sallnäs. 2006. Val av trädslag – Hur hanterar vi risken för stormskador?. In *Strategiska skogsbruksval – analys av två alternativ till trakthyggesbruk med gran: kontinuitetsskogsbruk med gran, självföryngrad björk och möjligheter till produktionshöjande åtgärder*, ed. B. Karlsson and L. Lönnstedt. Rapport 609, 82–91. Uppsala: Skogforsk.

Peltola, H., S. Kellomäki, A. Hassinen, and M. Granander. 2000. Mechanical stability of Scots pine, Norway spruce and birch: An analysis of tree-pulling experiments in Finland. *Forest Ecology and Management* 135: 143–53.

Persson, P. 1975. *Stormskador på skog – Uppkomstbetingelser och inverkan på skogliga åtgärder*. Research Notes 36. Stockholm: Department of Forest Yield Research, Royal College of Forestry.

Persson, J. 2002. De förbisedda objektsaspekterna av risk. In *Osäkerhetens horisonter*, ed. Å. Boholm, S.-O. Hansson, J. Persson, and M. Peterson, 15–28. Nora: Nya Doxa.

Persson, J. 2004. Riskkommunikation och tillit. In *Miljö och hållbar utveckling – samhällsvetenskapliga perspektiv från en lundahorisont*, ed. P. Wickenberg, A. Nilsson, and M.S. Sillén. Lund: Studentlitteratur.

Rogers, G.O. 1997. Dynamic risk perception in two communities: Risk events and changes in perceived risk. *Journal of Environmental Planning and Management* 40: 59–79.

Sahlin, N.-.E., and J. Persson. 1994. Epistemic risk: The significance of knowing what one does not know. In *Future risks and risk management*, ed. B. Brhemer and N.-E. Sahlin. Dordrecht: Kluwer.

Saville, P.S. 1983. Silviculture in windy climates. *Commonwealth Forestry Bureau* 44: 473–88.

Schelhaas, M.-J., G.-J. Nabuurs, and A. Schuck. 2003. Natural disturbances in the European forests in the 19th and 20th centuries. *Global Change Biology* 9: 1620–33.

Siegrist, M., and G. Cvetkovich. 2000. Perception of hazards: The role of social trust and knowledge. *Risk Analysis* 20: 713–19.

Skov, M.C. 2006. *Risk communication in forestry consulting – a survey among forestry professionals*. Examensarbete 73. Alnarp: Southern Swedish Forest Research Centre, Swedish University of Agricultural Sciences.

Slovic, P. 2000. Trust, emotion, sex, politics and science: Surveying the risk-assessment battlefield. In *The Perception of risk*, ed. P. Slovic, 390–412. London: Earthscan.

Slovic, P., M. Finucane, E. Peters, and D.G. MacGregor. 2002. The affect heuristic. In *Heuristics and biases. The psychology of intuitive judgment*, ed. T. Gilovich, D. Griffin, and D. Kahneman, 397–420. Cambridge: Cambridge University Press.

Sommerville, A. 1980. Wind stability: Forest layout and silviculture. *New Zealand Journal of Forestry Science* 10: 476–501.

Stjernquist, P. 1972. *Laws in the forest. A study of public direction of Swedish private forestry*. Lund: Gleerup.

———. 2001. Ordergivare eller rådgivare. Om skogsvårdsstyrelsernas strategi fram till 1990. In *Skogspolitisk historia*. Rapport 8A: ed. H. Ekelund and G. Hamilton. Jönköping: Swedish Forest Agency.

Swedish Forest Agency. 1983. *Skogsstyrelsens författningssamling 2*. Jönköping: Swedish Forest Agency.

———. 2003. *Handbok till skogsvårdslagen*. Jönköping: Swedish Forest Agency.

———. 2004. *Skogsstatistisk Årsbok*. Jönköping: Swedish Forest Agency.

———. 2005. *Skogsstatistisk Årsbok*. Jönköping: Swedish Forest Agency.

———. 2006. *Stormen 2005 – en skoglig analys,* Meddelande 1. Jönköping: Swedish Forest Agency.

Törnqvist, T. 1995. *Skogsrikets arvingar. En sociologisk studie av skogsägarskapet inom privat, enskilt skogsbruk,* Forskningsrapport 6. Uppsala: SAMU.

Valinger, E., M. Ottosson Lövenius, U. Johansson, J. Fridman, S. Claeson, and Å. Gustafsson. 2006. *Analys av riskfaktorer efter stormen Gudrun,* Rapport 8. Jönköping: Swedish Forest Agency.

The text and the tale: differences between scientific reports and scientists' reportings on the eruption of Mount Chance, Montserrat

Jonathan Skinner

The Queen's University Belfast, School of History and Anthropology, Belfast, Northern Ireland

This article looks at the difference between scientists' written reports and their oral accounts, explanations and stories. The subject of these discourses is the eruption of Mount Chance on Montserrat, a British Overseas Territory in the Eastern Caribbean, and its continued monitoring and reporting. Scientific notions of risk and uncertainty which feature in these texts and tales will subsequently be examined and critiqued. Further to this, this article will end by pointing out that, ironically, the latter – the tale – can in some cases be a more effective and approximate mode of communication with the public than the former – the text.

> As I understand it, the report was financed by the United Nations during the 1980s when there was a decade for the reduction of disaster in small territories. According to the paper itself, the report was handed to the Government of Montserrat at the end of it which would have been 1986–87. Hurricane Hugo struck Montserrat in 1989. If the Governor of the day had a copy of the report the Governor's office was blown totally into the sea in Hurricane Hugo, all the library was lost. A lot of the records of the Government of Montserrat were lost also. I can only assume the corporate memory on island had forgotten about this document. (Governor Savage's House of Commons testimony, cited in IDC 1998, 90)

> [T]he text is a weapon against time, oblivion and the trickery of speech. (Barthes 1981, 32)

Introduction

The first quotation above is from Frank Savage, former Governor of the British Overseas Territory Montserrat in the Eastern Caribbean, Foreign and Commonwealth Office (FCO) appointee, facing detailed cross-examination before the International Development Committee of the House of Commons concerning the handling of the eruption of Mount Chance which has been ongoing since 1995. At this stage in the proceedings, Savage was trying to explain why it was that post-hurricane Hugo development work had completely ignored a report by Wadge and Isaacs (1987) which warned of potential eruption and warned against extending development work in areas which they considered to be 'at risk'. He went on to wriggle before the House of Commons Committee by supposing that the Government of Montserrat's copies could have been blown into the sea by the hurricane! Unfortunately for Savage, his evasive words were recorded for posterity. The second quotation is from the post-structuralist Roland Barthes who is distinguishing between speech and text, between orality and literacy as part of his argument for a consideration of the reader's text – that special place where author and reader meet. The two quotations sit well, the (literary) criticism after the (bureaucratic) evasion. It is the intention of this article to use the example of the volcanic eruption on the British Overseas Territory of Montserrat to highlight the difference between the text and the tale, between the fixed report or document and the dynamic story or anecdote, or what Sharon Traweek (1992, 448) refers to as the difference between 'a message given in the authoritarian text and a contradictory one given in jest'. It goes without saying that the former, the text is considered by many to be exclusive, fixed and unchanging – bar its interpretation and the post-structuralist's musings; whereas the latter, the story or anecdote is dynamic, changing, off-the-cuff, immediate and inclusive.

Context and comprehension

Telling tales

It was Laura Bohannon (1952) who famously described how oral history changed to suit the circumstances of its telling, an unchanging but also changing 'genealogical charter' for the Tiv of Nigeria. She drew our attention to differences between orality and literacy in society, divisions and shifts also explored by Goody (1993), Goody and Watt (1981), Thomas (1992) and Ong (1989). The telling of tales and their enactments are in many cases examples of 'how societies remember' (Connerton 1992), how continuity is maintained across generations, how memory and the body are indoctrinated, socialized and made collective (Halbwachs 1992). The narrative contents, forms (see White 1987) and the processes of telling tales can all work to bring a group together and to bring an individual into a group – or so Gluckman's (1963) functionalist approach to gossip would have us believe, if the presumption of the group is to be allowed as transactionalist Robert Paine (1967) notes. Nevertheless, *there is* an inclusionary nature about telling tales, at least in the teller's attempts: 'sight isolates, sound incorporates' (Ong 1989, 72). This is particularly evident in the off-the-cuff remark during a lecture, the preamble before the text is read, the moment of eye contact with the audience before turning to the safety of the fixed lectern and printed or written lecture notes, or, as this article will go on to highlight, the difference between the technical written in-post science report and the analogous explanations, the public understanding of science products of a scientist in post and the revelatory talk anatomizing and exposing the production of scientific knowledge once free from work constraints.

Montserrat and my story

My illustrations come from Montserrat, a small British Overseas Territory dating back some 300 years to the seventeenth century when it was first settled by the British. It had the typical slave-plantation history common to the Caribbean islands (Fergus 1994; Skinner 2004). And it remained British, opting for financial reasons to retain colonial status in the twentieth century when its neighbours were turning independent (Fergus 1994, 212). I was working as a social anthropologist on Montserrat in the mid-1990s, studying ambivalent expressions of West Indian identity among the 10,000 Montserratian inhabitants and the some 300 white tourists, development and British government workers (Skinner 1997):

> June 1995, as my work on the island was drawing to a close, the extinct volcano which dominated the small island turned out to be dormant rather than extinct (for a full account, see Skinner 2000). The Dependent Territory Citizens [status to Overseas Territory Citizen only changed subsequent to the worst eruptions in 1997] – as Montserratians were known as then – reacted to the volcano by fitting it into their 'cognitive mazeway' (Oliver-Smith 1986, 267), by turning to prayer for the religious, by turning to voyeurism for the tourist, by turning to writing for the narrating and 'omniscient' anthropologist, and by turning to science for many of those with high school experience and memories. June 1995, with ash falling all around us and all over us, with the ground trembling, and sulphur burning, and an incessant rumbling of the volcano in the hills, we British Subjects and Dependent Territory Citizens waited for a team of geologists to fly in from Trinidad 'to give us the science' of this natural disaster developing around us.

The story continues – the one that I recount the most, and one which I recorded at the time (Skinner 2000):

> After a nervous wait of two days, a post-graduate student eventually arrived from Trinidad. Local radio broadcasts had advised us to sit tight and to wait for the scientists. Yet, this poor student had no scientist mentors and no scientific equipment with him, so he climbed one of the long narrow ghauts up Mount Chance and tested the ground, testing it with his tongue for volcanic materials – coarse and gritty for new material being thrown up, and fine and smooth for the usual surface material. The soil texture in his mouth felt smooth and so he judged the volcano to be safe until his scientific equipment arrived. Needless to say, the volcano grew ever more coarse and I and some 6000 Montserratians went on to leave the island in successive waves as curfews and evacuation zones were set up and our confidence in a post-grad's palette were queried and then substantiated. Though I left too, it was some months before I had my possessions shipped back to me, having fled with a rucksack of fieldnotes. And it was some months before my heart stopped surging into palpitations every time the washing machine went into spin-cycle and I was taken back to the shaking island. Such was the sudden change in environment that I had to catch up with.

Ten years on – and a number of volcano-related deaths later – and Montserratians are finally coming to terms with living with a live volcano above them, and the machinations, trials and tribulations of the British and Montserrat Governments, the agencies catering for islanders such as the Foreign and Commonwealth Office (FCO) and Department for International Development (DFID), disaster relief agencies, and the Montserrat Volcano Observatory (MVO). Elsewhere, I have referred to some of these agencies' 'social development' policies as versions of an '*anti-social* social development' (Skinner 2003a). In those publications I have explored the gathering and transmission of knowledge about the volcano and how the scientists' findings are often used politically on the island. I have shown how risk analysis is social and cultural and political and thus qualitative. I have worked from the premiss that 'we reason dramatically' (Dowie 1980) in this risk world of ours. And this stance has put me at odds with the likes of Giddens (1991) and Beck (1992a) who suggest that ours is a complex risk society, one characterized by reflexive modernization; that we live in a collective world of universal risk assessments despite pronouncements by the likes of anthropologist Robert Paine, or globalization and transnational scholars Jonathan Friedman (1990) and Ulf Hannerz (1990; see also Boholm 1996) to the contrary with their attention to individual meanings.

Analogy and anatomy

In the edited volume *Science as Practice and Culture*, Sharon Traweek looks at similar 'tales of objectivity' (1992, 429), assessing chains of conversation, the grand narrative stories of cause and effect. 'Narrative Leviathans' (1992, 430) are her description of her scientists' rhetorical constructions who enact a 'colonialist play' of knowledge and power in their research and their research encounters and recountings. Traweek's example is an ethnography of physicists in a Japanese facility and how they label and explain their facility in terms of operatic and pseudo-pornographic terms and how those labels frame and influence the scientists' behaviour. As such, the words are all terms of entitlement (1992, 461). My two examples which follow show by Analogy and Anatomy the gap between scientific report and scientist's jest: namely, my examples look at metaphor in a scientist's explanations and careless talk in a geologist's lecture on the anatomy of a volcano crisis.

Analogy

Reviewing the British Department for International Development's (DFID) work on Montserrat, the Overseas Development Institute (ODI), an external vetting organization working for the British Government, wrote reports exonerating DFID (Clay et al. 1999). I was a consultant brought in to listen to the vetters, to enable them to give them the impression of impartiality in their reports and it was interesting seeing the difference between the written reports and their rhetorical stories. The ODI evaluations, a whitewashing for the British government explained, in particular, how it was that the responses to the volcano were so slow, that the changes brought about by the volcano were so minor, in an accumulating fashion, that the overall changes were not felt by the public or development workers or organizations until disaster turned to catastrophe with 19 fatalities by pyroclastic mud flow on 25 June 1997. With every small change accommodated, with the expansion of the no-entry zones on the island, stasis occurred and a semblance of normality returned. No-go zones crept north on the island, particularly in 1997, settling for months at a time before moving – creeping – north again (for particular movement details see Druitt and Kokelaar 2002 ; Innanen 2006; MVO 2006). To give an example, ZJB radio station relocated six times in their slow creep north. Crucially, the local population grew accustomed to the new zones, to 'living with the unexpected' and to the monotony of the two daily MVO reports read out on the radio (Patullo 2000, 10). Explaining 'the creep' of the volcano – that 'beast' for the islanders – and thus how the natural disaster had developed 'un-opposed' and caught so many unawares, in an ODI dissemination of findings meeting, the members of the team verbally admitted to DFID shortcomings. This was contrary to the report findings they had convened to discuss. In this meeting, they went on to explain the volcano support operations, resorting to citing from the popular film 'Volcano' where the risk management problems are understood as like a frog being boiled in a pot: 'you put a frog into a pot of boiling water and it jumps out. But put a frog into a pot of cold water and then slowly heat up the pot and you'll boil the frog to death!' (Skinner 2003a, 108). Such was the fate of 19 Montserratians who ignored the incremental no-entry evacuation zones which they had become over-accustomed to. On Montserrat there was a degree of confusion between development agencies, and a lack of effective communication of risk to the public – particularly in the first two to three years of the volcano crisis, a matter which did not help to secure the entry zones. Such communication breakdown had resulted in a loss of colonial authority for the agencies, the creation of powerful island gossip networks, suspicion and mistrust of all bureaucracies on or associated with the island, and the ignoring of the situated indigenous knowledge retained by the locals (see Patullo 2000 for more detail). In terms of miscommunication the chain of 'intentionality' was weakened; miscommunication took place and the general condition of uncertainty came to be disputed by many and ignored by the fatal few.

This example of simile to explain the volcanic eruptions came at an awkward moment, not long after the death of Montserratians by pyroclastic mudflow. It is indicative, so I have argued elsewhere (Skinner 2003a), of a colonial mentality. It is also ribald in the sense that on Montserrat there is a species of giant frogs known locally as 'Mountain Chicken'. The use of analogy, such as this, is typically a means into the worldview of the other, a way of using the imagination to communicate with. Doctors connect with patients when explaining diagnoses, for instance (Rhodes 1984). We think figuratively and communicate through analogy according to Fernandez (1986, 6–7) who is continuing Lakoff and Johnson's thesis that 'human *thought processes* are largely metaphorical' (1981, 6, authors' emphasis). The Montserrat volcano, for example, is often described as 'a beast', alive with its pulsings, and sounding like 'a jet engine' with its roaring sounds; a volcano which between 1995 and 1997 had grown in volume by 65 million cubic metres ('the equivalent of a bonfire piled with 65 million sacks of coal' (Patullo 2000, 5)) and by 1998 had thrown out some 300 million cubic metres of rock, understood as 'enough to fill 3000 Royal Albert Halls' (Patullo 2000, 84), or equivalent to 2–3 fridges every second of at least the first two years of the eruption as one scientist once told me.

Anatomy

As mentioned earlier, Traweek (1992, 448) has drawn attention to the difference between written and verbal communication: the mode, style, longevity, inclusivity and authority all differ between the two. Traweek (1992, 448) goes further by using Gregory Bateson's idea of the double bind to explain this difference: one message is to be obeyed and the other message to be violated. This is the difference between the ODI reports and the ODI comments, jests and explanations through the use of popular media analogy. The text exonerates whilst the analogies account for the blame being attached to DFID. This difference was also very evident in the public lecture of a recently returned member of the British Geological Survey working on tours of duty at the Montserrat Volcano Observatory. Speaking in Oxford not so long ago, in a public/academic lecture tour, John Souter (name changed) chronicled, humorously and with a beautiful range of PowerPoint pictures, the early years of the Montserrat volcano crisis. His talk was billed as 'The anatomy of a crisis: the Montserrat Volcano emergency 1995–1999'. It discussed how the MVO monitored the volcano, predicted dome growth and collapse and liaised with the Montserrat and British governments as to how to divide up the island into safe and unsafe zones. In Souter's words, their work had a 'sub-agenda of dealing with uncertainty' on the island as independent evaluators; these members of the British Geological Survey team were more prepared than the Montserrat government or people for the volcano and changing situations, including the British development workers on the island who had had no emergency plans prior to the volcano eruption (this was a key criticism of DFID in the ODI evaluations: that the organization had no crisis management team or plans, and that development aid had never been used on an emergency basis before (IDC 1998, xx)). Writing about Montserrat and the volcano crisis, Patullo (2000, 54) quotes the former Governor's admission that their disaster preparedness manual re-written in early 1995 was severely lacking:

> When Frank Savage arrived in Montserrat as Governor in July 1993, he learned that the Caribbean Disaster and Emergency Response Agency [CDERA] rated Montserrat at the bottom of the disaster preparedness league. ... [S]o he got to work ... [and] asked a regional consultant to draw up a disaster preparedness manual for Montserrat. It was delivered in the first half of 1995. 'We opened V for volcanoes and there was nothing there', said Savage. No V for venting or V for viscosity, just V, apparently, for void.

At the start of his lecture, Souter outlined the history of the eruption of Mount Chance – 'history repeated itself' he remarked. The island was known to be volcanic, and yet, following the hurricane disaster of 1989 when Hugo swept all the trees and buildings from the island, rebuilding took place on the same land as landuse patterns were maintained from the past – a jetty at Plymouth Port, new Government Headquarters in the government district of the town and a new hospital on the slopes of Mount Chance near the city. This was all despite a now infamous report by Wadge and Isaacs (1987), two years prior to Hurricane Hugo, which had warned against any building or investment under the shadow of Mount Chance given that there was a 1% probability per century that the Soufrière Hills could erupt (cited in Patullo 2000, 52). But continuity of landuse will out, and the former British Governor of the island accounted for 17 million pounds worth of inappropriately placed development work (Skinner 2001) by suggesting that the Wadge and Isaacs report had left the island's corporate memory when it had blown away in the hurricane two years after it was produced (IDC 1998, 90).

Souter continued, in his lecture, by describing the procedures used by the MVO scientists to convert their readings and monitoring of the volcano into interpretations, speculations and predictions which could then be conveyed to the Montserrat and British government representatives on the island, those who subsequently set out the volcano policy on the island (evacuations and population zones, for example). Away from Montserrat and the Montserratians, Souter could be more candid about how the MVO dealt with uncertainty, risk and complexity on the island, and how those dealings played out politically and socially on the island and back home with the British Government. According to Souter, he and his colleagues constructed 'expert elicitations' to do with future and potential volcanic activity (Souter expressed a concern with how they were to quantify this 'expertise'). Souter admitted, in his talk, that there is an 'inherent uncertainty in the science of volcanicity'. The British Geological Survey strategy at the MVO, then, for constructing and communicating risk levels to the government, was by working with fuzzy numbers and the concept of 'fuzziness' (working from approximate rather than specific numbers). According to Souter, this communication avenue was taken by reading books on the subject rather than displaying any working knowledge of the topic, and by using fellow scientists' authority – the power of the integer – to convince the lay members of the public and members of the emergency committees. By constructing numbers, fuzzy and imaginary, the MVO staff had data which they could show the governments. From this 'hard' data, the MVO officers felt able to map zones of entry and living around the island – A, B, C ... – to which they then added 'styles of volcano activity, such as Amber, Orange and Red levels'. Their difficulties in these conversions for the public lay in the conjunction of letters and colours and how to get the public to understand and to abide by their measures. Elsewhere, Possekel (1999, 128) is more explicit about this system:

> There are four stages of alert: (1) initial alert, i.e. the volcano shows unusual activity, (2) yellow alert, i.e. volcanic activity is expected, (3) orange alert, i.e. volcanic activity is scientifically confirmed, (4) red alert, i.e. there are sufficient indicators prompting an evacuation of the area.

To continue with Souter from his lecture, behind the scenes, in front of their screens, scientists admitted to each other that it was 'difficult to understand' as a scheme, let alone work out and follow themselves. Whilst in Oxford, Souter described how the scientists would go around the room asking all who were present – including the cleaners – how they felt: if they felt certain that nothing was going to happen (in which case Orange was voted on as 'ok' state), or if they were not certain or confident of developments (in which they scored a '1' as in troubling). The participants in these exercises calibrated themselves, factored in medians and standard deviation from their responses and weighted the responses: '100% given to Sparks [Director of the MVO] and 10% to the cleaner', so Souter quipped, making the point that all had an input into the exercise though the volcanologists had the stronger say in this 'expert elicitation' and 'probabilistic forecasting'.

Unfortunately, if not unsurprisingly, the volcano did not behave according to statistics, and the work of the MVO became one more of 'retrospection than anticipation' as they sought to understand, to explain, to come to terms with the periodic ventings of the volcano, and to simplify and come to terms with the Montserrat complex system. It was only when far away from the MVO that Souter felt able to admit to the scientists' creative quest for certainty and confident numbers expressed in a language of approximation and with what Jim Fernandez (2001, 89) has referred to as 'the inflated professional rhetoric' of development workers. When 'on island', Souter and the other scientists presented their scientific facts and figures, exuding confidence in them and eschewing the process underpinning their pronouncements. In some cases, however, Souter noted that the government refused to accept their recommendations to evacuate swathes of Montserrat for political and socio-economic reasons. Finally, after recounting episodes in the volcano's history, particularly the eruptions and horrific deaths of 1997, Souter concluded his talk by reiterating the point that 'there is an inherent uncertainty in the analysis reflecting the nature of the current science of volcanology'. This was reflected in the changing zones, retrospective alterations to the zonation maps for events which were not predicted. 'This uncertainty will not disappear in the short term', though, with the Centre Hills of Montserrat acting as a buffer, island-wide evacuation has, to date, not been necessary, Souter concluded.

Uncertainty and the observatory

Ironically, Souter's descriptive tales now entertain scientists and members of the public in a ludic performance which contrasts in its candid risk communication with the communications and communiqués made when on Montserrat. Here is an example of a Press Release which would be publicized from the MVO. It contrasts with the verbal comments made by Souter or the analogy made by the ODI staff member.

PRESS RELEASE, 17-February-2006
FROM: Disaster Management Coordination Agency (DMCA) & Montserrat Volcano Observatory (MVO)
DATE: February 17, 2006
SUBJECT: Update on Soufrière Hills Volcano
The Volcano Executive Group (VEG) met this morning to discuss the recent developments at the Soufriere Hills Volcano.
Dr Sue Loughlin, the Director of the Montserrat Volcano Observatory (MVO) briefed VEG members on the recent change in activity at the volcano.
In addition, Dr Loughlin presented the Scientific Advisory Committee's (SAC) initial risk assessment of the renewed activity. The SAC reported that there has been no major change to the type of hazard presented in their last report but that the level of risk to individual workers in Plymouth had increased to a level of upper low to moderate on the UK Chief Medical Officer's scale.
VEG members felt that this risk was at an acceptable level and approved the resumption of commercial mining and shipping activities in Plymouth. This is on the understanding that the current activity remains at today's level and that those working in the area follow proper health & safety procedures.
Dr Loughlin reiterated that MVO staff were available to brief workers in Plymouth on the volcanic hazards and risks posed to workers in Plymouth.
Nevertheless, the VEG decided that there will be no tourist visits allowed into Plymouth during the current level of activity. (MVO 2006)

Souter's storytelling is a more effective communication of risk as he perceived it than when he was number crunching and report writing when he was out in the field. So too the frog analogy referred to earlier used to explain the creeping nature of the volcano proves to be an effective and memorable – and thus efficient – way of informing, conveying and persuading the committee audience, more so than the diplomatic review of DFID responses to the Government. Those in the audiences come on board with the scientists' way of thinking through these rhetorical narratives. In terms of risk analyses, here are examples of strategies used to unify the listeners, either those on Montserrat or those reading or hearing about Montserrat. Douglas (1990, 9) refers to this as the creation of 'solidarity' amongst disputants. For her (1990, 11), in the context of science, there are no certainties, 'even in science'; it is a morality play of conviction and persuasion where imprecise calculations of accountability, risk, forecasting and probability are constructed. It is thus the objective of the staff at the MVO to draw the Montserrat and British governments into their line of thinking.

This is the 'triumph of the calculus of risks' (Beck 1992b, 99) as society turns to mathematical ethics in our technological age, a modern uninsured risk society where the unforeseen is quantified – the incalculable made calculable (Beck 1992b, 100). The sociologist Ulrich Beck is describing our industrial and technological systems, drawing examples from litigation (the rise in the confusion of causation), accident settlement statistics (where blame is no longer paramount) and health and safety scales (where prediction comes close to production – to paraphrase (Beck 1992b, 106), in hazards there are no experts and in hazard prediction there is a 'double fuzziness' as acceptable risks become accepted risks and the successes of science sow new worries from new advancements). The Montserrat volcano is an excellent case in point and we are fortunate to have further information about the MVO activities from their staffs' publications. Writing an academic article about prediction, forecasting and risk assessment, Richard Sparks, former Director of the MVO and to whom – so Kokleaar notes above – the assessment team defers, along with William Aspinall, a colleague and consultant geologist (Sparks and Aspinall 2004), make a substantial elaboration upon Souter's verbal lecture points. It was Sparks who introduced the expert elicitation process at the MVO to assist with volcano forecasting on the island. Sparks and Aspinall (2004, 360) write about the volcano as an 'inherently unpredictable', non-linear complex system. They suggest that there are so many variables ranging from gas and magma pressure to geomorphology and weather and tidal systems that it is not possible to accurately 'predict' volcanic eruptions in the sense that a prediction is a statement about a specific event inevitably occurring within a specific time frame. Sparks and Aspinall (2004, 361) believe, however, that it is possible to 'forecast' and to give 'a probabilistic statement that a specific event might occur with a certain likelihood in a given time-frame' (the examples they give are from strain measurements forecasting the 2000 eruption of Mount Hekla in Iceland, and ground deformation measurements allowing forecasting of Mount St Helens volcanic activity in the US). These forecasts are best expressed in probabilistic terms and based on quantitative scientific evidence and analysis, so the authors (2004, 362) continue. This is because a qualitative statement that an eruption is 'very likely' is open to criticism if no eruption takes place, as in the 1976 mass evacuation of parts of Guadeloupe in fear of La Soufrière erupting. It did not. However, as Sparks and Aspinall note – corroborating my use of Douglas's argument about consensus building – public creditability is more likely to be maintained when a forecast is expressed as a percentage as in 'tomorrow there is an 80% chance of rain', for instance. The nature of couching such expressions, verbal for weather forecasters, but also, significantly, expressed as a percentage with a built in margin for error (20% in this case), is one which the public have become accustomed to. In their own words, they write:

> In the USA public and political decision-makers are used to weather forecasts. In general, weather forecasters have gained public creditability both by a record of improving the accuracy of forecasts and by getting the public accustomed to basic concepts of probability. The statement that there is an 80% chance of rain in a particular town is familiar, and most people will accept that if it doesn't rain then it was not necessarily a wrong forecast. (Sparks and Aspinall 2004, 362)

The scientists' creditability is paramount and the concern for constructing and maintaining that creditability is evident in examples such as these. This example also draws one's attention to the thought processes of the scientist and the public, showing by ratiocination that there is little difference between the thought processes involved in maintaining a belief system or conviction, whether religious or scientific (see Gluckman 1966; Evans-Pritchard 1987; Horton 1991). Yet, the scientist is to serve society and not to run it, and turning to the scientists' probabilistic prophecies might eradicate scientific error in the eyes of the public but it does little to resolve the typical tension between scientist as 'trustee and technician' (Freudenberg 1988, 47).

The shift to the use of probabilities and complex modelling involving fuzzy numbers is an example of the scientists' attempt to gain insight into an indeterminate system, one acknowledged to be 'inherently unstable' (Sparks and Aspinall 2004, 370). Sparks and Aspinall's (2004, 370) article is useful because it goes on to articulate how this 'quantification of uncertainty' is to be achieved. This returns us to the 'expert elicitation' rounds mentioned in Souter's lecture and again highlights the differences between orality and literacy, between the anatomy public lecture and the detailed academic publication, the text. When faced with continued dome growth in 2002, a Risk Assessment Panel was formed comprising staff of the MVO and external experts, chaired by Professor Sparks. To ascertain the level of risk of pyroclastic flow to lower Belham Valley on Montserrat, the panel complemented the modelling of flows with a 'structured discussion' (Sparks and Aspinall 2004, 365). The approach was designed to create both soft and hard information. In other words, the final quantitative risk estimate of flow probability was derived from the computer modelling *and* the pooled judgements of the panel members. This means that a space was created in these expert elicitations for the experts' intuition. Sparks and Aspinall (2004, 366) describe this process in detail:

> Each expert assesses their judgement of some parameter and his or her confidence limits on that assessment, based on shared, available scientific information. Experts are calibrated by a facilitator, so that the pooled results of the group are weighted according to the individual experts' ability to be informative and knowledgeable. This procedure is designed to give greater weight to those individuals with good judgement in urgent circumstances; that is, a ranking relevant to decision-making capabilities in crisis conditions, and not merely a metric of considered scholarship. The procedure has the advantage of reducing the influence of overconfident, vocal or highly opinionated individuals, while providing a neutral medium for the inclusive incorporation of a spectrum of views, and the outcome can be viewed as a mathematically rational consensus of the opinions of all participants.

Sparks and Aspinall (2004, 366) refer to the expert elicitation as a 'structured method'. Although the end result is a mathematical probability, expressed quantitatively, it incorporates qualitative assessments and so acknowledges the contribution qualitative data can have. Furthermore, the end-product is a blend or amalgam of two seemingly incommensurable data acquisition techniques. Statistical structuration? The final product could be either mathematical or descriptive, but the authors' leanings are to the power and authority invested in the number. This result is also less likely to be challenged or questioned by the public or Montserrat and British governments. It is a scientific consensus presented to the public to gain an even wider consensus. Whilst it is unlikely that the calibration of the views of the pool of scientists is evened out, and the influence of certain scientists is entirely neutralized, this written explanation of the expert elicitation process is more substantial and confidence-inspiring than Souter's amusing comments about round-the-room sampling which included the cleaner but always ultimately deferred to the Director of the MVO, namely, Richard Sparks.

Orality, literacy and uncertainty

Proportionally, the impact of a physical hazard disaster on a small island is going to be more comprehensive and 'total' than on a land-locked territory (Mossler 1996). The social impact is also heightened and more concentrated. On Montserrat there does seem to be an unfortunate culture of disaster recovery as Montserratians regularly struggle through strong annual hurricane seasons (Skinner 2006). Yet, the Montserrat volcano natural disaster cannot be seen to be ameliorated by this culture of resilience; certainly there was not a culture of preparedness about the island or instilled by the colonial government. The sense and concern for the future has been fostered by the work of the Montserrat Volcano Observatory. Whilst the insurance companies were withdrawing their policies and cover from the island – an end to this form of colonization and control of the future (see Giddens 1991) – the British Geological Survey (BGS) as well as Trinidad-based Seismic Research Unit (SRU) and the US Geological Survey teams were setting up a base for operation on the island. It was the BGS which came to dominate the MVO, an impartial scientific advisory from 1995 till 1999 when the MVO fell under the authority of the Government of Montserrat. From then it received large tranches of funding to move from rented premises behind some hills to a purpose built observatory with viewing platforms, lecture rooms and a shop for staff and visiting tourists and school children. There, you can experience ash clouds and small pyroclastic flows from a safe distance, take tours with the experts, and 'see the pulse of the volcano being recorded on seismographs' (Anon. 2001, 10; see also Skinner 2003b).

The MVO appears to be a success story in terms of their ability to respond to the volcano. They are adaptable in terms of their presentation methods and their positioning on the island, opening up to the tourist sector with the increasing stability of the volcano. Their shift to probabilistic forecasting given over the radio rather than qualitative prediction also means that they can always defend themselves against the accusation that they were wrong. Indeed, it is doubtful that such a criticism is now possible given the inbuilt reserve and hesitancy in their reports: the situation is now like Wadge and Isaacs with their 1% forecasting of an eruption in the twentieth century which resulted in the population holding them in high regard when the eruption occurred rather than looking at a 99% likelihood of there being no eruption/century translating to a more assertive and falsifiable statement such as 'it is extremely unlikely that there will be an eruption this century'. And yet, critiques of MVO practices are possible. To return to Freudenburg (1988, 46), scientists frequently introduce errors into their estimates, can be guilty of excessive confidence in their abilities, and are prone to making mistakes during calibration on a scale raising with the increased importance of the task – rather than intellectual or expertise differences. Freudenburg (1988, 46) goes on to note that he knows of only weather forecasters and experts who publish forecast prices for horse bets who have sufficient feedback to be able to accurately calibrate their estimates. Further to this, Patullo (2000, 73) makes the point that the SRU from Trinidad had the confidence of the island people because they were from the Caribbean region whereas the BGS, appointed and coming from Britain, struggled with the difficulty of being perceived to be complicit in the colonization of the islanders. The ways in which the BGS at the MVO communicated by report, by newspaper, by radio – their orality and their literacy – were alien to the people of Montserrat who were naturally suspicious of their statements. Montserratians were more likely to understand analogies associated with the volcano, but they were presented with probabilities. Ironically, the oral accounts of the volcano given by the ODI staff and Souter might just have been that much more effective in persuading Montserratians to abide by the evacuation zonings and so save the lives of a number of Montserratians.

Acknowledgements

I should like to thank Karen Fog Olwig, Vered Amit, Asa Boholm, Richard Sparks and Peter Phillimore for their support, encouragement and pointers in the writing of this article.

References

Anon. 2001. Nature tourism on Montserrat: The Soufriere Hills Volcano. In *Holiday Montserrat: The way the Caribbean used to be - an official tourist guide*, 10–11. St. John's, Antigua: West Indies Publishing Ltd.

Barthes, R. 1981. The theory of the text. In *Untying the text: A post-structuralist reader*, ed. R. Young, 31–47. London: Routledge.
Beck, U. 1992a. *Risk society: Towards a new modernity*. London: Sage Publications Ltd.
———. 1992b. From industrial society to risk society: Questions of survival, social structure and ecological enlightenment. *Theory, Culture and Society* 9: 97–123.
Bohannan, L. 1952. A genealogical charter. *Africa* XXII: 301–15.
Boholm, Å. 1996. Risk perception and social anthropology: Critique of cultural theory. *Ethnos* 61: 64–84.
Clay, E., C. Barrow, C. Benson, J. Dempster, P. Kokelaar, N. Pillai, and J. Seaman. 1999. *An evaluation of HMG's response to the Montserrat volcanic emergency - Volume 1*. Overseas Development Institute report for the Evaluation Department of DFID. London: DFID.
Connerton, P. 1992. *How societies remember*. Cambridge: Cambridge University Press.
Douglas, M. 1990. Risk as a forensic resource. *Daedalus* 199: 1–16.
Dowie, J. 1980. Bounds and biases. In *Risk and rationality*, ed. C. Wooldridge, N. Draper, and T. Hunter, 99–156. Milton Keynes: The Open University.
Druitt, T., and B.P. Kokelaar, eds. 2002. *The eruption of Soufrière Hills Volcano, Montserrat, from 1995 to 1999*. London: Geological Society.
Evans-Pritchard, E. 1987. *Witchcraft, oracles and magic among the Azande*. Oxford: Oxford University Press.
Fergus, H. 1994. *Montserrat: History of a Caribbean colony*. London: Macmillan Caribbean.
Fernandez, J. 1986. *Persuasions and performances: The play of tropes in culture*. Bloomington: Indiana University Press.
———. 2001. The irony of complicity and the complicity of irony in development discourse. In *Irony in action: Anthropology, practice, and the moral imagination*, ed. J. Fernandez and M. Huber, 84–102. London: The University of Chicago Press.
Freudenburg, W. 1988. Perceived risk, real risk: Social science and the art of probabilistic risk assessment. *Science* 242: 44–9.
Friedman, J. 1990. Being in the world: Localization and globalization. *Theory, Culture & Society* 7: 311–28.
Giddens, A. 1991. *Modernity and self-identity: Self and society in the late modern age*. Stanford: Stanford University Press.
Gluckman, M. 1963. Gossip and scandal. *Current Anthropology* 4: 307–15.
———. 1966. *Custom and conflict in Africa*. Oxford: Basil Blackwell.
Goody, J. 1993. *The interface between the written and the oral*. Cambridge: Cambridge University Press.
Goody, J., and I. Watt. 1981. The consequences of literacy. In *Literacy in traditional societies*, ed. J. Goody, 27–68. Cambridge: Cambridge University Press.
Halbwachs, M. 1992. *On collective memory*. London: University of Chicago Press.
Hannerz, U. 1990. Cosmopolitans and locals in world culture. In *Global culture: Nationalism, globalization and modernity*, ed. M. Featherstone, 237–51. London: Sage Publications.
Horton, R. 1991. African traditional thought and western science. In *Rationality*, ed. B. Wilson, 131–71. Oxford: Blackwell.
IDC. 1998. *Montserrat - First Report*. House of Commons. London: The Stationery Office.
Innanen, B. 2006. *Montserrat risk maps*. http://mni.ms/riskmaps/index.shtml.
Lakoff, G., and M. Johnson. 1981. *Metaphors we live by*. London: University of Chicago Press.
Mossler, M. 1996. Environmental hazard analysis and small island states: Rethinking academic approaches. *Geographische Zeitschrift* 84: 86–93.
MVO. 2006. 'Risk Assessments' and 'Press Releases'. *Montserrat Volcano Observatory Website*. http://www.mvo.ms/default.htm.
Oliver-Smith, A. 1986. *The martyred city: Death and rebirth in the Andes*. Albuquerque: University of New Mexico Press.

Ong, W. 1989. *Orality and literacy: The technologizing of the world.* London: Routledge.

Paine, R. 1967. What is gossip about? An alternative hypothesis. *Man* 2: 272–85.

Patullo, P. 2000. *Fire from the mountain: The tragedy of Montserrat and the betrayal of its people.* London: Constable.

Possekel, A. 1999. *Living with the unexpected: Linking disaster recovery to sustainable development on Montserrat.* London: Springer-Verlag.

Rhodes, L. 1984. 'This will clear your mind': The use of metaphors for medication in psychiatric settings. *Culture, Medicine and Psychiatry* 8: 49–70.

Skinner, J. 1997. Impressions of Montserrat: A partial account of contesting realities on a British Dependent Territory. PhD thesis, University of St. Andrews.

———. 2000. The eruption of Chances Peak, Montserrat, and the narrative containment of risk. In *Risk revisited*, ed. P. Caplan, 156–83. London: Pluto Press.

———. 2001. Licence revoked: When calypso goes too far. In *An anthropology of indirect communication*, ed. B. Watson and J. Hendry, 181–200. London: Routledge.

———. 2003a. Anti-social 'social development'? The DFID approach and the 'indigenous' of Montserrat. In *Negotiated development: Power and identity in development*, ed. J. Pottier, A. Bicker, and P. Sillitoe, 98–120. London: Pluto Press.

———. 2003b. Voyeurs, voyagers and disaster tourism from Mount Chance, Montserrat. In *Niche tourism and anthropology*, ed. D. Macleod, 129–44. Glasgow: University of Glasgow Press.

———. 2004. *Before the volcano: Reverberations of identity on Montserrat.* Kingston, Jamaica: Arawak Publications.

———. 2006. Disaster creation in the Caribbean and planning, policy and participation reconsidered. In *Environmental planning in the Caribbean: Context and case studies*, ed. J. Momsen and J. Pugh, 53–72. London: Ashgate Publishers.

Sparks, R., and W. Aspinall. 2004. Volcano activity: Frontiers and challenges in forecasting, prediction and risk assessment. *The State of the Planet: Frontiers and Challenges in Geophysics - Geophysical Monograph* 150: 359–73.

Thomas, R. 1992. *Oral tradition and written record in classical Athens.* Cambridge: Cambridge University Press.

Traweek, S. 1992. Border crossings: Narrative strategies in science studies and among physicists in Tsukuba Science City, Japan. In *Science as practice and culture*, ed. A. Pickering, 429–65. Chicago: University of Chicago Press.

Wadge, G., and M. Isaacs. 1987. *Volcanic hazards from Soufrière Hills Volcano Montserrat, West Indies.* Reading: University of Reading.

White, H. 1987. *The content of the form: Narrative discourse and historical representation.* Baltimore: Johns Hopkins University Press.

Governing the sea rescue service in Sweden: communicating in networks

Jenny Palm and Eva Törnqvist

Tema Institute, Department of Technology and Social Change, Linköping University, Linköping, Sweden

This paper discusses how various actors communicate about and coordinate sea rescue activities in networks. We combine a network approach with theories of inter-organization communication to understand how communication can facilitate or limit coordination in networks. Search and rescue officers retain the overall authority to direct rescue missions, and coordination with several other professions is central to this. When coordinating different professional actors in a network it is important to develop trust, legitimacy, and a shared and uniform understanding of the situation and of how to act. Communication deficiencies often result from the fact that involved actors belong to different organizations with different cultures or representing different professions. The greatest gap we found was between those working and not working at sea, and between those habitually or seldom involved in rescue activities. Communication comprises more than simply exchanging information; it also entails the forging of relationships, to facilitate future coordination and cooperation and to develop mutual trust and understanding. In any rescue operation it is important that the actors interpret communication in the same way and act according to a shared pattern. Joint training and follow-up are important conditions for continuous learning and development in this regard.

Introduction

In Scandinavia and elsewhere, people now have higher expectations regarding society's ability to handle emergencies. The occurrence of several natural disasters has drawn increased attention to society's emergency management abilities. Emergency management includes risk assessment as well as measures to prevent sudden, unexpected occurrences that entail serious consequences for fundamental functions of society. The seriousness of these consequences calls for preparation from authorities and other actors on all levels of society; it also calls for the engagement of volunteer and interest groups.

To structure and organize the contributions of these actors, a specific arrangement with the *frivillig resursgrupps* (FRGs) – or volunteer resource groups – has been introduced to the emergency management system. Such volunteer groups represent a resource available to municipal emergency management. FRGs are not regulated in law, and it is up to a municipality whether or not to use them. FRGs represent a structured form of cooperation between a municipality and volunteers. These groups comprise people with various competencies required by a municipality, giving municipalities relatively easy access to capable people in an emergency. FRG members should above all be able to take initiative and improvise in an emergency;

they should, for example, be able to administer first aid in an emergency or act as an extra resource in communicating with the public (Krisberedskapsmyndigheten 2005).

The involvement of such organizations calls for an ability to uphold communication between actors of different backgrounds and professions. The professional actor, working in emergency management on a daily basis, must be able to communicate with people who are seldom involved in such activities. In Sweden, the Maritime Search and Rescue Service (SAR) has long experience of coordinating organizations involving people with different professional skills. Essential for policy-making in changing situations is the actors' capacity to learn and handle new information, so a processes perspective expects actors to learn and change. This paper will focus on the Swedish Maritime Search and Rescue Service and their experience of communicating in networks, to find what lessons Swedish emergency management in general can learn from that sector.

Like Swedish emergency management in general, maritime SAR includes measures taken in response to sudden, unexpected events that are traumatic for those involved. Both maritime rescue and emergency management require the ability to coordinate cooperation between public and private actors. Designing and developing SAR services must be done in close cooperation with many different actors, often belonging to different professional groups. Since cooperation in interprofessional teams[1] is on the increase today, it is important to acquire competence in collaborating and communicating with other professions. We create, recreate, negotiate, and establish a shared and uniform understanding of phenomena through using language, and the language we use changes or confirms social relationships. The transfer of knowledge, meaning, and understanding via communication between people is neither neutral nor unproblematic. In civil defense, various professionals with different norms, knowledge bases, and technology use must communicate when cooperating to solve problems; this must be done without serious misunderstandings that lead to costly – in both human and material terms – consequences. The risk of misunderstanding in such communication increases in groups in which specialist knowledge is represented by professionals with different amounts of experience of collaborative work in networks of organizations with different structures, cultures, and geographical settings (Törnqvist 2004).

Purpose and methodology

In emergency management systems, a range of public and private organizations need to coordinate activities in emergencies. This paper examines how actors representing a range of different practices communicate with each other and coordinate their activities in networks. We combine a network approach with theories of inter-organization communication in attempting to understand how communication can facilitate or limit coordination in networks.

The empirical material cited here comes from case studies of the Swedish Maritime Search and Rescue Service, primarily its operations at the Gothenburg coordination center (Törnqvist 2004). These studies collected data using participant observation, document studies, and audio and video recordings. SAR officers were interviewed and documentation such as laws, regulations, manuals, and written communications were also examined to grasp the framework in which SAR activities occur.

The first part of the paper presents a conceptual discussion of network and governance steering, coordination, and cooperation and of the possibilities and limits of communication between different groups or organizations. The second part discusses the tradition of coordination and communication in maritime SAR in Sweden. We wrap up the paper with a concluding discussion that includes a brief consideration of some policy implications.

Communicating in networks

Emergency management and maritime SAR systems are designed and developed according to established institutions and legal frameworks. At the same time, this process requires coordination and cooperation between a range of public and private actors in networks. Current political science literature identifies two main theoretical approaches to organizational steering, namely government and governance. A traditional government approach highlights the formal chain of command of public organizations and decision-making by political actors. From a government perspective, the role of formal institutions and the state's monopoly on legitimate coercion is the focus. State institutions can thus make decisions and also have the capacity to enforce their implementation (Stoker 1998). The development of governance structures is regarded as a consequence of the state's increasing need to mobilize the resources of actors outside their formal control to be able to formulate and implement public policy. The governance approach identifies a network-oriented method of decision-making, including intricate interplay between public, private, and non-profit organizations. Peters and Pierre explain that governance '...refers to the process through which public and private actions and resources are coordinated and given a common direction and meaning' (Peters and Pierre 78).

The governance approach focuses on how and why actors continue to participate and interact in relatively stable networks. Networks are considered as non-hierarchical arrangements of mutually dependent actors, and are explained as arising from the needs of the involved actors to exchange resources and negotiate shared purposes. Cooperation and coordination in networks are regarded as the best ways to serve common interests (Börzel 1998). In this paper we mainly focus on how marine SAR has developed in networks of public and private actors.

Networks are considered as non-hierarchical arrangements of mutually dependent actors, characterized by uneven power relations and usually containing open exits. Interdependencies and resource exchange are regarded as central to networks and are also decisive for what actors are included in and excluded from a given network (Rhodes and Marsh 1992).

When cooperating in networks, actors need to communicate with others who may possess professional backgrounds different from their own. The particular form of communication used determines how problems will be handled and what policies will be developed. Kingdon (1995, 146–7) discusses the existence of a 'national mood' among central actors in a policy area. He is referring to how established actors involved in a discussion are comfortable with its contents and can sense when the mood shifts. There exists a key understanding, which, although facilitating communication, also constrains the development of policy and the introduction of new ideas. However, the existence of a national mood or shared understanding is crucial when coordinating actors in an emergency. To minimize risks and handle insecurity it is important to establish smoothly functioning communication.

High-reliability organizations (HROs) are organizations with a low tolerance for defects in the technical systems with which they work (Bea and Moore 1993; Bierley and Spender 1995; Roberts 1990; Rochlin 1993; Weick, Sutcliffe, and Obstfeld 1999). They thus constantly endeavor to eliminate defects from their systems (La Porte and Consolini 1991, 1998), as one mistake can cause a loss of public support. Communication in HROs is often not face-to-face but mediated, and the actors involved in the communication may not share the same values or professional background. Communication in large systems, such as marine SAR systems, serves two purposes: 1) to clarify roles, goals, relationships, and responsibilities, and 2) to develop trust and shared values (Grabowski and Roberts 1997).

In marine SAR, various public, private, and non-profit organizations must coordinate their activities in an emergency. According to Wenger (1998), an organization is a social construct consisting of various constellations of practices. These practices negotiate their own activities, even though they sometimes represent only responses to regulations. These practices are created, developed, and dissolved through learning. Practices also create their own boundaries, which may be—but are not necessarily—the same as the related institutional boundaries. These boundaries are especially interesting because they are a learning resource, giving rise to scope for new interplay between experience and competence. A practice includes elements such as language, tools, documents, symbols, well-defined roles, and regulations, all of which may be visible; equally important, however, are unspoken norms, embedded understandings, tacit agreements, etc. Wenger points out that a practice entails both acting and knowing, manual and mental effort, and concrete and abstract purposes; the process of being engaged in a community of practice thus involves the entire person. In networks, actors' relationships can be seen as communication channels, through which resources such as information, knowledge, and trust are exchanged (Kenis and Schneider 1991). The relationships differ, however, in their level of intensity, normalization, standardization, and interaction (Börzel 1998), and the actors are entangled with each other by webs of rumor, friendship, mutual interdependence, etc. Members of networks exchange resources, and cooperation and coordination are seen as the best ways to address common interests in networks (Börzel 1998). Relationships and resource exchange in networks take place over a long period of time through a general pattern of interaction, in which the members take part in mutual and common actions (Powell 1991; Kickert, Klijn, and Koppenjan 1997).

Emergency enforces cooperation between different organizations and relies on the existence of relatively open networks. In an open network no decision will be legitimate just because it is regarded as so by *one* organization or institution; rather, legitimacy emanates from interplay between legal interpretations, shared understanding, and trust in the network. Maritime SAR depends on these networks and on properly functioning communication between different organizations. When members of an organization talk with each other and with outside parties, quite apart from communicating, they are also constructing the organization itself through the process and substance of the communication. It can be said that an organization is created, maintained, and activated through communication, and if the communication is misunderstood the organization's existence is potentially weakened. If communication in a network is to be effective, without costly misunderstandings, network members must acquire communicative competence, meaning an ability to use language appropriately given the parties involved in a specific situation. Whether the communication is face-to-face or is in some way mediated must also be considered (Vaughan 1996; Walther and Burgoon 1992; Walther 1996).

SAR practice is formed from the interplay of participating actors and institutional arrangements. We will now discuss some examples of the communication involved in this and its implications for policy in the SAR area.

Swedish Maritime Search and Rescue Service

The Swedish Maritime Administration is the state authority responsible for maritime search and rescue (SAR). Their operations are managed and coordinated from the Maritime Rescue Co-ordination Centre (MRCC), located in Gothenburg. An SAR officer can use all the resources of society in attempting to save lives, meaning that s/he must cooperate closely with a wide range of actors. Actors from different professions and organizations and even speaking different languages frequently collaborate in performing a rescue mission.

The various actors involved cannot interrupt ongoing cooperation during a rescue mission even if they think it is not working well; rather, they must cooperate to rescue people in distress at sea. The actors also know they will meet again in the future and therefore need to surmount any differences of opinion. The SAR officer is the management officer who is actually posted to the field, and s/he needs to weigh the risks of an operation from the information received from many different places and actors, both lay and specialist.

SAR officer in charge of rescue operations

The SAR organization is a lean organization that most often receives alerts as they first come through the '112 alarm' system. The SAR officer interviews the caller, alerts all vessels in the area involved, decides what to do, and then contacts the appropriate rescue units. Throughout the operation, the SAR officer stays in direct contact with each unit, including the vessel in distress if possible. If the accident is a major one, a staff is created from the experts that come to the rescue center. If many operations are taking place at the same time, the SAR officer can request help from the other organizations working in the same room as the MRCC, namely, the navy, coast guard, and Aeronautical Co-ordination Centre (ARCC) . These organizations can also participate in everyday SAR work without the creation of any formal staff. This makes it uncomplicated, for example, to alert airborne units of search or rescue missions.

The SAR officer has access to a full range of communications equipment and to computer support, such as databases. One such database contains information about available resources such as aircraft and boats, detailing their technical specifications, the kinds of missions they are suited for, exactly where they are at a given time, and how they are equipped. As well, computer programs allow the estimation, for example, of how far and in what direction a boat may have drifted under different conditions.

To become a senior SAR officer one must possess a higher nautical degree, and most officers have many years of experience of working at sea, often in an organization involved in rescue missions. A newly hired officer begins in which theory and practice are mixed; s/he is 'the third person', meaning that s/he works together with the two senior officers on duty. After a while, the apprentice replaces one of the senior officers on the watch. Exactly when the apprentice/junior officer is ready to become a master/senior officer is decided by the other senior officers in consultation with the apprentice. The Swedish model is an example of what Lave and Wenger (1991) and others describe as a learning process running 'from the periphery to the centre', in which learning is situated and happens in a particular socio–cultural context.

Technology decides how to communicate

In earlier times if one was in distress, one assumed that help would arrive from nearby and that one would only be able to alert helpers by means of visual signaling, for example, using signal flags, arm movements, or light. The amount of transferable information was limited, and though only a few interpretations of such signals were possible, misunderstandings could occur. Newer technologies allowed more information to be transferred faster, over greater distances, and to more people. This in turn has increased expectations as to the number of people and units who will react to an alarm, and more people taking part in rescues has meant a greater need for coordination if such cooperation is to be effective. Thus, the person in charge of the rescue operation must be able to interpret the information and sift out what is essential in the information flow. This requires great communicative competence on the part of all those involved in a rescue operation.

The new Global Maritime Distress Safety System (GMDSS) consists of different parts which are sending out distress signals more or less automatically. With this system the emergency frequencies do not need to be constantly monitored, since the rescue center itself is directly and immediately alerted (Törnqvist 2004).

Up to approximately 99% of these GMDSS alarms are false, in the sense that someone has 'pressed the emergency button' by mistake. When this happens, however, there is no way to cancel the signal that has alerted the rescue center, even though one may think one knows what has actually happened. Sending a 'May Day' call by voice, however, demands more consideration and unintentional alarms are virtually never propagated in this way.

The spread of the cellular phone has made it easier for those at sea to maintain contact with friends and relatives, potentially reducing the number of missing person alarms. On the other hand, the technology has raised expectations as to contact frequency: people now expect to be able to call someone at any time, or that someone on holiday should report at a certain time every day. If these expectations are disappointed, anxiety and alarm can result. The 'missing' person may simply have forgotten that the batteries were flat or that the phone was out of range, which also means that the cellular phone cannot be used in a real emergency. The cellular phone can thus create a false sense of security, and SAR officers end up spending a lot of time simply calming down relatives.

Organizations can also contact each other using mobile phones, which, compared to the use of VHF radios, prevents communications from being monitored or overheard by uninvolved parties. Cellular phones can also be used to trace a person's movements and with whom s/he has been in contact and when. The cellular phone is used not only for alerting but also to coordinate rescue operations.

However, a disadvantage of the cellular phone for this purpose is that all the units involved in the mission cannot simultaneously follow what is being said. This can cause information gaps that could even delay the rescue or – in the worst case – result in misunderstandings or incorrect information going uncorrected. With the help of telecommunication companies, cellular phones also let the SAR officer pinpoint exactly where the cellular phone has been used and at what date and time. This is obviously a great advantage when an SAR officer needs information regarding the whereabouts of a person in perceived or real distress, or when a person in distress does not know where s/he is.

All this new technology means that more people can be reached by a distress signal and that many parties can now cooperate in Swedish maritime SAR efforts. Current technology also allows rescue operations to be coordinated and led from a great distance; consequently, there is only one rescue center for all of Sweden, located in Gothenburg.

Rescue operations in practice

In every rescue operation a log is kept in which operational activities are recorded in detail. Nowadays such logs are kept using a purpose-developed computer program; the program is connected to the SAR databases and automatically records information such as date, time, and phone numbers. If the SAR officer finds something particularly noteworthy in connection with the case, s/he can record it in the program as well. Often these records are used to highlight aspects of operations that need to be improved, for example, whether a distress signal was handled improperly due to misunderstanding or knowledge gaps. Problems can also arise because an SAR officer was deliberately given wrong information because of competition between rescue units. Of course, such records also include information about aspects of cooperation that went particularly well. These logs, complete with any appended notes, are the basis of discussions among SAR officers that form part of ongoing group competence development.

A not so obvious part of such continuous learning is the ongoing discussion between SAR officers of how best to use the computerized log. Such discussion arises spontaneously in relation to search and rescue operations, and can continue for an extended period. Superficially, the SAR officers may seem merely to be discussing the technology when reflecting on how the program works, i.e., how and where to record an event. Unawares, however, they are first and foremost discussing their own ways of getting the work done, and the dialogue can encompass both concrete work methods and ethics questions. The officers are both directly and indirectly sharing their experiences of how rescue operations can develop and their views of how 'skilled' rescue operations should be conducted and how 'skilled' SAR officers ought to act. At a deeper level, these discussions concern how they view their community of practice. This learning process can be said to exemplify how socio–cultural learning is also situated learning.

Much, if not all, of this enormously important communication is mediated, and most people involved in a rescue operation and mission have never met face-to-face. How is trust and confidence created between people who have only 'met' by phone or radio? Researchers such as Walther (1996) maintain that one can create a functional relationship through computerized contact if sufficient time is allotted, and we will give an example of this from the Gothenburg rescue center. Earlier we mentioned the database containing information on all available resources, including aircraft, boats, etc. The database is updated in real time, when the person monitoring a particular resource contacts the SAR officer (for example, to say that a boat is leaving its normal port for a few hours for an exercise) who then updates the database. You would suspect these calls to be relatively brief, but besides simply dealing with the matter at hand, the weather, current cases being worked on together, friends in common, etc., all come up – in short, small talk about anything or nothing, which can seem meaningless to an outside observer. Such personal communication does not serve to exchange information or convey knowledge, but rather to create or cement a mediated relationship, to facilitate upcoming cooperation in rescue operations and missions. This relationship, created via one or more communication channels, must then be recreated when the parties meet face-to-face. Diane Vaughan (1996) claims that one factor contributing to the disastrous decision to launch Challenger was the fact that the decision was made over the phone, something that had not happened before (prior launch decisions were made face-to-face). In this instance the actors misunderstood each other, despite their previous relationships, as this time they could not see gestures or other body language. The decision-making setting was thus different, and the parties to the decision could not correctly interpret the information they gave each other. The decision setting is also important in everyday work at the MRCC.

Example of communication problems

Swedish marine SAR is built on cooperation among many different organizations that are active in different geographical places. Many notes in the computerized log concern such cooperation, and mainly discuss mistakes occurring when the wrong actor was called on in an emergency or there was a lack of knowledge of how to act in a specific situation.

In one example taken from the SAR log we can sense the irritation between the SOS operator and the SAR officers:

> SOS … calls and wonders who had called a helicopter. Answered that it was the sea rescue that called and received a very sharp 'Why?' and 'On what grounds did you do that?' Answered that we called the helicopter and that we did not even need to contact the SOS center to do that. The person kept up his argument, and to avoid go crazy[2] I connected to ARCC[3] … Discussing this during an alarm and while on operative duty doesn't lead anywhere.

In this communication we can sense the irritation between the two parties, but also disagreement over who is responsible for what issue. They cannot even agree as to when discussion of the matter should take place.

What issues should be part of an organization's responsibility is also under discussion. The overall responsibility of marine SAR has not been clarified to all parties, as can be seen in the following note written by an SAR officer:

> Doctor … forcefully questioned why he should concern oneself with a pleasure sailor who is situated in a pleasure craft and who has fallen on a bare hillrock …

From the notes it is also obvious that as well as some units not knowing how to contact the MRCC, they are also unsure of who is responsible for what. In these situations they simply act the way they usually do in their organizations:

> Obvious routines must be updated with the fire boat. Communication works very badly and they do not know how to contact MRCC. MRCC only wishes to have the fire boat. SOS and the fire officer also alerted the fire engine and a minor boat …

Other notes from SAR officers describe how their decisions have been questioned by other units:

> Again a lot of unnecessary drivel from Mr. … whether he should go forward to the position or return. In this case MRCC[4] required him to go forward. [The reasons] why do not need to be discussed, but are obvious. No answer after a selective call, so MRCC had to use a telephone.

An SAR officer's decision is rarely questioned, though it could happen when someone is unaware of an SAR officer's formal authority. However, this was unlikely the case here, since it was noted that the party involved had questioned the officer before. These communication difficulties need not be attributed to the fact that the people involved represent different organizations, are in different locations, or belong to different communities; individual people could well have difficulties cooperating quite independently of the organizations to which they belong.

The notes do not only contain comments describing poor communication. There are also notes that single out other organizations and instances as examples of good work and cooperation.

The SAR organizations also evaluate the work done in a rescue case. Many comments in the log-book from these evaluations concern whether and how to report cases of drunkenness at sea. One explanation of why such reports are not generally forwarded to authorities could be the different goals of different organizations. For the police, for example, who must control sea drunkenness as one of their everyday tasks, it is important that SAR officers report any such suspicions. For SAR officers, however, an obligation to report such cases could actually impede their efforts to save lives, as a person involved in an emergency but who has used alcohol could, for fear of being reported, avoid calling for help.

Knowledge of the standard abbreviations in use is one small facet of communication competence in the marine SAR field. For one working with these abbreviations every day, such knowledge comes rather quickly. However, for a fire fighter or volunteer who seldom takes part in rescue operations at sea, ignorance of these terms can be a major hindrance to communication. These abbreviations are, however, standard worldwide, facilitating international communication in the SAR context.

Different communication abilities are required depending on whether one is communicating face-to-face or via telephone, radio, or computer but information and communication technology can create a substitute for physical space from a geographical context. A virtual pocket of local order can be created where projects are carried out based on the resources and restrictions available to the various organizations involved in the Search and Rescue Service and coordinated during a search and rescue mission.

Concluding discussion

SAR officers retain the overall authority to direct and make decisions pertaining to SAR missions. In this activity technology is important, having several different functions:

- Technology that *replaces humans*, for example, an emergency position indicating radio beacon (EPIRB) that replaces a wireless operator
- Technology that *activates humans*, for example, automatic alarms
- Technology that *complements* or aids humans, for example, various computer programs for making advanced calculations
- Technology that *is activated by* humans, for example, information from operators about where a specific cellular phone is located, when it was last used, etc.

In Sweden, the technology developed for the SAR field is mainly technology that can be said to be *controlled* by SAR officers, i.e., it is the officer who actively decides whether or not to use a certain computer program designed to support her or his work. Globally, standardized systems that *activate* the SAR officer are gaining ground, and the SAR officer is likely to become more of a system monitor.

At the same time, coordination and cooperation with various professionals is still central to the work of SAR officers. Sharing resources such as knowledge, information, and artifacts is essential to marine rescue operations. The actors must depend on each other if they are successfully to complete a sea rescue mission. As we have seen, marine SAR largely concerns the ability to coordinate cooperation among various organizations and units. Independently of whether the groups whose activities are being coordinated are paid or voluntary, they still need to deal with the same elementary problems concerning interplay, roles, and relationships. This aspect has been overlooked in most network literature, and research into it would build our understanding of how actors coordinate and cooperate in networks.

When coordinating different professional actors in networks it is important to develop trust, legitimacy, and a shared and uniform understanding of the situation and how to act in it, and communication is crucial to this process. Despite all the technology used in the marine SAR field, it is the people involved in search and rescue who matter the most; in the final analysis it is their experience, competence, and commitment that are decisive.

Most of the communication in SAR missions is mediated, i.e., the people communicating are not present in the same geographical place, so great communicative competence is demanded of SAR officers.

As exemplified in our case study, such communication is facilitated by the fact that most actors in the marine SAR network have established long-standing relationships and hence are familiar with each other. Although they may not always agree, they understand how to communicate with each other and when it is time to end a protracted discussion that is leading nowhere. These are important factors enhancing coordination and facilitating cooperation in the context of a rescue mission and are also central to a community of practice. Locally, rescue units have developed communities of practice in which the members share routines, a certain way of solving problems, and a special culture. It is worth noting that this community building is an ongoing process, taking place mainly during SAR mission operations. Not least, the log has proved to be important in helping actors improve their communication and cooperation routines.

In relation to other emergency management activities, one lesson drawn from marine SAR is obvious: one must develop a shared language, without too many abbreviations or acronyms. In the studied case, communication deficiencies often could not be attributed to the fact that people belonged to different organizations with different cultures or represented different professions. The greatest observed communication gaps were between those who did and did not work at sea, and between those who habitually and seldom worked on rescue activities. It is crucial to realize that communication represents more than just an exchange of information; it is also a way to forge relationships that facilitate future coordination and cooperation and to build mutual trust and understanding. In a rescue operation it is important that all the actors interpret communication in the same way and act according to a shared pattern. Joint training and follow-up are important for ongoing learning and development in this regard.

Investing in a computer program that is available to all actors and can be used for logging events and reporting good and bad experiences during a rescue mission is one way to facilitate communication among those involved and in the end deepen cooperation. In marine SAR, the mobile phone is the lowest common denominator, and was even regarded as a hindrance by the involved actors. At the same time, it is essential for communication to develop shared, uniform routines, such as taking notes during a mission for evaluation afterwards. In the marine SAR case, the log proved to be an important learning resource, even though not all actors had access to it; using it, past events could be recalled and analyzed in the interest of future improvement. As demonstrated, shared technology is crucial for success, and it could be used to create both virtual and physical arenas in which actors can meet, practice, and develop shared routines, language, and values.

Acknowledgements

The research undertaken in preparation for this article comprises part of the research program, 'Emergency management for technical infrastructure – regional and municipal strategies for coordination and implementation', funded by the Swedish Emergency Management Agency (SEMA).

Notes

1. We use Donnellon's definition of a team as, 'A group of people who are necessary to accomplish a task that requires the continuous integration of the expertise distributed among them' (Donnellon 1996, 10).
2. Original: 'gå i taket'.
3. ARCC=Aeronautical Rescue Coordination Centre.
4. Maritime Rescue Sub Centre.

References

Bea, R.G., and W.H. Moore. 1993. Operational reliability and marine systems. In *New challenges to understanding organizations*, ed. K.H. Roberts, 199–229. New York: Macmillan.

Bierly, P.E. III, and J.-C. Spender. 1995. Culture and high reliability organizations: The case of the nuclear submarine. *Journal of Management* 21: 639–756.

Börzel, T.A. 1998. Organizing Babylon—on the different conceptions of policy networks. *Public Administration* 76: 253–73.

Donnellon, A. 1996. *Team talk: The power of language in team dynamics*. Boston, MA: Harvard Business School Press.

Grabowski, M., and K.H. Roberts. 1997. Risk mitigation in large-scale systems: Lessons from high reliability organizations. *California Management Review* 39: 153–62.

Kenis, P., and V. Schneider. 1991. Policy network and policy analyses: Scrutinizing a new analytical toolbox. In *Policy networks, empirical evidence and theoretical considerations*, ed. B. Marin and R. Mayntz, 25–62. Frankfurt am Main: Campus Verlag.

Kickert, W., E.-H. Klijn, and J. Koppenjan. 1997. *Managing complex networks: Strategies for the public sector*. London: Sage Publications.

Kingdon, J.W. 1995. *Agendas, alternatives and public policies*. Boston, MA: Little Brown and Company.

Krisberedskapsmyndigheten. 2005. *Krisberedskapsboken—Förebygga och förbereda*. Stockholm: Krisberedskapsmyndigheten. http://www.krisberedskapsmyndigheten.se/templates/PublicationList.aspx?id=820.

La Porte, T., and P. Consolini. 1991. Working in practice but not in theory: Theoretical challenges of high reliability organizations. *Journal of Public Administration Research and Theory* 31: 19–47.

———. 1998. Theoretical and operational challenges of 'high-reliability organizations': Air-traffic control and aircraft carriers. *International Journal of Public Administration* 21: 847–52.

Lave, J., and E. Wenger. 1991. *Situated learning: Legitimate peripheral participation*. Cambridge: Cambridge University Press.

Peters, G., and J. Pierre. 2004. Multi-level governance and democracy: A Faustian bargain? In *Multi-level governance*, ed. I. Bache and M. Flinders, 75–91. New York: Oxford University Press.

Powell, W.W. 1991. Neither market nor hierarchy: Network forms of organization. In *Markets, hierarchies and networks: The coordination of life*, ed. G. Thompson and J. Frances, 265–76. London: Sage.

Rhodes, R.A.W., and D. Marsh. 1992. New directions in the study of policy networks. *European Journal of Political Research* 21: 181–205.

Roberts, K.H. 1990. Managing high-reliability organizations. *California Management Review* 32: 101–13.

Rochlin, G.I. 1993. Defining 'high reliability' organizations in practice: A taxonomic prologue. In *New challenges to understanding organizations*, ed. K.H. Robert, 11–32. New York: Macmillan.

Stoker, G. 1998. Governance as theory: Five propositions. *International Social Science Journal* 50: 17–28.

Törnqvist, E.K. 2004. *Bland grynnor och blindskär: Kommunikation, lärande och teknik i samarbetsprojektet Sjöräddning*. Linköping Studies in Arts and Science no. 297. Linköping: Linköpings Universitet.

Vaughan, D. 1996. *The Challenger launch decision: Risky technology, culture and deviance at NASA*. Chicago, IL: The University of Chicago Press.

Walther, J.B. 1996. Computer-mediated communication: Impersonal, interpersonal and hyper-personal interaction. *Communication Research* 28: 3–43.

Walther, J.B., and J.K. Burgoon. 1992. Relation communication in computer-mediated interaction. *Human Communication Research* 19: 50–88.

Weick, K.E., K.M. Sutcliffe, and D. Obstfeld. 1999. Organizing for high reliability: Processes of collective mindfulness. *Research in Organizational Behavior* 21: 81–123.

Wenger, E. 1998. *Communities of practice. Learning, meaning, and identity*. Cambridge: Cambridge University Press.

Wrestling with uncertain risks: EU regulation of GMOs and the uncertainty paradox

Marjolein B.A. van Asselt[a] and Ellen Vos[b]

[a]*Faculty of Arts and Social Science, Maastricht University, The Netherlands;* [b]*Faculty of Law, Maastricht University, The Netherlands*

Dealing with uncertain risks is an important challenge. Uncertain risks need to be sharply distinguished from traditional, simple risks which can be calculated by means of statistics. Arguably current governance of uncertain risks tends to erode into organized irresponsibility, i.e., society's ill-preparedness and inability to deal with surprises, negative consequences and/or long-term impacts which are associated with uncertain risks, notwithstanding all institutions and procedures in place. In such cases, a particular pattern in risk regulation can be identified, the uncertainty paradox, which is an umbrella term for situations in which uncertainty is acknowledged, but the role of science is framed as one of providing certainty. This article highlights the need to further understand the uncertainty paradox by investigating actual decision-making processes. Through case-study research on EU GMO regulation this article examines how various actors actually deal with science, knowledge and uncertainty. Four mechanisms sustaining the uncertainty paradox are inferred: 1) uncertainty intolerance; 2) boundary work; 3) the inclination to equate uncertainty with risk; and 4) technocratic provisions. In conclusion, some suggestions are advanced as to how to break through the uncertainty paradox.

Uncertain risks and organized irresponsibility

Prominent risk scholars have made very clear that dealing with uncertain risks[1] is a key challenge in current societies (e.g., Ravetz 2001; Wynne 1995; Beck 1986; Jasanoff 1990; Renn 2006; Löfstedt 2005; Nowotny, Scott, and Gibbons 2001; Harremoës et al. 2002; Wynne 1982). The notion uncertain risks refers to possible, new, imaginable hazards, with which society has no or limited experience. It is uncertain whether the particular activity, product or phenomenon constitutes a risk to humans and/or the environment, because causalities are complex, the possible multiple effects are heterogeneous and extend to the long-term and/or the global scale and risk perceptions clash. The use of genetically modified organisms is an example of uncertain risks. As Lang and Hallman (2005) phrase it: 'the potential for risk in using [GMO] (…) remains just that – potential. There has yet to be an event that would allow institutions and experts to move [it] (…) from an uncertain risk to a quantifiable hazard' (p. 1243).[2] Scientific or historic proofs of harmful consequences are lacking, but suspicions cannot be fully refuted either. Notwithstanding efforts to develop comprehensive frameworks (i.e., applicable to all kinds of risk) (e.g., Rosa 1998; Renn 2006; Renn and Walker 2008; Bouder, Slavin, and Löfstedt 2007;

MIDIR project: www.midir.eu), we are convinced that uncertain risks pose different governance challenges. Uncertain risks need to be sharply distinguished from traditional, simple risks which can be calculated by means of statistics on frequencies and actual impacts. Approaches, tools, routines, procedures and structures that work quite well in the regulation of simple risks are not just inadequate, but may even hamper responsibly dealing with uncertain risks. In the case of uncertain risks, basic, seemingly simple, questions as to whether there is a 'real' risk or whether there is 'enough' safety cannot be answered by science. The presence of uncertainty challenges, or at least complicates, the role of experts as risk assessors.

However, uncertainty is not equivalent to no knowledge at all.[3] Scientific experts, whether tacitly or explicitly, do possess what we have called 'uncertainty information' (van Asselt and Petersen 2003; van Asselt 2005, van Asselt and Vos 2005, 2006; van Asselt, Rooijackers, and Vos forthcoming; compare Funtowicz and Ravetz 1990; Krayer von Kraus 2005; Morgan 2003): i.e., insights on which uncertainties are (deemed) important, (attributed) sources of uncertainty, whether and how uncertainty is reducible and which interpretations seem valid in view of the current state of knowledge (i.e., the limits to interpretative flexibility). The question is whether such uncertainty information is welcomed and shared or whether 'plausibility proofs' are requested and constructed (van Asselt and Vos 2005, 2006). Plausibility proofs are statements by experts in which their risk judgements are cast as conclusive evidence on the existence or non-existence of an uncertain risk. There is no room for uncertainty in plausibility proofs: risks exist or they do not.

Without doubt, governance of uncertain risks is difficult, and inherently political. In this context, Beck (1986) provocatively coined the notion 'organized irresponsibility'. We understand organized irresponsibility as a situation in which society is ill-prepared for, and unable to adequately deal with, inevitable surprises, negative consequences and/or long-term impacts associated with uncertain risks, notwithstanding all institutions and procedures in place and the pretence of certainty and control. The sense of being organized and in control further contributes to society's ill-preparedness and inability, as it appeases people's awareness and alertness. As a consequence, a situation of organized irresponsibility has negative spiral characteristics. Organized irresponsibility is not just a luminous theoretical notion coined by a scholar critical of the social structure. For example, the BSE crisis of 1996 has been considered an example of organized irresponsibility (Hajer and Schwarz 2001). Another dramatic real-life example can be found in Mourik (2004), analyzing a case of polluted water and dead cows. Here, due to uncertainty in the mono-causal relationship between polluted water and dead cows, it turned out to be impossible to solve the actual problem of dying cows, even, or especially, in such an organized country as the Netherlands, due to the existing structures of allocating responsibility. This kind of real life case demonstrates that responsibly dealing with uncertainty is an important societal challenge.

In our earlier analyses, we have identified a particular pattern in risk regulation that we refer to as the uncertainty paradox: an umbrella term for situations in which uncertainty is present and acknowledged, but the role of science is framed as one of providing certainty (van Asselt and Vos 2005, 2006). In instances of this uncertainty paradox, policy-makers and judicial authorities resort to experts for conclusive evidence and definite answers, despite uncertainty precluding both conclusiveness and definitiveness. Hence, on the one hand, it is recognized that science cannot provide decisive evidence on uncertain risks, while on the other, policy-makers and judicial authorities increasingly turn to science for certainty (Weingart 1999; Wynne 2001). In uncertainty paradox situations, a very high level of scepticism as to what science can deliver, goes hand in hand with a very optimistic level of confidence regarding what science should be able to deliver (Forrester and Hanekamp 2006).

Yet, before even considering sensible and effective alternative ways of dealing with uncertain risks, we need to comprehend the essence of the challenge faced. To this end, we propose to examine various decision-making processes on uncertain risks so as to understand which mechanisms bring about, contribute to, and sustain the uncertainty paradox. In case-study research on EU GMO regulation, we examine how various actors actually deal with science, knowledge and uncertainty. With this, we aim to infer patterns and mechanisms illustrating the current governance of uncertain risks. In this article, we identify and discuss four mechanisms: 1) uncertainty intolerance; 2) boundary work; 3) the inclination to equate uncertainty with risk; and 4) technocratic provisions.

EU risk regulation of GMOs[4]

We analyzed three cases pertaining to the import of genetically modified organisms (GMOs), i.e.:

- NK603, a genetically modified maize made resistant to a particular herbicide to increase farmers' control of weed
- GT73, a genetically modified maize made resistant to the same herbicide as NK603
- MON 863 X MON 810, a maize composed of two genetically modified maize variants containing insecticidal proteins.

Our analysis is informed by analysis of all kinds of documents and views produced by the relevant actors and we benefited from complementary research on the same cases and/or EFSA, which involved interviews and participant observation (e.g., Chalmers 2004; Vos and Wendler 2006) as well as earlier analyses on EU regulation of GMOs.[5] All three cases developed around applications for authorization for import in the EU by the US-based biotechnology company Monsanto. The cases are particularly interesting because of the complex regulatory procedures and decision-making and societal controversies involved. In our analysis, we use the following notions inspired by Ravetz (2001) to refer to specific main actors, i.e.:

- *risk producers* – those pursuing potentially hazardous activities or technologies: Monsanto.
- *risk assessors* – experts seeking to analyze risks, e.g., the GMO-panel of the European Food Safety Authority (EFSA).
- *risk managers* – those charged with deciding on and controlling risks: the European Council and Commission and the Member States.
- *risk protesters* – those objecting new technologies or activities on behalf of potential risk victims: Greenpeace and Friends of the Earth. These risk protestors do not have a formal role, but they aim to influence the regulation process through lobbying, protesting and critical reports.

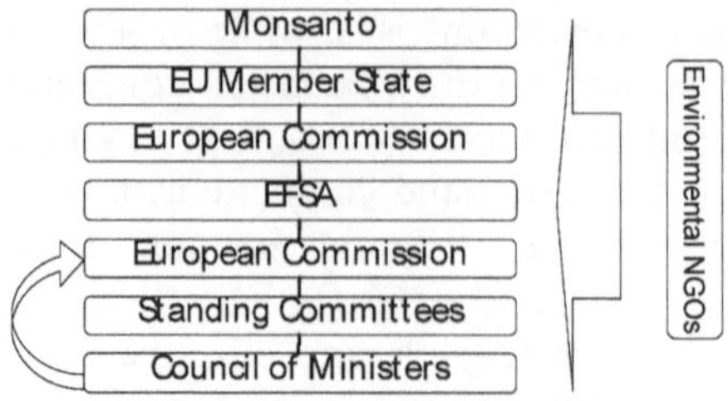

Figure 1. Schematic overview of the risk regulation process.

The risk regulation process (see Figure 1 for a schematic overview) began with an application by the risk producer Monsanto to one of the EU Member States to market a GM product in that country. Not only did Monsanto initiate the risk regulation process, it is legally required to carry out an assessment. Monsanto is thus not only the risk producer, but is also the first to assess the risks. Their assessment had been submitted to the relevant competent authority of the Member State. When the national assessment was completed, it was forwarded to the European Commission. The Commission then asked the GMO Panel of the European Food Safety Authority, established in 2002 as an independent body for science-based risk assessment, to consider the risks of the GM product in question. The GMO Panel has to operate within strict time limits[6] and it lacks the resources and facilities to test products. Our analysis demonstrates that this means that EFSA's risk assessments are *de facto* meta-reviews of Monsanto's assessments instead of an independent examination (van Asselt, Rooijackers, and Vos forthcoming). The claimed independency of EFSA is compromised by reliance on Monsanto's research facilities, tests, knowledge and willingness to disclose information.

The Panel's advice was forwarded to the relevant standing committees comprised of representatives from each Member State. Yet, they could not agree with the proposed authorization of the products and failed to give an opinion. This meant that the Commission had to refer the draft decision to the Council of Ministers of the EU. Similarly divided on these issues, the Council too failed to reach agreement and could not adopt a decision. This meant that in turn the final decision was up to the Commission who functions in these cases as 'deal breaker'. This procedure has been intended to serve as a provision for incidental deadlocks. Yet, in the three GMO cases that we analyzed, this turned out to be the *de facto* standard operating procedure.

Uncertainty intolerance

Our case study research on EU regulation of GMOs revealed that both the risk producers and the risk assessors expressed an intolerant attitude towards uncertainty. By uncertainty intolerance we mean that uncertainties are not acknowledged, deemed irrelevant, or are simply evaded, instead of genuinely and systematically investigated.[7] Uncertainty intolerance is associated with an unwillingness to demand and produce uncertainty information. In this section, we will detail how Monsanto and the EFSA GMO Panel framed uncertainty as well as their assessment behaviour.

Monsanto's framing

Our impression is that Monsanto's safety assessments are deliberate efforts to transform any uncertainty about risk into absolute certainty about safety. To this end, they avoid uncertainty in their communication, they define it away in their reports and they seem to attempt to suppress tests that may question the absolute certainty. This behaviour will be detailed below.

The risk producer seems to deliberately avoid the vocabulary of risk and uncertainty. In the case of NK 603, uncertainty is not even mentioned once in their 37-page report. Monsanto only used the term 'risk' in contexts not directly related to NK 603, or in contexts in which Monsanto claims that the risks are similar to or even less than those of conventional maize, implying that their product is as safe or even safer. In this way, Monsanto claimed that its GMO does not pose any risks to humans and the environment, albeit indirectly and implicitly, as it refused to frame its assessment in those terms. Instead of risk assessments, Monsanto conducts what are called 'safety assessments'. This 'safety framing' is not just a matter of wording. We would like to argue that the choice of the 'safety' instead of 'risk' symbolizes Monsanto's uncertainty intolerance. The notion of 'framing' refers to 'schemata of interpretation' (Goffman 1974), which enables to organize experience and to set boundaries, and which serves as 'bias for action' (6 2005), i.e., they push particular styles of decision-making and/or behavioural responses (compare Benford and Snow 2000). Framings impact on risk assessment, as 'frames define what counts as relevant for attention and assessment' (6 2005, 94).[8] Monsanto's safety framing reinforces a focus in the risk assessment on demonstration of no effect, which pre-empts, hampers or even derails any attempt to systematically analyze uncertainty.

Monsanto's assessment behaviour

Monsanto also displayed uncertainty intolerance in its assessment behaviour. Informed by an interview with Mr Morley,[9] UK environment minister, the *Guardian* journalist Brown stated that Monsanto did not immediately and voluntarily share test results indicating that MON 863 caused unexplained kidney damage to rats. A German law court in Münster ordered Monsanto (20 June 2005) to disclose the suppressed report on rat feeding trials related to MON 863. The court case was inspired by a Greenpeace campaign (Greenpeace 2005). The full rat study report (1100 pages) and a summary document were subsequently made available on Monsanto's website.[10] The rat study was dated December 2002, while the MON 863 assessment was released in August 2003. In the MON 863 assessment, these rat study results were not reported: it was suggested that these tests were not performed and no anomalous results had been found. Viewed against the background of this chronology, it is clear that the test results, which could have been interpreted as uncertainties about effects, were indeed suppressed.

Surprisingly this obvious violation by Monsanto of the obligation to disclose all relevant research seemed to have no consequences for EFSA's attitude towards Monsanto's applications. The authorization procedure prescribes companies to carry out the risk assessment which need to be reviewed by EFSA. EFSA is thus dependent on Monsanto's assessments. In such a situation of dependency, trust and trustworthiness are critical, especially in a so-called post-trust era (Löfstedt 2005). Trust is not a matter of belief, but it involves a system of checks and balances. Trust should not be given naively or relied on blindly. Any failure to adhere to an atmosphere of mutual trust and openness should at the very least have negative consequences for the risk producer. At it stands now, the regulators' attitude towards Monsanto was not affected by the affair concerning the missing rat study. Whether this missing data was a lack of due diligence or a deliberate attempt to hide uncertainty information was apparently unimportant to the risk regulators. How can one expect trustworthy behaviour when abuse is so easy?

Apart from the fact that the rat study report was held back, it is also worthwhile to look to the assessment behaviour in the contested study. According to the rat study report, the tested rats had less mineralized kidney tubules than average. Without motivation or justification, Monsanto claimed that this finding is of 'no biological significance' (11). The observed effects are dismissed by Monsanto as irrelevant. Van Asselt (2005) observed that in many practices claims of irrelevance enable experts to suggest certainty. Uncertainty intolerant assessors seem inclined to claim irrelevance, while uncertainty tolerant assessors would have investigated the uncertainty and would have sought to share the uncertainty information. Monsanto furthermore stated that the effect cannot be considered 'test related' (11). This is a quite puzzling statement. So there is an adverse effect; however, Monsanto claims that this is not caused by the intake of GM maize. It is not explained how this is possible in a control study set-up: the only difference between the two groups of rats was that the rats with altered kidneys were fed GM maize. Here, uncertainty intolerant assessment behaviour entailed creating a smoke screen with the aim to keep the uncertainty out of sight. It is intriguing that although Monsanto neutralized the uncertainty in the report itself through claims of irrelevance and a smoke screen, it still decided to conceal the rat study report. The combination of these assessment behaviours – claiming irrelevance, creating a smoke screen and suppression of the report – provide further evidence for our evaluation of Monsanto's stance: uncertainty intolerance is both manifest in the framing and the assessment behaviour.

EFSA's stance

In our cases, we also observed uncertainty intolerance on the part of EFSA's GMO panel, which is in line with the earlier observations of Levidow, Carr and Wield (2005): the opinions of EFSA's GMO Panel 'generally indicate no uncertainty' that might trigger extra risk management measures (270) and they 'have framed scientific uncertainties in such a way that they can be resolved by extra information, or can be readily manageable, or can be deemed irrelevant to any risk' (273). The risk assessor partly inherited Monsanto's uncertainty intolerance, as EFSA's risk assessments were in fact meta-reviews of Monsanto's assessment. This arrangement, which is laid down in the relevant legislation, did not only introduce dependency on the willingness of the applicant to disclose all relevant information, but also in terms of framing and the willingness to disclose uncertainty information. EFSA's room of manoeuvre is determined by Monsanto's framing and information about uncertainties. The behaviour of Monsanto with regard to the tests with adverse effects indicates that this dependency is not a theoretical issue, but a practical difficulty. Another illustration is that during the assessment of MON 863 × MON 810 (EFSA 2004a), the applicant had to repeatedly provide EFSA with additional data for a variety of reasons. Though Monsanto apparently satisfied the Panel in the end, it is clear that the company lingered with the information.

However, the uncertainty intolerance of EFSA's GMO Panel cannot only be explained by inheritance of uncertainty tolerance from Monsanto. It is noticeable that EFSA's GMO Panel often sides with Monsanto's evaluations and interpretations of data, and even explicitly 'agrees with the applicant', without further explanation or critical discussion of uncertainties that might have been overlooked by Monsanto. We did not find any signs that the rat study affair prompted more explicit requests for uncertainty information. EFSA did frame its assessment activities as risk assessment, but its actual assessment behaviour is very consistent with a safety, i.e., an uncertainty intolerant, orientation. Chalmers (2004) characterizes EFSA's style of reasoning as providing 'its own corpus of proof' to find no evidence of risk.

In all cases, the Commission asked the EFSA's panel to 'consider whether there is any scientific reason to believe' that the placing on the market of the GMOs 'is likely to cause any adverse effects on human health and the environment' (EFSA 2003, 2004a, b). The particular formulation 'whether there is any scientific reason to believe' is relevant if we compare it with the closed question (i.e., whether or not the activity constitutes a risk) the Commission asked in the virginiamycin case (van Asselt and Vos 2005, 2006). It can be argued that the terms of reference to EFSA were more uncertainty tolerant in the light of the following two elements: 1) instead of asking for a decisive answer and proof on whether the risk is a hazard, the Commission asked for indications that hint at adverse effects; and 2) instead of referring to science as the source of absolute truth and certainty, with the notion 'to believe' the Commission seemed to accept that science cannot provide certainty about uncertain risks.[11] The terms of reference could therefore have been read as invitation to systematically discuss the uncertainties involved and to provide uncertainty information. Interestingly, however, in all three cases, EFSA provided clear, 'certain' answers phrased in the following terms: the GMO is *as safe as*[12] the conventional counterpart and it is 'unlikely to have an adverse effect on human and animal health and (...) the environment'. 'Unlikely' can be read as 'low risk'. Combined with the claim 'as safe as', it can be argued that the effective message of EFSA's assessment is a plausibility proof, although such non risk contention was arguably not requested by the Commission. This production of plausibility proofs can be read as an uncertainty intolerant interpretation of the goal of the assessment. Our observations are consistent with findings of Stahl and Cimorelli (2005) about policy analysts involved in ozone risk assessment. They argue that:

> Even when uncertainty is discussed, policy analysts attempt to limit its consideration, address only uncertainties in specific pieces of data, indicate that uncertainties may be very large and produce impacts to the final conclusion that are not part of the analysis, or, in other cases, list the assumptions of the analysis so as to avoid specifically addressing some troubling uncertainties. (Stahl and Cimorelli 2005, 1110)

Wynne (2001) attributes to experts involved in risk assessment 'the inability (...) to recognize the limits of their knowledge'. In his observations, they instead forward 'claims that the risks and consequences are (or soon will be) adequately known' (447). These observations and analyses suggest that EFSA's uncertainty intolerance is not unique, but rather widespread among risk assessors institutionalized in the policy domain.

Monsanto's safety assessment and EFSA's risk assessment breathed a similar spirit due to shared uncertainty intolerance. This resemblance, which was also noted by others, led to accusations of bias and relationships with biotech industry, complot theories, etc. Friends of the Earth (2004), for example, argued that EFSA is used by the Commission to force GMOs onto the market. We do not agree with such views. Building upon our analysis, we argue that EFSA was as uncertainty intolerant and safety-oriented as Monsanto, with the consequence that it is understandable that EFSA felt at ease with Monsanto's arguments and that the assessments were very similar. We want to emphasize that we do neither suggest nor argue that EFSA is deliberately defending or advancing Monsanto's stakes. But, because of the shared uncertainty intolerance, their lines of reasoning were very similar. This similarity is not the point of departure, but in part the *effect* of EFSA's attitude towards uncertainty.

Boundary work and plausibility proofs

How did EFSA construct plausibility proofs about uncertain risks? Evading uncertainty, which allowed GM products to be declared safe, was made possible through boundary work. Boundary work, a notion coined by Gieryn (1983, 1999), is a strategic and purposeful act in which boundaries are drawn between realms, for example, between science and non-science and between science and politics. Boundary work involves drawing and maintaining contrasts through selective attributions, which effectively demarcate in order to construct 'self-evident justification' and 'superiority in designated terrains' (Gieryn 1999). It has been convincingly demonstrated that boundary work is not just a matter of formal responsibilities, but that it is an ongoing negotiation process on roles and tasks and how these are portrayed to others. Tactics and intentions are involved and boundary work has particular effects (Bal, Bijker, and Hendriks 2002; Halffman 2003, 2005; Hoppe and Huijs 2002; Jasanoff 1990, 2005).

EFSA explicitly stated to have been requested to consider scientific objections and not to assess non-scientific ones (EFSA 2003, 4). The GMO panel used the constructed boundary between science and non-science to argue that possible uncertainties about interference with the European environment and regular maize crops are non-scientific concerns. In this way, the scope of the risk assessment could be minimized. In a similar vein, EFSA constructed a boundary between risk assessment and risk management. Such boundary work enabled the GMO Panel to disqualify and dismiss Member States' concerns, which could have been read as uncertainty issues (compare Levidow, Carr, and Wield 2005), as 'non-scientific' and 'issues of risk management'.

The GMO Panel's responses to concerns regarding the risk of cross-contamination illustrate the dynamic nature of boundary work. With NK 603, it refused to consider the possibility that the GMO could contaminate regular maize crops. It argued that the question of contamination was one of risk management, not of risk assessment and thus felt beyond the limits of the Panel remit (EFSA 2003). Landfried (1999) (quoted in Levidow, Carr and Wield 2005, 263) argued: 'the difficulty of distinguishing between political and technical questions also provides an opportunity to those who might wish to reduce political questions to technical ones'. In our case, issues of uncertainty were recast as political topics. In this way, boundary work helped EFSA to minimize the scope of the assessment. The same point concerning cross-contamination was raised with regard to GT 73. This time, however, the Panel did engage in what it had previously argued to be questions of risk management. Concerns regarding contamination were still quickly dismissed with the reasoning that spillage of the GT 73 plant would likely only occur in ports located in industrial areas. Since these industrial areas offer little room for agricultural cultivation, contact with, and contamination of, other plants was portrayed as highly unlikely. Similar reasoning was employed when contamination concerns were raised with regard to MON 810 × MON 863. Also in that case, the Panel appeared more willing to address issues it dismissed in the NK 603 case. Chalmers (2004) argues that in the case of NK 603 the Panel interpreted its role in the narrowest and most formal sense and thus carried out a very minimal risk assessment. We would like to argue that there is not just a matter of interpreting roles. The interpretation of roles is an important dimension of boundary work, which facilitates a particular dealing with uncertainty. The different stances in the three cases illustrate the dynamic nature of boundary work.

Boundary work enabled EFSA to construct superiority, which facilitated the production of authority claims. The Panel often mobilized 'the scientific literature', but without specific references or they just agreed with some particular findings or conclusions, without providing further justification. Through this *ex cathedra* style, EFSA presented itself as an authoritative voice. 'Believe us, we are scientists', is the implicit message. The constructed authority is then used as an anchor to evade uncertainty. In the GT73 assessment, the Panel investigated the potential allergenicity of two genetically modified proteins (EFSA 2004b). Concerning one of these proteins, EFSA stated that no indications of allergenicity could be found in Monsanto's data (EFSA 2004b, 9), nor in the previous Panel assessment for NK 603 (EFSA 2004b, 9). Yet, neither references nor details were provided. Our analysis of the NK 603 assessment reveals that in that assessment EFSA merely stated that an allergy risk evaluation had been performed previously by the national competent authorities and the EC Scientific Committees. Again, neither references nor details were provided. In the GT 73 assessment, the Panel argued that new information on allergenicity would be required to change the Panel's opinion (EFSA 2004b, 9). The boundary work brought about self-evident justification for EFSA's role in deciding what counts as 'new science'. The 'new'-requirement (compare Chalmers 2004) in combination with the self-acclaimed authority in this domain, enabled the Panel to be satisfied with a reference to earlier assessments. In sum, boundary work facilitated the production of certainty about non-allergenicity.

Some further illustration of boundary work and its crucial role in constructing plausibility proofs can be found in the second opinion on MON 863 × MON 810 (EFSA 2005). On the basis of new rat study results, the Panel straightforwardly concluded that there are no safety concerns. However, that particular rat study yielded some anomalous results in organ weights of the rats fed the hybrid maize: the weights were statistically significantly lower. The significance of these results was immediately disarmed by the unmotivated assertion that 'this difference did not exhibit a dose-response relationship' (22). This is another example of authority claims used by EFSA to evade uncertainty.

Dynamic boundary work between risk assessment and risk management (or as Levidow, Carr and Wield (2005) phrase it: 'judgements in the grey area between risk assessment and risk management' (273)) as well as between science and non-science facilitated the construction of claims of irrelevance and authority claims, which were used as building blocks in creating certainty. EFSA's boundary work is an example of what Jasanoff (2005) refers to as 'less transparent, politically significant boundary work'.[13] It is beyond the scope of our expertise to assess the content of the GMO panel's claims: we only want to describe the assessment behaviour. Our analysis demonstrates that in instances that could have been read as uncertainty (Member States' concerns, adverse effects, open questions), the GMO panel actively evaded uncertainty through boundary work, instead of discussing these uncertainties and exploring whether and how they may matter. Let alone that the possibility of other and 'unknown' uncertainties was systematically considered with an open mind (compare Wynne 2001; Levidow, Carr, and Wield 2005).[14]

An 'Uncertainty = Risk' tendency

Our case-study research reveals another mechanism contributing to the uncertainty paradox. We discovered that various actors tend to equate uncertainty with risks. Although the notion 'uncertain risks' points to situations where risks are highly uncertain, this does not mean that *all* risks are highly uncertain and, importantly, not all uncertainties inhibit dangers. Not all uncertainties are by definition relevant or critical for every aspect in a risk assessment. Equating uncertainty with risks implies that any uncertainty is interpreted as a signal of risk.

We observed that both the risk producer (Monsanto) and the risk protestors (Greenpeace and Friends of the Earth) tend to equate uncertainty with risk, notwithstanding the fact that they have opposite positions in the GMO debate. Monsanto's uncertainty intolerance and the associated desire to prove no effect, seem to be grounded in the following reasoning:

$$\text{zero uncertainty} = \text{zero risk} = 100\% \text{ safe}$$

As a consequence they seem to fear uncertainty, as it threatens the idea of absolute safety. The risk protestors also tend to equate uncertainty with risk, but in a different way. Their reasoning seems to be:

$$\text{uncertainty} = \text{risk} = 100\% \text{ hazard}$$

The actual behaviour of the risk protestors is consistent with this logic, which is illustrated by the following example. Shortly after EFSA's opinion on GT 73 became public, Greenpeace published a short critique, in which it concluded that irregularities in the molecular characterization of GT 73 oilseed rape had not been investigated satisfactorily (Greenpeace 2004, 2). Greenpeace highlighted that in EFSA's report, it was stated that besides the genes added by Monsanto, there was a small fragment of DNA present in GT 73 that was not present in its unmodified counterpart, and it also lacked some DNA normally found in oilseed rape (EFSA 2004b, 5). Greenpeace argued that not investigating these differences was an error on the part of the Panel. The risk protestor seems to believe that insufficient molecular characterization implies uncertainty which means risk. Greenpeace apparently did not consider the possibility that the uncertainties pertaining to the altered DNA structure could also be 'safe uncertainties' (RMNO 2004; van Asselt 2004). The same logic seems to motivate the stance of Friends of the Earth. These risk protestors explicitly attacked the lack of uncertainty information in EFSA's assessment:

> By failing to frame its opinion within the context of continuing scientific debate and uncertainty about fundamental issues relating to its conclusions, the Panel has failed to provide decision makers with an adequate analysis of scientific uncertainty. (Friends of the Earth 2004, 15–16)

This argument is used to discredit EFSA. Such equating of uncertainty with risk may be interpreted as a way to politicize uncertainty. Politicization of uncertainty refers to the phenomenon of politicians, stakeholders, special interests groups and/or the public highlighting, amplifying or attenuating uncertainties in order to serve other interests (e.g., Funtowicz and Ravetz 1990; Pollack 2003; Smithson 1993; Stocking and Holstein 1993). Levidow, Carr, and Wield (2005) suggest that in the GMO case such politicization of uncertainty is taking place: 'Policy actors play down or emphasize various uncertainties – to challenge evidence of risk or safety, to justify their stances (...), to pursue greater rigour in demonstrating safety, to mediate among conflicting views and/or to delay (...) decisions' (273).[14] That is not the point we want to raise here. We would like to point to the fact that the risk producer tried to avoid uncertainty in order to demonstrate safety, while the risk protestors highlighted uncertainty to demonstrate risk. This tendency to equate uncertainty with risk sustains the uncertainty paradox as it hampers the production and sharing of uncertainty information. Those who adhere to the absolute safety logic (i.e., zero uncertainty = zero risk), do not welcome uncertainty information. On the other hand, actors who consider uncertainty as a sign of risk tend to politicize uncertainty information, which is, according to Frewer et al. (2003), one of the reasons why experts hesitate to communicate uncertainty.[15]

Technocratic provisions

We, furthermore, observed how technocratic regulatory provisions put the risk assessor, i.e., EFSA, centre stage. EFSA's role in the decision process is magnified, with the consequence that the role of EFSA's uncertainty intolerance further increased. In all three cases, deadlocks[16] occurred because no agreement pro or against import was reached in the relevant Committees and/or the Council. To avoid procedural standstill, in such situations the Council delegates political powers to the Commission: the relevant legislation stipulates that it is the Commission that ultimately is empowered to adopt a decision.[17] Despite resistance of several Member States, the Commission could issue consents for the placing on the market of NK 603[18] and GT 73 oilseed[19] under Directive 2001/18, taking EFSA's opinion as its justification. This provision designed for extraordinary circumstances was the *de facto* standard operating procedure and it gave leeway to adopt authorization in line with EFSA's opinion. The Commission's decisions were a matter of rubberstamping EFSA's plausibility proofs. As a consequence, EFSA was the *de facto* risk manager in the NK 603 and the GT 73 cases.

In the MON 810 × MON 863 case, the decision-making process was more complex, because of disagreement within EFSA's GMO panel. In 2004, the Member States could not reach a majority in favour or against authorization, but this time the Commission could not rely on a plausibility proof. So the decision-making process temporarily evaporated, partly due to changes in the regulatory regime. In 2005,

EFSA issued a new opinion; this time it succeeded in producing a plausibility proof, which was again the basis for a Commission proposal to be discussed in the Committee, but again the Member States could not reach a majority, so it was referred back to the Council. However, the Council did not react in the appropriate period, with the consequence that also in this case, the Commission authorized MON 863 × MON 810 in line with EFSA's plausibility proof.[20]

In view of the societal and political controversies in Europe pertaining to genetic modification (e.g., Durant, Bauer, and Gaskell 1998; Gaskell et al. 2000, 2004; Horlick-Jones et al. 2007; Jasanoff 2005; Levidow 2001; Levidow, Carr, and Wield 2000, 2005; Levidow, Murphy, and Carr 2007; Tait 2001; Wynne 2001), it is striking that notwithstanding the political turbulence about possible authorization of these three GMOs, in the end the decisions about the import were taken in a technocratic manner: the Commission could simply authorize these products. Our analysis provides further evidence for Borrás' conclusion (Borrás 2006) that scientific experts continue to have a central, undisputed position in actual EU regulation of GMOs, notwithstanding rhetoric and institutional changes after the BSE crisis. So there is a clear political deficit. It is exactly such technocratic division of responsibility, and associated credibility and legitimacy questions, which were the main political concern after the BSE crisis and the *raison d'être* of EFSA (Vos 1999, 2000). Our analysis adds another dimension to the old concerns: technocratic provisions sustain or even reinforce the uncertainty paradox in situations of uncertainty intolerance on behalf of the risk producer and the risk assessor.

Conclusions and challenges for risk regulation

Our findings reveal that the uncertainty paradox is also manifest in the case of EU regulation of GMO import. Our analysis indicates that in the case of EU regulation of NK 603, GT 73 and MON 810 × MON 863, the uncertainty paradox is sustained and reinforced by the interplay of four mechanisms (see Figure 2).

The first pattern is *uncertainty intolerance* on the part of Monsanto and EFSA (1). EFSA's uncertainty intolerance was accompanied by *boundary work* which enabled to claim irrelevance and to construct authority claims which served as building blocks in creating plausibility proofs (2). The third mechanism is the *tendency to equate uncertainty with risk*, which (further) confined risk producers and risk assessors to the role of uncertainty-intolerant producers of plausibility proofs

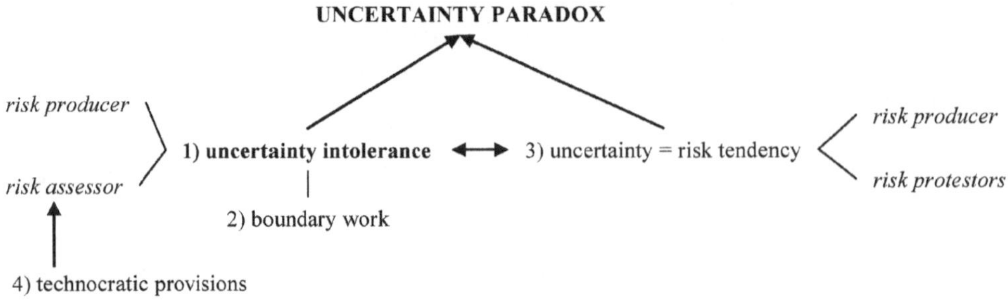

Figure 2. A schematic overview of the uncertainty paradox mechanisms manifest in the cases of EU regulation of GMOs as described in this article.

(3). *Technocratic provisions* (4) resulted in an even stronger, and *de facto* political, role for EFSA, with the consequence that its uncertainty intolerance became a critical and decisive factor in the interplay. Our analysis also demonstrated that in this way the political responsibility for a highly sensitive risk dossier got lost, notwithstanding all institutions and procedures in place and the pretence of certainty and control. Only the future will teach us whether as a consequence the European society is ill-prepared and unable to deal with possible surprises, negative consequences and long-term impacts in the event that GMOs should turn out to have such effects. Yet, it must be emphasized that it is too early to conclude that the current situation is indeed one of organized irresponsibility.

The identification of different, although interacting, uncertainty paradox mechanisms contributes to a further understanding of the complexities of the governance of uncertain risks and as such it provides input for theory development in the field of risk governance. The key question from a policy perspective is whether these insights can be translated into recommendations for risk regulation. The main challenge for risk governance is to break (out of) the uncertainty paradox and to rethink the role of science and expertise in risk regulation (compare Wynne (2001), Levidow, Carr, and Wield (2005) and Jasanoff (2005)). Our analysis provides insights into the ways in which actors wrestle with uncertainty. What would be more responsible ways of dealing with uncertainty in the context of risk regulation?

The case studies discussed in this article lend further importance to the issues of the dependence on the risk producer, which is also raised in the current political debate on the regulatory procedures in GM authorisation.[21] The current arrangement does not only introduce dependency in the willingness of the applicant to disclose all relevant information, but also in terms of framing and the willingness to disclose uncertainty information. Uncertainty intolerance on behalf of the risk producer is easily inherited, which in its turn contributes to sustaining the uncertainty paradox. Especially the rat study affair illustrates that dependence is not just a theoretical issue, but an actual concern. We would recommend policy actors to develop mechanisms that enable to investigate and punish violations of trustworthy behaviour. It should not be so easy to hide (uncertainty) information. Furthermore, it seems needed to develop incentives that invite risk producers to be more uncertainty tolerant.

Our analysis suggests that uncertainty intolerance is a core problem. Recognizing and addressing uncertainty can inform knowledge-generation (Levidow, Carr, and Wield 2005). To counteract uncertainty intolerance, uncertainty training is needed. In current education programs, science is presented as a body of certainty (Pollack 2003; see also Collins 1987), which nourishes uncertainty intolerance. In all scientific education, more explicit attention should be paid to uncertainty aspects of science (van Asselt and Petersen 2003). Our own experience with uncertainty training suggests that it changes the attitude towards uncertainty and creates openness for communicating uncertainty, without hampering the willingness to bear responsibility and take decisions. Another aim of uncertainty training is to facilitate awareness of uncertainty paradox mechanisms among people involved in risk regulation.

Furthermore, risk assessment bodies, such as EFSA, should not only include experts on issues relevant with regard to specific uncertain risks, but should also welcome uncertainty tolerant experts who are aware of mechanisms associated with the uncertainty paradox in order to organize resistance to the production of plausibility proofs. In addition to this, it might prove fruitful to organise realms where risk producers, risk assessors, risk managers and risk protestors meet, next to the indirect communication through formal reports. Such two-way exchanges might help to discuss uncertainties in a different way, and might enable risk regulators to gain an understanding of what science can and cannot provide.

Informed by our research, we are convinced that the uncertainty paradox is deeply ingrained in current risk regulation arrangements and the broad socio-political order. So it will be difficult to break (out of) the uncertainty paradox. Insights into, and broader awareness of mechanisms that bring about and sustain the uncertainty paradox are only first, but necessary steps. When looking at the current debate on EU GMO regulation, we observe that most critics continue to frame EFSA's role in terms of the uncertainty paradox: EFSA's GMO Panel is expected to resolve diverging scientific opinions and is still cast as 'a body responsible for supplying *unimpeachable* scientific advice and guidance' (Randall 2006, 413, emphasis added). It is not recognized that the problems arise from the great burden placed on science as the basis for decisions (Levidow, Carr, and Wield 2005). In this way, an artificial certainty that glosses over any uncertainty regarding risks of GMOs is asked for, which only encourages and reinforces 'the self-elusions of institutional science' (Wynne 2001, 457) and the 'lack of reflexivity about the quality of the knowledge it provides' (Wynne 2001, 458). This direction further reifies the uncertainty paradox. We would like to argue that any regulatory reform is doomed to fail if the uncertainty paradox is not recognized, let alone addressed.

Acknowledgements

The research reported in this paper is part of our collaborative research, which was partly carried out with the career grants (the so-called Innovation Research Incentives Scheme), the Dutch National Science Foundation (NWO) granted to both Marjolein van Asselt and Ellen Vos (2001–2006). We would like to thank NWO for the necessary financial support and the freedom to follow one's own avenues. An earlier version of this paper was presented at the research colloquium of Science and Society Studies at Maastricht University and the RiskCom 2006 conference in Gothenborg, Sweden. We benefited from the questions and feedback. Furthermore, we would like to thank our former student-assistant Bram Rooijackers for his contribution to the case-study research, our colleague Elisavetta Radulova for useful comments and our student-assistant Elise Gielisse for her support in additional literature review. Last but not least, we thank Ragnar Lofstedt and the anonymous reviewer for their constructive criticism. A chronological description of each of the cases (NK603, GT 73, MON 810 × MON 863) can be found in: van Asselt, Rooijackers and Vos (forthcoming).

Notes

1. Notwithstanding Knight's famous distinction between uncertainty and risk, it has been convincingly argued that risk implies uncertainty (Vercelli 1995; RMNO 2004; van Asselt 2000; van Asselt and Vos 2005, 2006; van Asselt, Rooijackers, and Vos forthcoming). However, in the case of many risks there is certainty that a particular activity yields detrimental effects (such as smoking and lung cancer), notwithstanding uncertainties pertaining to when health effects will manifest themselves, by whom, to which degree and/ or where and to which extent environmental impacts will materialize. In such cases, uncertainty is limited to the contingencies of the occurrence. But there is another category

of risks in which there is not only uncertainty with regard to the contingencies of the occurrence. In many controversial risk dossiers there is also uncertainty with regard to potentially complex causal relationships as well as uncertainty pertaining to kind of negative impacts, if any. In such cases, usually ambiguity also reigns, for example, with regard to the evaluation of the activity, benefits and risks. To emphasize the impossibility to calculate such risks with classical risk assessment methods, we prefer to refer to the latter class of risks as 'uncertain risks'. The notion can be considered a tautology, but tautologies do have meaning: like 'a large giant' emphasizes that even for a giant this particular one is very large, the notion 'uncertain risks' emphasizes that uncertainty is a prominent feature of this category of risks.

2. Compare Hunter (1999), Siegrist (2003) and Levidow, Carr, and Wield (2005) who also explicitly refer to gene technology in terms of 'unknown risks'/'unknown, unquantifiable and (…) undiscoverable risks', 'a very unknown risk' and 'uncertain risks' respectively.
3. The relationship between uncertainty and knowledge is complex (van Asselt 2005). More knowledge may also imply more uncertainty, as more scientific unknowns are highlighted (compare van Asselt (2000) and Levidow, Carr, and Wield (2005)).
4. For in-depth treatment and detailed overviews of EU regulation of GMOs, see Hunter (1999); Grossman and Endres (2000); Guehlstorf and Hallstrom (2005); Vos and Wendler (2006); Borrás (2006); and Levidow, Carr, and Wield (2007).
5. There many scholars studying the European GMO debate and regulation of GMOs in Europe. Many analyses of the GMO debate focus on risk perceptions among the public (see, for example, Tulloch and Lupton 2002; Siegrist 2003; Gaskell et al. 2004). Other studies discuss GMO regulation with an emphasis on the precautionary principle on the role of the precaution/the precautionary principle (see, for example, Tait 2001; Levidow, Carr, and Wield 2005). For comparative studies on regulatory regimes, e.g., EU versus the United States, see e.g., Jasanoff (2005) and Guehlstorf and Hallstrom (2005). Some of these colleagues, e.g., Brain Wynne, Sheila Jasanoff and Les Levidow and his co-authors, share our special interest in how actors deal with science, knowledge and uncertainty in the GMO debate and/or regulation of GMOs. See also de Wilde and Reithler (2006).
6. As per Directive 2001/18/EC, the EFSA has to provide its opinion within 90 days of the European Commission's request.
7. See Wynne (2001) who argues that already the framing of the issue at stake in terms of 'what are the risks?' tends to narrow down to the questions 'what are the scientifically demonstrable risks?', which implies excluding 'unknown' and unanticipated effects, i.e., the uncertain risks.
8. This is the reason that Renn (2006; Renn and Walker 2008) includes framing as an important element in the pre-assessment phase.
9. Further sources were not specified.
10. At the time of our research, the rat study report and its summary were available on Monsanto's website at http://www.monsanto.com/monsanto/content/sci_tech/prod_safety/ratstudy.pdf. Since this website had been revamped in June 2006, this link is no longer functional and the reports are no longer available.
11. Also in the regulatory documents related to the directives relevant for the regulation of GMOs and in additional guidelines, the Commission expresses openness to uncertainty or even explicitly calls for the description and explicit treatment of uncertainty (Levidow, Carr, and Wield 2005).
12. It is noteworthy that this phrasing is very much in line with what Jasanoff (2005) describes as the 'product' perspective on biotechnology (i.e., whether the GM product is different from the natural product) as opposed to the 'process' perspective on biotechnology (i.e., whether biotechnology as such is acceptable). Jasanoff (2005) argues that the product perspective dominates in the United States, while the process perspective dominates in Europe. Our analysis suggests that the product perspective is promoted in the EU

regulatory practice, which provides another angle to explain the discontent among risk protestors with the regulatory state of affairs. In the EU, the product and the process perspectives seem to collide. It is beyond the scope of the current paper to substantiate this view on EU regulation of GMOs.

13. Compare also Cranor (1990) who argued that 'the supposedly objective scientific studies used for estimating risks to human health for regulatory purposes can be considerably (...) political' (126). It can furthermore be argued that earlier boundary work in the processes of setting up the regulatory structures, enabled EU policy to define 'agribiotechnology as an expert scientific issues (...) kept separate from socio-ethical issues' (Levidow, Carr, and Wield 2005, 266) which facilitated EFSA's boundary work.
14. See also Levidow, Murphy, and Carr (2007) in which they argue that '[c]ompeting policy agendas framed scientific uncertainty in different ways' (26).
15. We did not include members of the public in our analysis. Frewer et al. (2003) reveal that risk experts assume that information about uncertainty will cause panic. It is beyond the scope of this study to investigate this hypothesis. However, informed by an analysis of the GMO debate in the UK and more broadly in Europe, Wynne (2001) qualifies such portrayal of the public in terms of 'a naïve "demand for zero uncertainty"' as 'grotesque' (476–7). He argues that the public recognizes that uncertainty is inevitable, that it is not as intolerant as suggested and that 'they actually recognize a more radical uncertainty (indeed indeterminacy) than that admitted by science' (477). Wynne's claim is further supported by observations from focus group research: 'not only did focus groups recognize the inherent uncertainties present in the GM-agriculture and food issue, they were particularly troubled by the failure of expert institutions to acknowledge [these] (...) uncertainties' (PABE report 2001, 7 as cited in Guehlstorf and Hallstrom (2005, 338). These observations suggest uncertainty information does not cause problems, but the experts' inability to acknowledge uncertainty.
16. Interestingly enough, these 'regulatory delays have been often called "the de facto EU council moratorium" – a misnomer for a period of intense activity' (Levidow, Carr, and Wield 2005, 262).
17. Article 58 (2) Regulation 2002/178/EC *OJ* 2002, L31/1 refers to the procedure laid down in Decision 1999/468/EC *OJ* L 184, establishing the procedure to be followed.
18. Commission Decision 2004/643/ec.
19. Commission Decision 2005/635/ec.
20. Decision 2006/47/EC, 16 January 2006.
21. EU Food Law Weekly, no. 247, 10 March 2006.

References

6, Perry. 2005. What's in a frame? Social organization, risk perception and the sociology of knowledge. *Journal of Risk Research* 8, no. 2: 91–118.

Bal, R., W.E. Bijker, and R. Hendriks. 2002. *Paradox of scientific authority: On the societal impact of the Health Council's advice* (in Dutch). Den Haag, the Netherlands: Gezondheidsraad.

Beck, U. 1986. *Risk society: Towards a new modernity*. London: Sage.

Benford, R.D., and D.A. Snow. 2000. Framing processes and social movements: An overview and assessment. *Annual Review of Sociology* 26: 611–39.

Borrás, S. 2006. Legitimate governance of risk at the EU level? The case of genetically modified organisms. *Technological Forecasting and Social Change* 73, no. 1: 61–75.

Bouder, F., D. Slavin, and R. Löfstedt, eds. 2007. *The tolerability of risk: A new framework for risk management*. London: Earthscan.

Chalmers, D. 2004. Risk, anxiety and the European mediation of the politics of life: The European food safety authority and the government of biotechnology. http://www.law.harvard.edu/programs/elrc/events/2004-2005/socialregulation.pdf.

Collins, H.M. 1987. Certainty and the public understanding of science: Science on television. *Social Studies of Science* 17, no. 4: 689–713.

Cranor, C.F. 1990. Some moral issues in risk assessment. *Ethics* 101, no. 1: 123–43.

de Wilde, R., and M. Reithler. 2006. Post-normal science in action: A case-study on the role of scientific risk assessment in a political controversy (in Dutch). *Filosofie en Praktijk* 27, no. 1: 52–64.

Durant, J., M.W. Bauer, and G. Gaskell. 1998. *Biotechnology in the public sphere: A European source book*. London: Cromwell Press.

EFSA. 2003. Opinion of the Scientific Panel on Genetically Modified Organisms on a request from the Commission related to the safety of foods and food ingredients derived from herbicide-tolerant genetically modified maize NK603, for which a request for placing on the market was submitted under Article 4 of the Novel Food Regulation (EC) No 258/97 by Monsanto. *The EFSA Journal* 9: 1–14.

———. 2004a. Opinion of the Scientific Panel on Genetically Modified Organisms on a request from the Commission related to the Notification (Reference C/DE/02/9) for the placing on the market of insect protected genetically modified maize MON 863 and MON 863 × MON 810, for import and processing, under Part C of Directive 2001/18/EC from Monsanto. *The EFSA Journal* 49: 1–25.

———. 2004b. Opinion of the Scientific Panel on Genetically Modified Organisms on a request from the Commission related to the Notification (Reference C/NL/98/11) for the placing on the market of glyphosatetolerant oilseed rape event GT73, for import and processing, under Part C of Directive 2001/18/EC from Monsanto. *The EFSA Journal* 29: 1–19.

———. 2005. Opinion of the Scientific Panel on Genetically Modified Organisms on a request from the Commission related to the Notification (Reference C/DE/02/9) for the placing on the market of insect-protected genetically modified maize MON 863 × MON 810, for import and processing, under Part C of Directive 2001/18/EC from Monsanto. *The EFSA Journal* 251: 122.

Forrester, I., and J.C. Hanekamp. 2006. Precaution, science and jurisprudence: A test case. *Journal of Risk Research* 9, no. 4: 297–311.

Frewer, L.J., S. Hunt, M. Brennan, S. Kuznesof, M. Ness, and C. Ritson. 2003. The views of scientific experts on how the public conceptualize uncertainty. *Journal of Risk Research* 2, no. 6: 75–85.

Friends of the Earth. 2004 Throwing caution to the wind: A review of the European Food Safety Authority and its work on genetically modified foods and crops. http://www.foeeurope.org/GMOs/publications/EFSAreport.pdf.

Funtowicz, S.O., and J.R. Ravetz. 1990. *Uncertainty and quality in science for policy*. Dordrecht, The Netherlands: Kluwer.

Gaskell, G., N. Allum, M. Bauer, J. Durant, A. Allansdottir, H. Bonfadelli, D. Boy, S. de Cheveigné, B. Fjaestad, J.M. Gutteling, J. Hampel, E. Jelsøe, J. Correia Jesuino, M. Kohring, N. Kronberger, C. Midden, T.H. Nielsen, A. Przestalski, T. Rusanen, G. Sakellaris, H. Torgersen, T. Twardowski, and W. Wagner. 2000. Biotechnology and the European public. *Nature Biotechnology* 18: 935–8.

Gaskell, G., N. Allum, W. Wagner, G. Gaskell, N. Kronberger, H. Torgersen, J. Hampel, and J. Bardes. 2004. GM foods and the misperception of risk perception. *Risk Analysis* 24: 185–94.

Gieryn, T.F. 1983. Boundary-work and the demarcation of science from non-science: Strains and interests in professional ideologies of scientists. *American Sociological Review* 48: 781–95.

———. 1999. *Cultural boundaries of science: Credibility on the line*. Chicago, IL: University of Chicago Press.

Goffman, E. 1974. *Frame analysis: An essay on the organization of experience*. New York: Harper Colofon.

Greenpeace. 2004. Greenpeace critique of Monsanto's Roundup Ready Oilseed Rape, GT73. http://eu.greenpeace.org/downloads/gmo/GPTechCritiqueOfEFSAOpinion.pdf.

———. 2005. Monsanto ordered to make secret study public. http://eu.greenpeace.org/downloads/gmo/MON863GermanCourt050620.pdf.

Grossman, M.R., and B. Endres. 2000. Regulation of genetically modified organisms in the European Union. *American Behavioral Scientist* 44, no. 2: 378–434.

Guehlstorf, N.P., and L.K. Hallstrom. 2005. The role of culture in risk regulations: A comparative case study of genetically modified corn in the United Stated of America and European Union. *Environmental Science & Policy* 8: 327–42.

Hajer, M., and M. Schwarz. 2001. Inleiding. Contouren van de risicomaatschappij. In *De wereld als risicomaatschappij*, ed. U. Beck, 7–22. Amsterdam: De Balie.

Halffman, W. 2003. *Boundaries of regulatory science: Ecoltoxicology and aquatic hazards of chemicals in the US, England, and the Netherlands, 1970–1995*. Amsterdam: Universiteit van Amsterdam (UvA).

———. 2005. Science-policy boundaries: National styles? *Science and Public Policy* 32, no. 6: 457–67.

Harremoës, P., D. Gee, M. MacGarvin, A. Stirling, J. Keys, B. Wynne, and S. Guedes Vaz. 2002. *The precautionary principle in the 20th century: Late lessons from early warnings*. London: Earthscan.

Hoppe, R., and S. Huijs. 2002. *Boundary work between science and policy: Paradoxes and dilemmas* (in Dutch). Den Haag, The Netherlands: RMNO.

Horlick-Jones, T., J. Walls, G. Rowe, N. Pidgeon, W. Poortinga, G. Murdock, and T. O'Riordan. 2007. *The GM debate: Risk, politics and public engagement*. New York: Routledge.

Hunter, R. 1999. European regulation of genetically modified organisms. In *Fearing food: Risk, health and environment*, ed. J. Morris and R. Bate, 189–230. Newton, MA: Butterworth-Heinemann.

Jasanoff, S. 1990. *The fifth branch: Science advisers as policy makers*. Cambridge, MA: Harvard University Press.

———. 2005. *Designs on nature: Science and democracy in Europe and the United States*. Princeton, NJ: Princeton University Press.

Krayer von Kraus, M.P. 2005. Uncertainty in policy relevant sciences. PhD thesis, Technical University of Denmark.

Landfried, C. 1999. The European regulation of biotechnology by polycratic governance. In *EU committees: Social regulation, law and politics*, ed. C. Joerges and E. Vos, 173–94. Portland, OR: Hart Publishers.

Lang, J.T., and W.K. Hallman. 2005. Who does the public trust? The case of genetically modified food in the United States. *Risk Analysis* 25, no. 5: 1241–52.

Levidow, L. 2001. Genetically modified crops: What transboundary harmonization in Europe? In *Transboundary risk management*, ed. J. Linnerooth-Bayer, R.E. Löfstedt, and G. Sjöstedt. London: Earthscan.

Levidow, L., S. Carr, and D. Wield. 2000. Genetically modified crops in the European Union: Regulatory conflicts as precautionary opportunities. *Journal of Risk Research* 3, no. 3: 189–208.

———. 2005. European Union regulation of agri-biotechnology: Precautionary links between science, expertise and policy. *Science and Public Policy* 32, no. 4: 261–76.

Levidow, L., J. Murphy, and S. Carr. 2007. Recasting 'substantial equivalence': Transatlantic governance of GM food. *Science, Technology & Human Values* 32, no. 1: 26–64.

Löfstedt, R.E. 2005. *Risk management in post-trust societies*. New York: Palgrave Macmillan.

Morgan, M.G. 2003. Characterizing and dealing with uncertainty: Insights from the integrated assessment of climate change. *Integrated Assessment* 4, no. 1: 46–55.

Mourik, R. 2004. Did water kill the cows? The distribution and democratisation of risk, responsibility and liability in a Dutch agricultural controversy on water pollution and cattle sickness. PhD thesis, Maastricht University.

Nowotny, H., P. Scott, and M. Gibbons. 2001. *Re-thinking science: Knowledge and the public in an age of uncertainty*. Cambridge, UK: Polity Press in association with Blackwell Publishers.

Pollack, H.N. 2003. *Uncertain science… Uncertain world.* Cambridge, UK: Cambridge University Press.

Randall, E. 2006. Not that soft or informal: A response to Eberlein and Grande's account of regulatory governance in the EU with special reference to the European Food Safety Authority (EFSA). *Journal of European Public Policy* 13, no. 3: 402–19.

Ravetz, J.R. 2001. Models of risk: An exploration. In *Knowledge, power and participation in environmental policy analysis*, ed. M. Hisschemöller, R. Hoppe, W.N. Dunn, and J.R. Ravetz. London: Transaction Books.

Renn, O. 2006. *Risk governance: Towards an integrative approach. IRGC White Paper 1.* Geneva, Switzerland: International Risk Governance Council (IRGC).

Renn, O., and K. Walker. 2008. Global risk governance: Concept and practice using the IRGC framework. In *International risk governance council book series*, ed. C. Bunting. Dordrecht, the Netherlands: Springer.

RMNO. 2004. *Spotting new risks?* (in Dutch). The Hague, the Netherlands: Lemma.

Rosa, E.A. 1998. Metatheoretical foundations for post-normal risk. *Journal of Risk Research* 1, no. 1: 15–44.

Siegrist, M. 2003. Perception of gene technology and food risks: Results of a survey in Switzerland. *Journal of Risk Research* 6, no. 1: 45–60.

Smithson, M. 1993. Ignorance and science: Dilemmas, perspectives, and prospects. *Knowledge: Creation, Diffusion, Utilization* 15, no. 2: 133–56.

Stahl, C.A., and A.J. Cimorelli. 2005. How much uncertainty is too much and how do we know?: A case example of the assessment of ozone monitoring network options. *Risk Analysis* 25, no. 5: 1109–20.

Stocking, S.H., and L.W. Holstein. 1993. Constructing and reconstructing scientific ignorance: Ignorance claims in science and journalism. *Knowledge: Creation, Diffusion, Utilization* 15, no. 2: 186–210.

Tait, J. 2001. More faust than Frankenstein: The European debate about the precautionary principle and risk regulation for genetically modified crops. *Journal of Risk Research* 4, no. 2: 175–89.

Tulloch, J., and D. Lupton. 2002. Consuming risk, consuming science: The case of GM foods. *Journal of Consumer Culture* 2, no. 3: 363–83.

van Asselt, M.B.A. 2000. *Perspectives on uncertainty and risk: The PRIMA approach to decision support*. Dordrecht, The Netherlands: Kluwer Academic Publishers.

———. 2004. Onzekere risico's en riskante onzekerheden. In *Risico en voorzorg in de rechtsmaatschappij*, ed. G. van Calster and E. Vos. Maastricht, The Netherlands: IUS Commune Seminarreeks "Europaeum".

———. 2005. The complex significance of uncertainty in a risk era: Logics, manners and strategies in use. *International Journal for Risk Assessment and Management* 5, nos. 2–4: 125–58.

van Asselt, M.B.A., and A.C. Petersen. 2003. *Not afraid of uncertainty* (in Dutch). Den Haag, The Netherlands: Lemma/RMNO.

van Asselt, M.B.A., B. Rooijackers, and E. Vos. Forthcoming. Science, knowledge and uncertainty in EU risk regulation. In *Uncertain risks regulated*, ed. E. Vos, M. Everson, and J. Scott. London: UCL Press/Cavendish Publishing.

van Asselt, M.B.A., and E. Vos. 2005. The precautionary principle in times of intermingled uncertainty and risk: Some regulatory complexities. *Water, Science and Technology* 52, no. 6: 35–41.

———. 2006. The precautionary principle and the uncertainty paradox. *Journal of Risk Research* 9, no. 4: 313–36.

Vercelli, A. 1995. From soft uncertainty to hard environmental uncertainty. *Economie appliquée* 48, no. 2: 251–69.

Vos, E. 1999. *Institutional frameworks of community health and safety regulation: Committees, agencies and private bodies*. Oxford: Hart Publishing.

———. 2000. EU food safety regulation in the aftermath of the BSE crisis. *Journal of Consumer Policy* 23: 227–55.

Vos, E., and F. Wendler. 2006. Food safety regulation at the EU level. In *Food safety regulation in Europe. A comparative institutional analysis*, ed. E. Vos and F. Wendler, 65–138. Mortsel, Belgium: Intersentia Publishing.

Weingart, P. 1999. Scientific expertise and political accountability: Paradoxes of science in politics. *Science and Public Policy* 26, no. 3: 151–61.

Wynne, B. 1982. *Rationality and ritual: The windscale inquiry and nuclear decisions in Britain*. Chalfont, St Giles: British Society for the History of Science.

———. 1995. Technology assessment and reflexive social learning: Observations from the risk field. In *Managing technology in society: The approach of constructive technology assessment*, ed. A. Rip, T.J. Misa, and J. Schot. London: Pinter Publishers.

———. 2001. Creating public alienation: Expert cultures of risk and ethics. *Science as Culture* 10, no. 4: 445–81.

Index

Page numbers in *Italics* represent tables.
Page numbers in **Bold** represent figures.